AQA Chemistry

2nd Edition

A LEVEL YEAR 2

Ted Lister
Janet Renshaw

UNIVERSITY PRESS

OXFORD
UNIVERSITY PRESS

Great Clarendon Street, Oxford, OX2 6DP, United Kingdom

Oxford University Press is a department of the University of Oxford. It furthers the University's objective of excellence in research, scholarship, and education by publishing worldwide. Oxford is a registered trade mark of Oxford University Press in the UK and in certain other countries

British Library Cataloguing in Publication Data
Data available

978-0-19-835771-1

10 9 8 7 6 5 4 3

Paper used in the production of this book is a natural, recyclable product made from wood grown in sustainable forests. The manufacturing process conforms to the environmental regulations of the country of origin.

Printed in China by Golden Cup

Message from AQA

This textbook has been approved by AQA for use with our qualification.
This means that we have checked that it broadly covers the specification and we are satisfied with the overall quality. Full details of our approval process can be found on our website.

We approve textbooks because we know how important it is for teachers and students to have the right resources to support their teaching and learning. However, the publisher is ultimately responsible for the editorial control and quality of this book.

Please note that when teaching the *AQA AS or A-Level Chemistry* course, you must refer to AQA's specification as your definitive source of information. While this book has been written to match the specification, it cannot provide complete coverage of every aspect of the course.

A wide range of other useful resources can be found on the relevant subject pages of our website: www.aqa.org.uk.

The authors would like to acknowledge Colin Chambers and Lawrie Ryan, as well as the contribution of their editors – Sophie Ladden, Alison Schrecker, Sadie Ann Garratt, and Sarah Ryan.

The authors and publisher are grateful to the following for permission to reproduce photographs and other copyright material in this book.

Acknowledgements COVER: PHIL BOORMAN/ SCIENCE PHOTO LIBRARY **p2-3**: Rynio Productions/ Shutterstock; **p11**: Photolab / Lawrence Berkeley National Laboratory; **p49**: Andrew Lambert Photography/Science Photo Library; **p56**(T): Erik Khalitov/iStockphoto; **p56**(B): cjp/iStockphoto; **p58**: Colin Cuthbert/Science Photo Library; **p59**: D Burke/ Alamy; **p62**: Andrew Lambert Photography/Science Photo Library; **p64**: Public Health England/Science Photo Library; **p65**: Science Photos/Alamy; **p68**: Auscape/Getty Images; **p71**(R): Martyn F. Chillmaid/ Science Photo Library; **p71**(L): Andrew Lambert Photography/Science Photo Library; **p78**(T): Belmonte/ BSIP/Science Photo Library; **p78**(BL): urbanbuzz/ Shutterstock; **p89**: Rynio Productions/Shutterstock; **p90-91**: Kuttelvaserova Stuchelova/Shutterstock; **p93**(T): Andrew Lambert Photography/Science Photo Library; **p93**(B): Andrew Lambert Photography/Science Photo Library; **p94**: Charles D. Winters/Science Photo Library; **p105**: Martyn F. Chillmaid/Science Photo Library; **p110**: Andrew Lambert Photography/Science Photo Library; **p112**: SunChan/iStockphoto; **p128**(L): Andrew Lambert Photography/Science Photo Library; **p128**(R): Andrew Lambert Photography/Science Photo Library; **p130**: Andrew Lambert Photography/Science Photo Library; **p132**: Andrew Lambert Photography/ Science Photo Library; **p141**: Kuttelvaserova Stuchelova/Shutterstock; **p142-143**: Sergio Stakhnyk/ Shutterstock; **p151**(B): Martyn F. Chillmaid/Science Photo Library; **p151**(T): Martyn F. Chillmaid/Science Photo Library; **p151**(C): wareham.nl (sport)/Alamy; **p159**(CL): Martyn F. Chillmaid/Science Photo Library; **p159**(CR): Science Photo Library; **p162**: Andrew Lambert Photography/Science Photo Library; **p165**: Martyn F. Chillmaid/Science Photo Library; **p167**: Ria Novosti/Science Photo Library; **p174**(L): Molekuul/ Science Photo Library; **p174**(R): Science Photo Library; **p179**: Martyn F. Chillmaid/Science Photo Library; **p180**: Press Illustrating Service/National Geographic Creative/Corbis; **p186**: Adventure_Photo/iStockphoto; **p187**: MeePoohyaPhoto/Shutterstock; **p189**: evemilla/ iStockphoto; **p191**: Varipics/iStockphoto; **p192**: urbanbuzz/Shutterstock; **p196**(T): pidjoe/iStockphoto; **p192**(B): dt03mbb/iStockphoto; **p198**(T): Charles D. Winters/Science Photo Library; **p198**(B): Michael Stokes/Shutterstock; **p200**: FLPA/Alamy; **p221**: Argonaut technologies; **p229**: OUP; **p238**(T): Ezz Mika Elya/Shutterstock; **p238**(B): Kondor83/Shutterstock; **p242**: Charles D. Winters/Science Photo Library; **p244**: Mark Sykes/Science Photo Library; **p253**: Sergio Stakhnyk/Shutterstock; **p260**: Science Photo Library;

Artwork by Q2A Media

Although we have made every effort to trace and contact all copyright holders before publication this has not been possible in all cases. If notified, the publisher will rectify any errors or omissions at the earliest opportunity.

Links to third party websites are provided by Oxford in good faith and for information only. Oxford disclaims any responsibility for the materials contained in any third party website referenced in this work.

This book has been written to support students studying for AQA A Level Chemistry. The sections covered are shown in the contents list, which shows you the page numbers for the main topics within each section. There is also an index at the back to help you find what you are looking for.

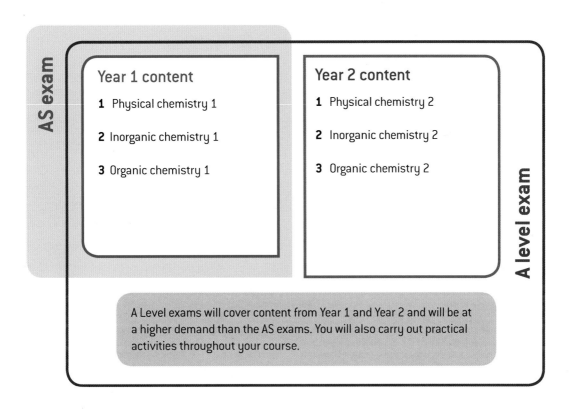

AS exam

A level exam

Year 1 content

1 Physical chemistry 1

2 Inorganic chemistry 1

3 Organic chemistry 1

Year 2 content

1 Physical chemistry 2

2 Inorganic chemistry 2

3 Organic chemistry 2

A Level exams will cover content from Year 1 and Year 2 and will be at a higher demand than the AS exams. You will also carry out practical activities throughout your course.

How to use this book   vi
Kerboodle   ix

Section 1 Physical chemistry   2

17 Thermodynamics   4
17.1 Enthalpy change   4
17.2 Born–Haber cycles   7
17.3 More enthalpy changes   13
17.4 Why do chemistry reactions take place?   15
     Practice questions   22

18 Kinetics   24
18.1 The rate of chemical reactions   24
18.2 The rate expression and order of reaction   27
18.3 Determining the rate equation   30
18.4 The Arrhenius equation   35
18.5 The rate-determining step   39
     Practice questions   42

19 Equilibrium constant K_p   44
19.1 Equilibrium constant K_p for
     homogeneous systems   44
     Practice questions   47

**20 Electrode potentials and
electrochemical cells**   48
20.1 The electrochemical series   48
20.2 Predicting the direction of redox
     reactions   52
20.3 Electrochemical cells   56
     Practice questions   60

21 Acids, bases, and buffers   62
21.1 Defining an acid   62
21.2 The pH scale   64
21.3 Weak acids and bases   68
21.4 Acid–base titrations   71
21.5 Choice of indicators for titrations   74
21.6 Buffer solutions   77
     Practice questions   82
Section 1 practice questions   84
Section 1 summary   88

Section 2 Inorganic chemistry 2   90

22 Periodicity   92
22.1 Reactions of Period 3 elements   92
22.2 The oxides of elements in Period 3   95
22.3 The acidic/basic nature of the
     Period 3 oxides   99
     Practice questions   101

23 The transition metals   104
23.1 The general properties of transition
     metals   104
23.2 Complex formation and the shape of
     complex ions   106
23.3 Coloured ions   110
23.4 Variable oxidation states of transition
     elements   113
23.5 Catalysts   119
     Practice questions   123

**24 Reactions of inorganic compounds in
aqueous solutions**   124
24.1 The acid–base chemistry of aqueous
     transition metal ions   124
24.2 Ligand substitution reactions   129
24.3 A summary of acid–base and substitution
     reactions of some metal ions   132
     Practice questions   134
Section 2 practice questions   136
Section 2 summary   140

Section 3 Organic chemistry 2   142

25 Nomenclature and isomerism   144
25.1 Naming organic compounds   144
25.2 Reactions of the carbonyl group in
     aldehydes and ketones   147
25.3 Synthesis of optically active compounds   150
     Practice questions   154

26 Compounds containing the carbonyl group 156
26.1 Introduction to aldehydes and ketones 156
26.2 Reactions of the carbonyl group in aldehydes and ketones 158
26.3 Carboxylic acids and esters 161
26.4 Reactions of carboxylic acids and esters 164
26.5 Acylation 168
Practice questions 173

27 Aromatic chemistry 174
27.1 Introduction to arenes 174
27.2 Arenes – physical properties, naming, and reactivity 177
27.3 Reactions of arenes 179
Practice questions 183

28 Amines 186
28.1 Introduction to amines 186
28.2 The properties of amines as bases 188
28.3 Amines as nucleophiles and their synthesis 190
Practice questions 194

29 Polymerisation 196
29.1 Condensation polymers 196
Practice questions 202

30 Amino acids, proteins, and DNA 204
30.1 Introduction to amino acids 204
30.2 Peptides, polypeptides, and proteins 206
30.3 Enzymes 212
30.4 DNA 213
30.5 The action of anti-cancer drugs 217
Practice questions 218

31 Organic synthesis and analysis 220
31.1 Synthetic routes 220
31.2 Organic analysis 225
Practice questions 226

32 Structure determination 228
32.1 Nuclear magnetic resonance (NMR) spectroscopy 228

32.2 Proton NMR 232
32.3 Interpreting proton, 1H, NMR spectra 234
Practice questions 240

33 Chromatography 242
33.1 Chromatography 242
Practice questions 246
Section 3 practice questions 248
Section 3 summary 252

Additional practice questions 254
Section 4 Practical skills 260
Section 5 Mathematical skills 264

Reference
  Data table 272
  Periodic Table 274
  Glossary 275
  Answers 276
  Index 285
  Acknowledgements 289

How to use this book

Learning objectives

→ At the beginning of each topic, there is a list of learning objectives.

→ These are matched to the specification and allow you to monitor your progress.

→ A specification reference is also included.
Specification reference: 3.1.1

This book contains many different features. Each feature is designed to foster and stimulate your interest in chemistry, as well as supporting and developing the skills you will need for your examinations.

Terms that you will need to be able to define and understand are highlighted in **bold orange text**. You can look these words up in the glossary.

Sometimes a word appears in **bold**. These are words that are useful to know but are not used on the specification. They therefore do not have to be learnt for examination purposes.

Synoptic link

These highlight how the sections relate to each other. Linking different areas of chemistry together is important, and you will need to be able to do this.

There are also links to the mathematical skills on the specification. More detail can be found in the maths section.

Application features

These features contain important and interesting applications of chemistry in order to emphasise how scientists and engineers have used their scientific knowledge and understanding to develop new applications and technologies. There are also application features to develop your maths skills, with the icon √𝑥, and to develop your practical skills, with the icon ⚗.

Extension features

These features contain material that is beyond the specification designed to stretch and provide you with a broader knowledge and understanding and lead the way into the types of thinking and areas you might study in further education. As such, neither the detail nor the depth of questioning will be required for the examinations. But this book is about more than getting through the examinations.

1 Extension and application features have questions that link the material with concepts that are covered in the specification. Short answers are inverted at the bottom of the feature, whilst longer answers can be found in the answers section at the back of the book.

Study tips

Study tips contain prompts to help you with your revision. They can also support the development of your practical skills (with the practical symbol ⚗) and your mathematical skills (with the math symbol √𝑥).

Hint

Hint features give other information or ways of thinking about a concept to support your understanding. They can also relate to practical or mathematical skills and use the symbols ⚗ and √𝑥.

Summary questions

1 These are short questions that test your understanding of the topic and allow you to apply the knowledge and skills you have acquired. The questions are ramped in order of difficulty.

2 Questions that will test and develop your mathematical and practical skills are labelled with the mathematical symbol (√𝑥) and the practical symbol (⚗).

Section 1

Introduction at the opening of each section summarises what you need to know.

18 Kinetics
19 The equilibrium constant K_p
20 Electrode potentials
21 Acids, bases, and buffers

...mics builds on the ideas introduced in AS Chemistry ...ow you can calculate enthalpy changes that are hard, ...to measure directly. It applies this idea to the enthalpy ...ed in forming ionic compounds. It introduces the idea ...measure of disorder – that drives chemical reactions and ...dissociation energy, a way of predicting whether a reaction will take place at a particular temperature.

Kinetics is about rates of reactions. The rate equation is an expression that links the rate of a reaction to the concentrations of different species in the reaction mixture. The idea of a reaction mechanism as a series of simple steps is introduced along with the concept of the rate – determining step.

Equilibria shows how to apply the equilibrium law and Le Châtelier's principle to reversible reactions that take place in the gas phase. The idea of partial pressure is introduced as well as the gaseous equilibrium constant K_p, which is expressed in terms of partial pressures.

In **Electrode potentials and electrochemical cells**, the idea of half cells which can be joined to generate an electrical potential difference is introduced. This leads on to a method of predicting the course of redox reactions and also to a description of how a number of types of batteries work.

Acids, bases, and buffers extends the definition of acids and bases and gives an expression to find the pH of a solution. The idea of strong and weak acids and bases is introduced and applied quantitatively. Titrations between strong and weak acids and bases are discussed and the operation of buffer solutions, which resist changes of pH, is explained.

What you already know:

The material in this unit builds on knowledge and understanding that you have built up at AS level. In particular the following:

- ☐ It is possible to measure energy (enthalpy) changes in chemical reactions.
- ☐ The rates of chemical reactions are governed by collisions between particles that occur with sufficient energy.
- ☐ Reversible reactions may reach equilibrium.
- ☐ Le Chatelier's Principle can be used to make predictions about the position of an equilibrium.
- ☐ The equilibrium law expression for K_c allows you to make calculations about the position of equilibrium for reactions in solution.
- ☐ Redox reactions involve the transfer of electrons and can be kept track of using the idea of oxidation state (oxidation numbers).

A checklist to help you assess your knowledge from KS4, before starting work on the section.

Visual summaries of each section show how some of the key concepts of that section interlink.

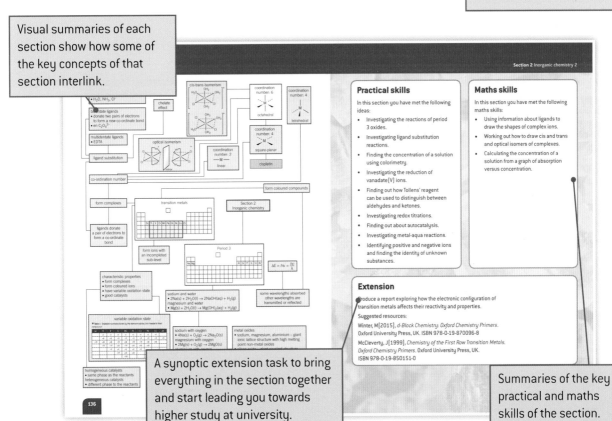

Section 2 Inorganic chemistry 2

Practical skills

In this section you have met the following ideas:

- Investigating the reactions of period 3 oxides.
- Investigating ligand substitution reactions.
- Finding the concentration of a solution using colorimetry.
- Investigating the reduction of vanadate(V) ions.
- Finding out how Tollens' reagent can be used to distinguish between aldehydes and ketones.
- Investigating redox titrations.
- Finding out about autocatalysis.
- Investigating metal-aqua reactions.
- Identifying positive and negative ions and finding the identity of unknown substances.

Maths skills

In this section you have met the following maths skills:

- Using information about ligands to draw the shapes of complex ions.
- Working out how to draw cis and trans and optical isomers of complexes.
- Calculating the concentration of a solution from a graph of absorption versus concentration.

Extension

Produce a report exploring how the electronic configuration of transition metals affects their reactivity and properties.

Suggested resources:

Winter, M[2015], *d-Block Chemistry. Oxford Chemistry Primers*. Oxford University Press, UK. ISBN 978-0-19-870096-8

McCleverty, J[1999], *Chemistry of the First Row Transition Metals. Oxford Chemistry Primers*. Oxford University Press, UK. ISBN 978-0-19-850151-0

A synoptic extension task to bring everything in the section together and start leading you towards higher study at university.

Summaries of the key practical and maths skills of the section.

Section 5
Mathematical skills

Mathematical section to support and develop the mathematical skills required for your course. Remember, at least 20% of your exam will involve mathematical skills.

Practical skills section with questions for each suggested practical on the specification. Remember, at least 15% of your exam will be based on practical skills.

Section 4
Practical skills

Section 3 practice questions

Practice questions at the end of each chapter and each section, including questions that cover practical and maths skills. There are also additional practice questions at the end of the book.

This book is supported by next generation Kerboodle, offering unrivalled digital support for independent study, differentiation, assessment, and the new practical endorsement.

If your school subscribes to Kerboodle, you will also find a wealth of additional resources to help you with your studies and with revision:

- Study guides
- Maths skills boosters and calculation worksheets
- On your marks activities to help you achieve your best
- Practicals and follow up activities to support the practical endorsement
- Interactive objective tests that give question-by-question feedback
- Animations and revision podcasts
- Self-assessment checklists.

Revise with ease using the study guides to guide you through each chapter and direct you towards the resources you need.

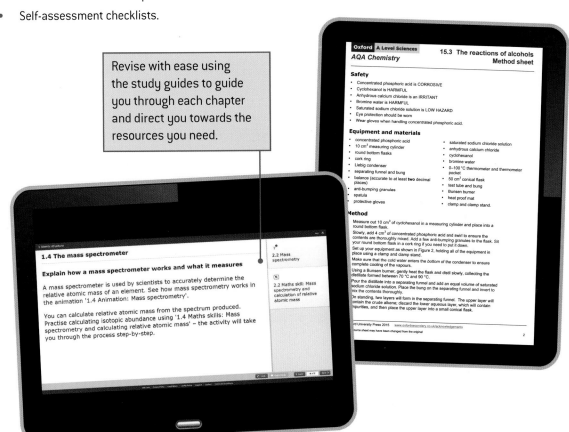

If you are a teacher reading this, Kerboodle also has plenty of further assessment resources, answers to the questions in the book, and a digital markbook along with full teacher support for practicals and the worksheets, which include suggestions on how to support and stretch your students. All of the resources are pulled together into teacher guides that suggest a route through each chapter.

Section 1
Physical chemistry 2

Chapters in this section

17 Thermodynamics

18 Kinetics

19 The equilibrium constant K_p

20 Electrode potentials

21 Acids, bases, and buffers

Thermodynamics builds on the ideas introduced in AS Chemistry and looks at how you can calculate enthalpy changes that are hard, or impossible, to measure directly. It applies this idea to the enthalpy changes involved in forming ionic compounds. It introduces the idea of entropy – a measure of disorder – that drives chemical reactions and the idea of free energy, a way of predicting whether a reaction will take place at a particular temperature.

Kinetics is about rates of reactions. The rate equation is an expression that links the rate of a reaction to the concentrations of different species in the reaction mixture. The idea of a reaction mechanism as a series of simple steps is introduced along with the concept of the rate – determining step.

Equilibria shows how to apply the equilibrium law and Le Châtelier's principle to reversible reactions that take place in the gas phase. The idea of partial pressure is introduced as well as the gaseous equilibrium constant K_p, which is expressed in terms of partial pressures.

In **Electrode potentials and electrochemical cells**, the idea of half cells which can be joined to generate an electrical potential difference is introduced. This leads on to a method of predicting the course of redox reactions and also to a description of how a number of types of batteries work.

Acids, bases, and buffers extends the definition of acids and bases and gives an expression to find the pH of a solution. The idea of strong and weak acids and bases is introduced and applied quantitatively. Titrations between strong and weak acids and bases are discussed and the operation of buffer solutions, which resist changes of pH, is explained.

What you already know:

The material in this unit builds on knowledge and understanding that you have built up at AS level. In particular the following:

- ☐ It is possible to measure energy (enthalpy) changes in chemical reactions.
- ☐ The rates of chemical reactions are governed by collisions between particles that occur with sufficient energy.
- ☐ Reversible reactions may reach equilibrium.
- ☐ Le Chatelier's Principle can be used to make predictions about the position of an equilibrium.
- ☐ The equilibrium law expression for K_c allows you to make calculations about the position of equilibrium for reactions in solution.
- ☐ Redox reactions involve the transfer of electrons and can be kept track of using the idea of oxidation state (oxidation numbers).

Learning objective:

→ List the enthalpy changes that are relevant to the formation of ionic compounds.

Specification reference: 3.1.8

Synoptic link

You will need to know the energetics, states of matter, ionic bonding, and change of state studied in Chapter 3, Bonding, and Chapter 4, Energetics.

Hint

You may also refer to the enthalpy of atomisation of a *compound*.

Study tip

It is important that you know these definitions.

Hint

Ionisation enthalpies are *always* positive because energy has to be put in to pull an electron away from the attraction of the positively charged nucleus of the atom.

Study tip

The second ionisation energy of sodium is *not* the energy change for

$$Na(g) \rightarrow Na^{2+}(g) + 2e^-$$

It is the energy change for

$$Na^+(g) \rightarrow Na^{2+}(g) + e^-$$

Hint

Hess' law states that the enthalpy change for a chemical reaction is always the same, whatever route is taken from reactants to products.

Hess's law

You have already seen how to use Hess's law to construct enthalpy cycles and enthalpy diagrams. In this chapter you will return to Hess's law and use it to investigate the enthalpy changes when an ionic compound is formed.

Definition of terms

When you measure a heat change at constant pressure, you call it an enthalpy change.

Standard conditions chosen are 100 kPa and a stated temperature, usually 298 K.

The **standard molar enthalpy of formation** $\Delta_f H^\ominus$ is the enthalpy change when one mole of a compound is formed from its constituent elements under standard conditions, all reactants and products in their standard states.

For example: $H_2(g) + \frac{1}{2}O_2(g) \rightarrow H_2O(l)$ $\Delta_f H^\ominus = -286\,kJ\,mol^{-1}$

The standard enthalpy of formation of an element is, by definition, zero.

The **standard molar enthalpy change of combustion** $\Delta_c H^\ominus$ is the enthalpy change when one mole of substance is completely burnt in oxygen.

For example: $CH_4(g) + 2O_2(g)$ $CO_2(g) + 2H_2O(l)$

$$\Delta_c H^\ominus = -890\,kJ\,mol^{-1}$$

The **standard enthalpy of atomisation** $\Delta_{at} H^\ominus$ is the enthalpy change which accompanies the formation of one mole of gaseous atoms from the element in its standard state under standard conditions.

For example: $Mg(s) \rightarrow Mg(g)$ $\Delta_{at} H^\ominus = +147.7\,kJ\,mol^{-1}$

$\frac{1}{2}Br_2(l) \rightarrow Br(g)$ $\Delta_{at} H^\ominus = +111.9\,kJ\,mol^{-1}$

$\frac{1}{2}Cl_2(g) \rightarrow Cl(g)$ $\Delta_{at} H^\ominus = +121.7\,kJ\,mol^{-1}$

This is given per mole of chlorine or bromine *atoms* and not per mole of chlorine or bromine *molecules*.

First ionisation energy (first IE) $\Delta_i H^\ominus$ is the standard enthalpy change when one mole of gaseous atoms is converted into a mole of gaseous ions each with a single positive charge.

For example: $Mg(g) \rightarrow Mg^+(g) + e^-$ $\Delta_i H^\ominus = +738\,kJ\,mol^{-1}$

or first IE $= +738\,kJ\,mol^{-1}$

The **second ionisation energy** (second IE) refers to the loss of a mole of electrons from a mole of singly positively charged ions.

For example: $Mg^+(g) \rightarrow Mg^{2+}(g) + e^-$ $\Delta_i H^\ominus = +1451\,kJ\,mol^{-1}$

or second IE $= +1451\,kJ\,mol^{-1}$

The **first electron affinity** $\Delta_{ea} H^\ominus$ is the standard enthalpy change when a mole of gaseous atoms is converted to a mole of gaseous ions, each with a single negative charge.

For example: $O(g) + e^- \rightarrow O^-(g)$ $\qquad \Delta_{ea}H^{\ominus} = -141.1\,\text{kJ mol}^{-1}$

or first EA $= -141.1\,\text{kJ mol}^{-1}$

This refers to single atoms, not to oxygen molecules O_2.

The second electron affinity $\Delta_{ea}H^{\ominus}$ is the enthalpy change when a mole of electrons is added to a mole of gaseous ions each with a single negative charge to form ions each with two negative charges.

For example: $O^-(g) + e^- \rightarrow O^{2-}(g)$ $\qquad \Delta_{ea}H^{\ominus} = +798\,\text{kJ mol}^{-1}$

or second electron affinity $= +798\,\text{kJ mol}^{-1}$

Lattice enthalpy of formation $\Delta_L H^{\ominus}$ is the standard enthalpy change when one mole of solid ionic compound is formed from its gaseous ions.

For example: $Na^+(g) + Cl^-(g) \rightarrow NaCl(s)$ $\qquad \Delta_L H^{\ominus} = -788\,\text{kJ mol}^{-1}$

When a lattice forms, new bonds are formed, resulting in energy being given out, so $\Delta H^{\ominus}$ is always negative for this process.

The opposite process, when one mole of ionic compound separates into its gaseous ions, is called the **enthalpy of lattice dissociation**.

Lattice enthalpy of dissociation is the standard enthalpy change when one mole of solid ionic compound dissociated into its gaseous ions.

Lattice enthalpies cannot be measured directly – they need to be calculated (Topic 17.2). The enthalpy of lattice dissociation has the same numerical value as the lattice enthalpy, but $\Delta H^{\ominus}$ is always positive for this process.

Enthalpy of hydration $\Delta_{hyd}H^{\ominus}$ is the standard enthalpy change when water molecules surround one mole of gaseous ions.

For example: $Na^+(g) + aq \rightarrow Na^+(aq)$ $\qquad \Delta_{hyd}H^{\ominus} = -406\,\text{kJ mol}^{-1}$

or $\qquad Cl^-(g) + aq \rightarrow Cl^-(aq)$ $\qquad \Delta_{hyd}H^{\ominus} = -363\,\text{kJ mol}^{-1}$

Enthalpy of solution $\Delta_{sol}H^{\ominus}$ is the standard enthalpy change when one mole of solute dissolves completely in sufficient solvent to form a solution in which the molecules or ions are far enough apart not to interact with each other.

For example: $NaCl(s) + aq \rightarrow Na^+(aq) + Cl^-(aq)$ $\quad \Delta_{sol}H^{\ominus} = +19\,\text{kJ mol}^{-1}$

Mean bond enthalpy is the enthalpy change when one mole of gaseous molecules each breaks a covalent bond to form two free radicals, averaged over a range of compounds.

For example: $CH_4(g) \rightarrow C(g) + 4H(g)$ $\qquad \Delta_{diss}H^{\ominus} = +1664\,\text{kJ mol}^{-1}$

So the mean (or average) C—H bond energy in methane is:

$$\frac{1664}{4} = +416\,\text{kJ mol}^{-1}$$

It is important to have an equation to refer to for enthalpy changes.

Ionic bonding

In a simple model of ionic bonding, electrons are transferred from metal atoms to non-metal atoms. Positively charged metal ions and negatively charged non-metal ions are formed that all have stable outer shells of electrons. These ions arrange themselves into a lattice so that ions of opposite charge are next to one another (Figure 1).

Synoptic link

Successive ionisation energies can help us to understand the arrangement of electrons in atoms – see Topic 1.6, Electron arrangements and ionisation energy.

Hint

First electron affinities are *always* negative as energy is given out when an electron is attracted to the positively charged nucleus of an atom. However, second electron affinities are *always* positive as energy must be put in to overcome the repulsion between an electron and a negatively charged ion.

Hint

Standard enthalpy changes are written in full as, for example, $\Delta_f H^{\ominus}_{298}$, but the 298 is often omitted – the symbol $\ominus$ being taken to imply that the figure refers to a temperature of 298 K (25 °C), which is around room temperature.

Synoptic link

Ionic bonding is discussed in Topic 3.1, The nature of ionic bonding.

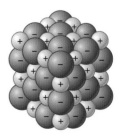

▲ **Figure 1** *Part of an ionic lattice*

Hint

Do not confuse ionisation energy that refers to electron *loss* with electron affinity, which refers to electron *gain*.

Enthalpy changes on forming ionic compounds

If a cleaned piece of solid sodium is placed in a gas jar containing chlorine gas, an exothermic reaction takes place, forming solid sodium chloride:

$$Na(s) + \frac{1}{2}Cl_2(g) \rightarrow NaCl(s) \qquad \Delta_f H^\ominus = -411\,kJ\,mol^{-1}$$

You can think of it as taking place in several steps:

- The reaction involves solid sodium, not gaseous, and chlorine *molecules*, not separate atoms, so you must start with the enthalpy changes for atomisation:

$$Na(s) \rightarrow Na(g) \qquad \Delta_{at} H^\ominus = +108\,kJ\,mol^{-1}$$
$$\frac{1}{2}Cl_2(g) \rightarrow Cl(g) \qquad \Delta_{at} H^\ominus = +122\,kJ\,mol^{-1}$$

Energy has to be *put in* to pull apart the atoms ($\Delta_{at} H^\ominus$ is positive in both cases).

- The gaseous sodium atoms must each give up an electron to form gaseous Na^+ ions:

$$Na(g) \rightarrow Na^+(g) + e^-$$

The enthalpy change for this process is the enthalpy change of first ionisation (ionisation energy, first IE) of sodium and is $+496\,kJ\,mol^{-1}$.

- The chlorine atoms must gain an electron to form gaseous Cl^- ions:

$$Cl(g) + e^- \rightarrow Cl^-(g)$$

The enthalpy change for this process of electron *gain* is the first electron affinity. The first electron affinity for the chlorine atom is $-349\,kJ\,mol^{-1}$ (i.e., energy is given out when this process occurs).

There is a further energy change. At room temperature sodium chloride exists as a solid lattice of alternating positive and negative ions, and not as separate gaseous ions. If positively charged ions come together with negatively charged ions, they form a solid lattice and energy is given out due to the attraction between the oppositely charged ions. This is called the lattice formation enthalpy $\Delta_L H^\ominus$ and it refers to the process:

$$Na^+(g) + Cl^-(g) \rightarrow NaCl(s) \qquad \Delta_L H^\ominus = -788\,kJ\,mol^{-1}$$

The following five processes lead to the formation of NaCl(s) from its elements.

- Atomisation of Na:

$$Na(s) \rightarrow Na(g) \qquad \Delta_{at} H^\ominus = +108\,kJ\,mol^{-1}$$

- Atomisation of chlorine:

$$\frac{1}{2}Cl(g) \rightarrow Cl(g) \qquad \Delta_{at} H^\ominus = +122\,kJ\,mol^{-1}$$

- Ionisation (e^- loss) of Na:

$$Na(g) \rightarrow Na^+(g) + e^- \qquad \text{first IE} = +496\,kJ\,mol^{-1}$$

- Electron affinity of Cl:

$$Cl(g) + e^- \rightarrow Cl^-(g) \qquad \text{first electron affinity} = -349\,kJ\,mol^{-1}$$

- Formation of lattice:

$$Na^+(g) + Cl^-(g) \rightarrow NaCl(s) \qquad \Delta_L H^\ominus = -788\,kJ\,mol^{-1}$$

Hess's law tells us that the total energy (or enthalpy) change for a chemical reaction is the same *whatever route is taken*, provided that the initial and final conditions are the same. It does not matter whether the reaction actually takes place *via* these steps or not.

So the sum of the first five energy changes (taking the signs into account) is equal to the enthalpy change of formation of sodium chloride. You can calculate any one of the quantities, provided all the others are known. You do this by using a thermochemical cycle, called a Born–Haber cycle.

Summary questions

1 **a** 🔢 What is the value of ΔH for this process?
$NaCl(s) \rightarrow Na^+(g) + Cl^-(g)$

   b Explain your answer.

   c What is the term that describes this process?

2 Explain why:

   a Loss of an electron from a sodium atom (ionisation) is an endothermic process.

   b Gain of an electron by a chlorine atom is an exothermic process.

3 **a** Write the equation to represent:

   i the first ioinsation energy of aluminium

   ii the second ionisation energy of aluminium.

   b In terms of the enthalpy changes for the two processes in **a**, what is the enthalpy change when a $Al^{2+}(g)$ ion is formed from a $Al(g)$ atom?

17.2 Born–Haber cycles

A **Born–Haber cycle** is a thermochemical cycle that includes all the enthalpy changes involved in the formation of an ionic compound. Born–Haber cycles are constructed by starting with the elements in their standard states. All elements in their standard states have zero enthalpy by definition.

The Born–Haber cycle for sodium chloride

There are six steps in the Born–Haber cycle for the formation of sodium chloride. Here you will use the cycle to calculate the lattice formation enthalpy ($\Delta_L H$ or LE). The other five steps are shown in Figure 1. (Remember that if you know any five, you can calculate the other). Figure 1 shows you how each step is added to the one before, starting from the elements in their standard state. Positive (endothermic changes) are shown upwards, and negative exothermic changes downwards.

$$Na(s) \rightarrow Na(g) \qquad \Delta_{at}H^\ominus \, Na = +108 \, kJ \, mol^{-1}$$

$$\tfrac{1}{2}Cl_2(g) \rightarrow Cl(g) \qquad \Delta_{at}H^\ominus \, Cl = +122 \, kJ \, mol^{-1}$$

$$Na(g) \rightarrow Na^+(g) + e^- \qquad \text{first IE} = +496 \, kJ \, mol^{-1}$$

$$Cl(g) + e^- \rightarrow Cl^-(g) \qquad \text{first electron affinity} = -349 \, kJ \, mol^{-1}$$

$$Na(s) + \tfrac{1}{2}Cl_2(g) \rightarrow NaCl(s) \qquad \Delta_f H^\ominus = -411 \, kJ \, mol^{-1}$$

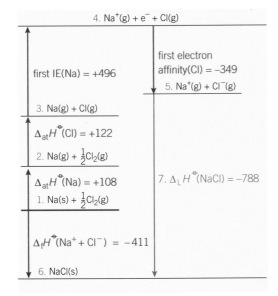

▲ **Figure 1** *Stages in the construction of the Born–Haber cycle for sodium chloride, NaCl, to find the lattice enthalpy. All enthalpies are in kJ mol⁻¹*

Using a Born–Haber cycle you can see why the formation of an ionic compound from its elements is an exothermic process. This is mainly due to the large amount of energy given out when the lattice forms.

1   Start with elements in their standard states. This is the energy zero of the diagram.

Learning objectives:

→ Illustrate how a Born–Haber cycle is constructed for a simple ionic compound.

→ Describe how Born–Haber cycles can be used to predict enthalpy changes of formation of theoretical compounds.

Specification reference: 3.1.8

Study tip

Remember, the standard enthalpy of atomisation is the enthalpy change which accompanies the *formation* of one mole of gaseous atoms.

To form one mole of Cl(g) you need $\tfrac{1}{2}Cl_2(g)$.

Study tip √x

Most errors in Born–Haber cycle calculations result either from lack of knowledge of the enthalpy change definitions, or lack of care with signs.

2   Add in the atomisation of sodium. This is positive, so it is drawn 'uphill'.

3   Add in the atomisation of chlorine. This too is positive, so draw 'uphill'.

4   Add in the ionisation of sodium, also positive and so drawn 'uphill'.

5   Add in the electron affinity of chlorine. This is a negative energy change and so it is drawn 'downhill'.

6   Add in the enthalpy of formation of sodium chloride, also negative and drawn 'downhill'.

7   The final unknown quantity is the lattice formation enthalpy of sodium chloride. The size of this is $788 \, kJ \, mol^{-1}$ from the diagram. Lattice formation enthalpy is the change from separate ions to solid lattice and you must therefore go 'downhill', so $LE(Na^+ + Cl^-)(s) = -788 \, kJ \, mol^{-1}$.

When drawing Born–Haber cycles:

*   make up a rough scale, for example, one line of lined paper to $100 \, kJ \, mol^{-1}$

*   plan out roughly first to avoid going off the top or bottom of the paper. (The zero line representing elements in their standard state will need to be in the middle of the paper.)

*   remember to put in the sign of each enthalpy change and an arrow to show its direction. Positive enthalpy changes go up, negative enthalpy changes go down.

Worked example: The Born–Haber cycle for magnesium chloride

Figure 2 shows the complete Born–Haber cycle for the formation of magnesium chloride, $MgCl_2$, from its elements, together with notes on how it is constructed.

Since two chlorine atoms are involved all the quantities related to chlorine are doubled, that is, $\Delta_{at}H^{\ominus}(Cl)$ and the first electron affinity are both *multiplied* by two.

Also notice that the ionisation of magnesium, $Mg \rightarrow Mg^{2+}$, is the first ionisation enthalpy *plus* the second ionisation enthalpy. The second ionisation enthalpy is larger because it is more difficult to lose an electron from a positively charged ion than from a neutral atom.

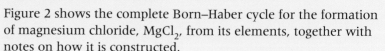

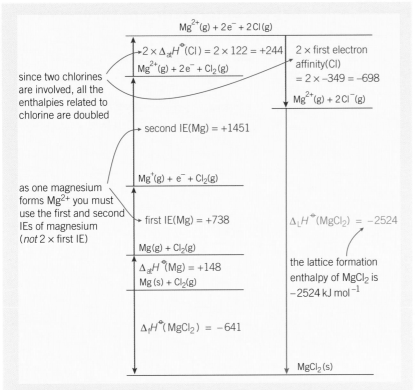

Hint

For chlorine, you multiply the first electron affinity and $\Delta_{at}H^{\ominus}$ by two.

For magnesium, you add together the first and second ionisation energy.

This is because there are two chlorine atoms gaining *one* electron (so two first electron affinities), but *one* magnesium atom loosing *two* electrons (so the first *and* second ionisation energy).

▲ **Figure 2** *The Born–Haber cycle for magnesium chloride, MgCl$_2$. All enthalpies are in kJ mol^{-1}*

Trends in lattice enthalpies

The lattice formation enthalpies of some simple ionic compounds of formula M^+X^- are given in Table 1.

▼ **Table 1** *Some values of lattice formation enthalpies in kJ mol^{-1} for compounds M^+X^-*

		Larger negative ions (anions)			
		F^-	Cl^-	Br^-	I^-
Larger positive ions (cations)	Li^+	−1031	−848	−803	−759
	Na^+	−918	−788	−742	−705
	K^+	−817	−711	−679	−651
	Rb^+	−783	−685	−656	−628
	Cs^+	−747	−661	−635	−613

Larger ions lead to smaller lattice enthalpies. This is because the opposite charges do not approach each other as closely when the ions are larger.

Table 2 shows lattice enthalpies for some compounds $M^{2+}X^{2-}$. You can see the same trend related to size of ions as before in Table 1.

Comparing Table 1 with Table 2 shows that for ions of approximately similar size (i.e., formed from elements in the same group of the Periodic Table, such as Na^+ and Mg^{2+} or F^- and O^{2-}) the lattice enthalpy increases with the size of the charge. This is because ions with double the charge give out roughly twice as much energy when they come together.

▼ **Table 2** *Some values of lattice formation enthalpies for compounds $M^{2+}X^{2-}$*

		Larger anions	
		O^{2-}	S^{2-}
Larger cations	Be^{2+}	−4443	−3832
	Mg^{2+}	−3791	−3299
	Ca^{2+}	−3401	−3013
	Sr^{2+}	−3223	−2848
	Ba^{2+}	−3054	−2725

Predicting enthalpies of formation of theoretical compounds – worked example

Born–Haber cycles can be used to investigate the enthalpy of formation of theoretical compounds to see if they might be expected to exist. The cycles in Figure 3 are for CaF, CaF_2, and CaF_3. They use lattice enthalpies that have been calculated using sensible assumptions about the crystal structures of these compounds and the sizes of the Ca^+ and Ca^{3+} ions.

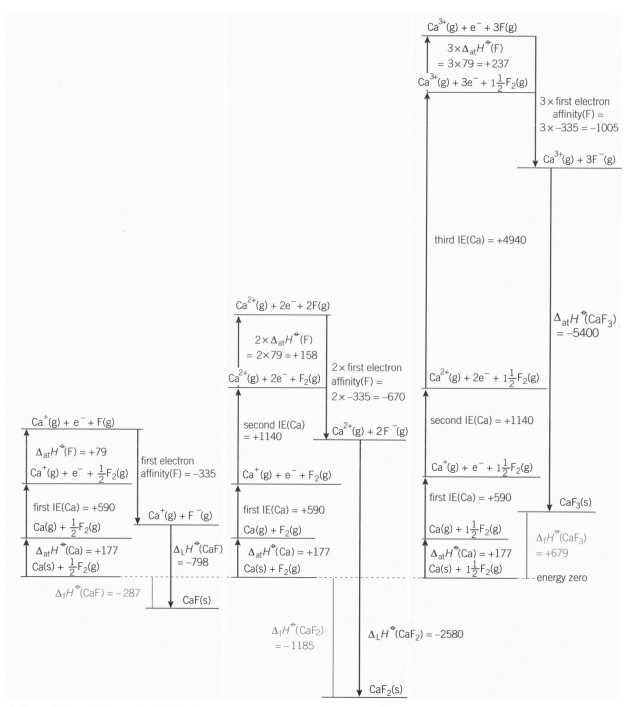

▲ **Figure 3** *Born–Haber cycles for CaF, CaF₂, and CaF₃. All enthalpies in kJ mol⁻¹*

Look at the enthalpies of formation. A large amount of energy would have to be put in to form CaF_3. The formation of CaF would give out energy but not as much as CaF_2. This explains why only CaF_2 has been prepared as a stable compound. CaF has indeed been made but readily turns into CaF_2 and + Ca. It is unstable with respect to CaF_2 + and Ca.

You can see from the relative enthalpy levels of CaF and CaF_2 on Figure 3 that CaF_2 is $1185 - 287 = 898\,kJ\,mol^{-1}$ below CaF.

You can draw a thermochemical cycle to calculate ΔH for the reaction:

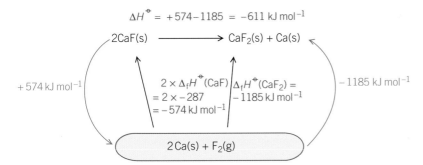

$$\Delta H^{\ominus} = +574 - 1185 = -611\ kJ\,mol^{-1}$$

$2CaF(s) \longrightarrow CaF_2(s) + Ca(s)$

$+574\ kJ\,mol^{-1}$

$2 \times \Delta_f H^{\ominus}(CaF)$
$= 2 \times -287$
$= -574\ kJ\,mol^{-1}$

$\Delta_f H^{\ominus}(CaF_2) =$
$-1185\ kJ\,mol^{-1}$

$-1185\ kJ\,mol^{-1}$

$2\,Ca(s) + F_2(g)$

The first noble gas compound

The noble gases are often called the inert gases and, until 1962, they seemed to be just that – inert. There were no known compounds of them at all. This was explained on the basis of their stable electron arrangements. It had been predicted that there might be compounds of krypton and xenon with fluorine, but no one took much notice. However, in 1962 British chemist Neil Bartlett created a chemical sensation when he announced that he had prepared the first noble gas compound, xenon hexafluoroplatinate (V). Although the name may seem exotic, Bartlett predicted that the compound had a good chance of existing by using a very simple piece of chemical theory.

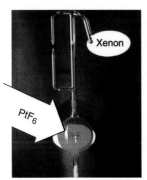
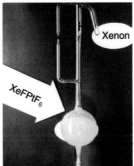

▲ **Figure 4** *Formation of XeFPtF$_5$*

He had previously found that the powerful oxidising agent platinum (VI) fluoride gas, PtF_6, would oxidise oxygen molecules to form the compound dioxygenyl hexfluoroplatinate (V), $O_2{}^+PtF_6{}^-$, in which the oxidising agent has removed an electron from an oxygen molecule.

He then realised that the first ionisation energy of xenon (the energy required to remove an electron from an atom of it) was a little less positive than that of the oxygen molecule, so that if platinum(VI) fluoride could remove an electron from oxygen, it should also be able to remove one from xenon. The values are:

$$Xe(g) \rightarrow Xe^+(g) + e^- \quad \Delta H = +1170 \text{ kJ mol}^{-1} \text{ (first IE of xenon)}$$

$$O_2(g) \rightarrow O_2^+(g) + e^- \quad \Delta H = +1183 \text{ kJ mol}^{-1} \text{ (first IE of an oxygen molecule)}$$

This is not the same as the first ionisation energy of an oxygen *atom*.

The experiment itself was surprisingly simple, as soon as the two gases came into contact, the compound was formed immediately – no heat or catalyst was required. In Bartlett's own words 'When I broke the seal between the red PtF_6 gas and the colourless xenon gas, there was an immediate interaction, causing an orange-yellow solid to precipitate. At once I tried to find someone with whom to share the exciting finding, but it appeared that everyone had left for dinner!'

This was one of those moments when all the textbooks had to be re-written.

The reaction can be represented:

$$Xe(g) + PtF_6(g) \rightarrow Xe^+PtF_6^-(s)$$

More recently it has been realised that the formula of the product may be a little more complex than this.

There are now over 100 noble gas compounds known, although most are highly unstable.

1 Write an equation to represent the first ionisation energy of an oxygen *atom*.
2 If you assume that noble gas compounds are formed with positive noble gas ions, suggest why compounds of xenon and krypton were predicted rather than ones of helium or neon.
3 Why might you *not* expect platinum(VI) fluoride to be a gas?
4 Explain the oxidation states of the elements in $Xe^+PtF_6^-$.

Summary questions

1 a Draw a Born–Haber cycle to find the lattice formation enthalpy for sodium fluoride, NaF. The values for the relevant enthalpy terms are given below.

b What is the lattice formation enthalpy of NaF, given these values?

$Na(s) \rightarrow Na(g)$	$\Delta_{at}H^\ominus = +108 \text{ kJ mol}^{-1}$
$\frac{1}{2}F_2(g) \rightarrow F(g)$	$\Delta_{at}H^\ominus = +79 \text{ kJ mol}^{-1}$
$Na(g) \rightarrow Na^+(g) + e^-$	first IE $= +496 \text{ kJ mol}^{-1}$
$F(g) + e^- \rightarrow F^-(g)$	first electron affinity $= -328 \text{ kJ mol}^{-1}$
$Na(s) + \frac{1}{2}F_2(g) \rightarrow NaF(s)$	$\Delta_f H^\ominus = -574 \text{ kJ mol}^{-1}$

Finding the enthalpy of solution

Ionic solids can only dissolve well in polar solvents. In order to dissolve an ionic compound the lattice must be broken up. This requires an input of energy – the lattice enthalpy. The separate ions are then solvated by the solvent molecules, usually water. These cluster round the ions so that the positive ions are surrounded by the negative ends of the dipole of the water molecules and the negative ions are surrounded by the positive ends of the dipoles of the water molecules. This is called **hydration** when the solvent is water (Figure 1).

The enthalpy change of hydration shows the same trends as lattice enthalpy – it is more negative for more highly charged ions and less negative for bigger ions.

You can think of dissolving an ionic compound in water as the sum of three processes:

1 Breaking the ionic lattice to give separate gaseous ions – the lattice dissociation enthalpy has to be put in.
2 Hydrating the positive ions (cations) – the enthalpy of hydration is given out.
3 Hydrating the negative ions (anions) – the enthalpy of hydration is given out.

For ionic compounds the enthalpy change of hydration has rather a small value and may be positive or negative. For example, the enthalpy of hydration $\Delta_{hyd}H^\ominus$ for sodium chloride is given by the equation:

$$NaCl(s) + aq \rightarrow Na^+(aq) + Cl^-(aq)$$

It may be calculated via a thermochemical cycle as shown below.

These are the steps that are needed:

4 $NaCl(s) \rightarrow Na^+(g) + Cl^-(g)$ $\qquad \Delta_L H^\ominus = +788\,kJ\,mol^{-1}$
This is the enthalpy change for lattice dissociation.

5 $Na^+(g) + aq + Cl^-(g) \rightarrow Na^+(aq) + Cl^-(g)$ $\Delta_{hyd}H^\ominus = -406\,kJ\,mol^{-1}$
This is the enthalpy change for the hydration of the sodium ion.

6 $Na^+(aq) + Cl^-(g) + aq \rightarrow Na^+(aq) + Cl^-(aq)$ $\Delta_{hyd}H^\ominus = -363\,kJ\,mol^{-1}$
This is the enthalpy change for the hydration of the chloride ion.

7 So $\Delta_{hyd}H^\ominus(NaCl) = \Delta_L H^\ominus(NaCl) + \Delta_{hyd}H^\ominus(Na^+) + \Delta_{hyd}H^\ominus(Cl^-)$
$\qquad\qquad +788 \qquad\quad -406 \qquad -363$ you $= +19\,kJ\,mol^{-1}$

The process of dissolving can be represented on an enthalpy diagram (Figure 2) or calculated directly as above. Either method is equally acceptable.

Lattice enthalpies and bonding

It is possible to work out a theoretical value for the lattice formation enthalpy of an ionic compound if you know the charge on the ions, their distance apart, and the geometry of its structure.

Learning objectives:

→ Describe how to find the enthalpy change of solution.

→ Describe the evidence that theoretical calculations for lattice enthalpies provide about bonding.

Specification reference: 3.1.8

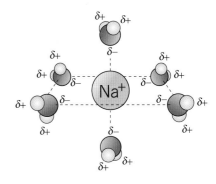

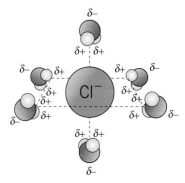

▲ **Figure 1** *The hydration of sodium and chloride ions by water molecules*

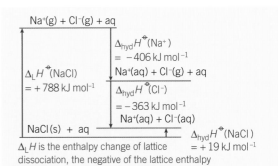

$\Delta_L H$ is the enthalpy change of lattice dissociation, the negative of the lattice enthalpy

▲ **Figure 2** *Thermochemical cycle for the enthalpy of hydration of sodium chloride*

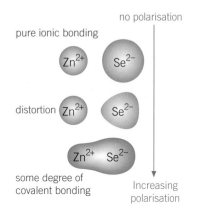

pure ionic bonding

Zn^{2+} Se^{2-}

distortion Zn^{2+} Se^{2-}

Zn^{2+} Se^{2-}

some degree of covalent bonding

no polarisation

Increasing polarisation

▲ **Figure 3** *Polarisation in zinc selenide*

Study tip

It is important that you know why lattice enthalpies obtained by theoretical calculation and from a Born–Haber cycle can differ.

Summary questions

1 $\boxed{\sqrt{x}}$ Draw a diagram to calculate the enthalpy of hydration of potassium bromide using the experimental value of lattice enthalpy in Table 1.
$\Delta_{hyd}H^{\ominus}$ $(K^+) = -322\,kJ\,mol^{-1}$
$\Delta_{hyd}H^{\ominus}$ $(Br^-) = -335\,kJ\,mol^{-1}$
Why is the value relatively small?

2 Explain why compounds of beryllium and aluminium are nearly all significantly covalent. Would you expect the calculated value of lattice enthalpy to be greater or smaller than the experimental value? Explain your answer.

For many ionic compounds, the lattice formation enthalpy determined from experimental values via a Born–Haber cycle agrees very well with that calculated theoretically, and this confirms that you have the correct model of ionic bonding in that compound. However, there are some compounds where there is a large discrepancy between the two values for lattice formation enthalpy because the bond in question has some covalent character.

For example, zinc selenide, ZnSe, has an experimental lattice formation enthalpy of $-3611\,kJ\,mol^{-1}$. The theoretical value, based on the model of complete ionisation ($Zn^{2+} + Se^{2-}$) is $-3305\,kJ\,mol^{-1}$, some 10% lower. The greater experimental value implies some extra bonding is present.

This can be explained as follows. The ion Zn^{2+} is relatively small and has a high positive charge, whilst Se^{2-} is relatively large and has a high negative charge. The small Zn^{2+} can approach closely to the electron clouds of the Se^{2-} and distort them by attracting them towards it (Figure 3). The Se^{2-} is fairly easy to distort, because its large size means the electrons are far from the nucleus and its double charge means there is plenty of negative charge to distort. This distortion means there are more electrons than expected concentrated *between* the Zn and Se nuclei, and represents a degree of electron sharing or covalency which accounts for the lattice enthalpy discrepancy. The Se^{2-} ion is said to be **polarised**.

The factors which increase polarisation are:

- positive ion (cation) – small size, high charge
- negative ion (anion) – large size, high charge.

Table 1 shows some values of experimentally determined lattice enthalpies, compared with those calculated assuming pure ionic bonding. The biggest discrepancy (the most extra covalent-type bonding) is cadmium iodide. The cadmium ion is small and doubly charged, whilst the iodide ion is large and easily polarised.

▼ **Table 1** *Some values of experimental and calculated lattice enthalpies*

Compound	Experimental value of LE / kJ mol^{-1}	Calculated value of LE assuming ionic bonding / kJ mol^{-1}
LiF	−1031	−1021
NaCl	−780	−777
KBr	−679	−667
CaF$_2$	−2611	−2586
CdI$_2$	−2435	−1986
AgCl	−890	−769

So all ionic and covalent bonds can be seen as part of a continuum from purely ionic to purely covalent. For example, caesium fluoride, Cs^+F^-, which has a large singly charged positive ion and a small singly charged negative ion, is hardly polarised at all and is almost completely ionic, whereas a bond between two identical atoms *must* be 100% covalent.

Chemists use the terms **feasible** or **spontaneous** to describe reactions which could take place of their own accord. The terms take no account of the rate of the reaction, which could be so slow as to be unmeasurable at room temperature.

You may have noticed that many of the reactions that occur of their own accord are exothermic (ΔH is negative). For example, if you add magnesium to copper sulfate solution, the reaction to form copper and magnesium sulfate takes place and the solution gets hot. Negative ΔH is a factor in whether a reaction is spontaneous, but it does not explain why a number of endothermic reactions are spontaneous.

For example, both the following reactions, which occur spontaneously, are endothermic (ΔH is positive):

$$C_6H_8O_7(aq) + 3NaHCO_3(aq) \rightarrow Na_3C_6H_5O_7(aq) + 3H_2O(l) + 3CO_2(g)$$

citric acid　　sodium hydrogencarbonate　　　sodium citrate　　　water　　carbon dioxide

$$NH_4NO_3(s) + aq \rightarrow NH_4NO_3(aq)$$

ammonium nitrate　　　　　aqueous ammonium nitrate

Entropy or randomness

Many processes which take place spontaneously involve mixing or spreading out, for example, liquids evaporating, solids dissolving to form solutions, or gases mixing.

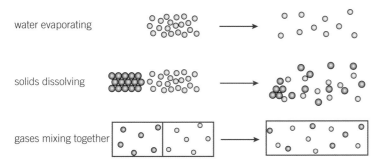

water evaporating

solids dissolving

gases mixing together

▲ **Figure 1** *Spontaneous processes*

This is the clue to the second factor which drives chemical processes – a tendency towards randomising or disordering, that is, towards chaos. Gases are more random than liquids, and liquids are more random than solids, because of the arrangement of their particles.

So endothermic reactions may be spontaneous if they involve spreading out, randomising, or disordering. This is true of the two reactions above – the arrangement of the particles in the products is more random than in the reactants.

The randomness of a system, expressed mathematically, is called the entropy of the system and is given the symbol S. A reaction like the two above, in which the products are more disordered than the reactants, will have positive values for the entropy change ΔS.

Learning objectives:

→ Explain why endothermic reactions occur.

→ Explain how a temperature change affects feasibility.

Specification reference: 3.1.8

Hint

The reaction between citric acid and sodium hydrogencarbonate takes place on your tongue when you eat sherbet – you can feel your tongue getting cold.

Study tip

In chemistry the words spontaneous and feasible mean exactly the same thing – that a reaction has a tendency to happen.

Entropies have been determined for a vast range of substances and can be looked up in databases. They are usually quoted for standard conditions: 298 K and 100 kPa pressure. Table 1 gives some examples.

Study tip

Take care – the units of entropy are $J\,K^{-1}\,mol^{-1}$, *not* $kJ\,K^{-1}\,mol^{-1}$.

▼ **Table 1** *Some values of entropy*

Substance	State at standard conditions	Entropy S / $J\,K^{-1}\,mol^{-1}$
carbon (diamond)	solid	2.4
carbon (graphite)	solid	5.7
copper	solid	33.0
iron	solid	27.0
ammonium chloride	solid	95.0
calcium carbonate	solid	93.0
calcium oxide	solid	40.0
iron(III) oxide	solid	88.0
water (ice)	solid	48.0
water (liquid)	liquid	70.0
mercury	liquid	76.0
water (steam)	gas	189.0
hydrogen chloride	gas	187.0
ammonia	gas	192.0
carbon dioxide	gas	214.0

Study tip

Unlike enthalpies, the entropies of elements in their standard states are *not* zero. All entropies are positive numbers, but it is possible to have a negative entropy *change* for a reaction.

In general, gases have larger values than liquids, which have larger values than solids.

Table 1 shows that the entropy increases when water turns to steam. Entropies increase with temperature, largely because at higher temperatures particles spread out and randomness increases.

Calculating entropy changes

The entropy change for a reaction can be calculated by adding all the entropies of the products and subtracting the sum of the entropies of the reactants. For example:

$$CaCO_3(s) \rightarrow CaO(s) + CO_2(g)$$

Using the values from Table 1:

$$\text{entropy of products} = 40 + 214 = 254\,J\,K^{-1}\,mol^{-1}$$

$$\text{entropy of reactant} = 93\,J\,K^{-1}\,mol^{-1}$$

$$\Delta S = 254 - 93 = +161\,J\,K^{-1}\,mol^{-1}$$

This is a large positive value – a gas is formed from a solid.

The Gibbs free energy change ΔG

You have seen above that a combination of *two* factors govern the feasibility of a chemical reaction:

- the enthalpy change
- the entropy change.

These two factors are combined in a quantity called the **Gibbs free energy G**. If the change in G, ΔG, for a reaction is negative, then this reaction is feasible. If ΔG is positive, the reaction is not feasible.

ΔG combines the enthalpy change ΔH and entropy change ΔS factors as follows:

$$\Delta G = \Delta H - T\Delta S$$

ΔG depends on temperature, because of the term $T\Delta S$. This means that some reactions may be feasible at one temperature and not at another. So an endothermic reaction can become feasible when temperature is increased if there is a large enough positive entropy change. (A *positive* value for ΔS will make ΔG more *negative* because of the negative sign in the $T\Delta S$ term.)

Here are some examples of how this works.

Take the reaction:

$$CaCO_3(s) \rightarrow CaO(s) + CO_2(g) \qquad \Delta H = +178\,kJ\,mol^{-1}$$

You have seen above that $\Delta S = +161\,J\,K^{-1}\,mol^{-1} = 0.161\,kJ\,K^{-1}\,mol^{-1}$

So at room temperature (298 K):

$$\Delta G = \Delta H - T\Delta S$$
$$\Delta G = 178 - (298 \times 0.161) = +130\,kJ\,mol^{-1}$$

This positive value means that the reaction is not feasible at room temperature. The reverse reaction will have $\Delta G = -130\,kJ\,mol^{-1}$ and will be feasible:

$$CaO(s) + CO_2(g) \rightarrow CaCO_3(s)$$

This is the reaction that occurs in desiccators to absorb carbon dioxide.

However, if you do the calculation for a temperature of 1500 K, you get a different result:

At 1500 K:

$$\Delta G = \Delta H - T\Delta S$$
$$\Delta G = -178 - (1500 \times 0.161)$$
$$\Delta G = -178 - 242$$
$$\Delta G = -64\,kJ\,mol^{-1}$$

ΔG is negative and the reaction is feasible at this temperature. This is the reaction that occurs in a lime kiln to make lime (calcium oxide) from limestone (calcium carbonate).

What happens when $\Delta G = 0$?

There is a temperature at which $\Delta G = 0$ for this reaction. This is the point at which the reaction is just feasible. You can calculate this temperature for the reaction above:

$$\Delta G = \Delta H - T\Delta S$$
$$0 = \Delta H - T\Delta S$$
$$\Delta H = T\Delta S \text{ where } \Delta H = +178\,kJ\,mol^{-1} \text{ and } \Delta S = 0.161\,kJ\,K^{-1}$$
$$\text{So } T = \frac{178}{0.161}$$
$$= 1105.6\,K$$

> **Study tip** $\sqrt{x}$
>
> Remember to convert the entropy units by dividing by 1000 because enthalpy is measured in $kJ\,mol^{-1}$ and entropy in $J\,K^{-1}\,mol^{-1}$.

In fact, the reaction does not suddenly flip from feasible to non-feasible. In a closed system an equilibrium exists around this temperature in which both products and reactants are present.

Calculating an entropy change

You can use the temperature at which $\Delta G = 0$ to calculate an entropy change. For example, a solid at its melting point is equally likely to exist as a solid or a liquid – an equilibrium exists between solid and liquid. So ΔG for the melting process must be zero and:

$$0 = \Delta H - T\Delta S$$

For example, the melting point for water is $273\,\text{K}$ and the enthalpy change for melting is $6.0\,\text{kJ mol}^{-1}$. Putting these values into the equation:

$$0 = 6.0 - 273 \times \Delta S$$

$$\Delta S = \frac{6.0}{273} = 0.022\,\text{kJ K}^{-1}\,\text{mol}^{-1} = +22\ \text{J K}^{-1}\,\text{mol}^{-1}$$

This is the entropy change that occurs when ice changes to water. It is positive, which you would expect as the molecules in water are more disordered than those in ice.

> **Hint**
>
> At the melting point of a substance:
> $\Delta H = T\Delta S$

  Determining an entropy change

You can measure the abstract quantity of an entropy change using kitchen equipment.

A chemistry teacher set out to find the entropy change for the vaporisation of water at home using the household kettle and a top pan balance.

At its boiling point, water is equally likely to exist as liquid or vapour (water or steam), so for vaporisation, $\Delta G = 0$.

Inserting the ΔG value into $\Delta G = \Delta H - T\Delta S$ gives:

$0 = \Delta H - T\Delta S$

So $\Delta H = T\Delta S$

Rearrange to $\Delta S = \dfrac{\Delta H}{T}$

The boiling point of water (at atmospheric pressure) is $100\,^\circ\text{C}$ ($373\,\text{K}$), $T = 373\,\text{K}$ so all they needed to measure was ΔH.

The kettle had a power rating of 2.4 kW, which means it supplies 2.4 kJ of energy per second.

They brought some water to the boil in an ordinary kitchen kettle, and weighed the kettle and its contents on a top pan balance that read to the nearest gram. They switched on the kettle again and allowed it to boil for 100 seconds holding down the automatic switch. They then reweighed the kettle to find how much water had boiled away. They found that 100 g of water had boiled away, that is, turned from water to steam (vaporised).

Calculate $\Delta_{vap}S$ by the following steps:

1 Calculate how many kilojoules of heat were supplied to the water in 100 s.
2 Calculate the value of M_r for water and hence find how many moles of water were vaporised.

3 Calculate $\Delta_{vap}H$ for the process in kJ mol^{-1}and convert this into J mol^{-1}.

4 Use $\Delta_{vap}S = \dfrac{\Delta_{vap}H}{T_b}$ to calculate the entropy change of vaporisation.

5 To how many significant figures can you quote your answer?

6 What systematic error (experimental design error) is there in this experiment? How could you reduce it?

7 If the top pan balance weighs to the nearest gram, what is the percentage error in the weighing?

8 The value of $\Delta_{vap}S$ for water is higher than for most liquids. Suggest why. Hint – you are measuring the increase in *disorder* between the liquid and vapour phases. Think about what causes order in the liquid state of water.

Extracting metals

A good way of extracting metals from their oxide ores is to heat them with carbon, which removes the oxygen as carbon dioxide and leaves the metal. This has the advantage that carbon (in the form of coke) is cheap. The gaseous carbon dioxide simply diffuses away, so there is no problem separating it from the metal (although it does contribute to global warming as it is a greenhouse gas). You can use ΔG to investigate under what conditions the reaction might be feasible for different metals.

One of the most important metals is iron and its ore is largely iron(III) oxide, Fe_2O_3. You can calculate $\Delta G^\ominus$ from a thermochemical cycle.

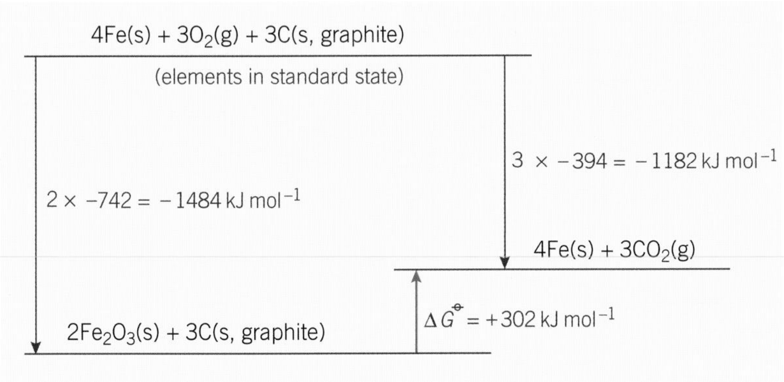

$4Fe(s) + 3O_2(g) + 3C(s, graphite)$

(elements in standard state)

$3 \times -394 = -1182\,kJ\,mol^{-1}$

$2 \times -742 = -1484\,kJ\,mol^{-1}$

$4Fe(s) + 3CO_2(g)$

$2Fe_2O_3(s) + 3C(s, graphite)$

$\Delta G^\ominus = +302\,kJ\,mol^{-1}$

▲ **Figure 2** *Free energy diagram for the reduction of iron(III) oxide by graphite*

$2Fe_2O_3(s) + 3C(s, graphite) \rightarrow 4Fe(s) + 3CO_2(g)\ \Delta G^\ominus = +302\,kJ\,mol^{-1}$

So this reaction is not feasible under standard conditions (298 K).

Will the reaction take place at a higher temperature?

You can work out the temperature at which the reaction just becomes feasible (this is when $\Delta G = 0$) using:

$$\Delta G = \Delta H - T\Delta S$$

Synoptic link

Look back at Chapter 4, Energetics to revise thermochemical cycles.

Calculating ΔH

ΔH for the reaction can be calculated from the following thermochemical cycle.

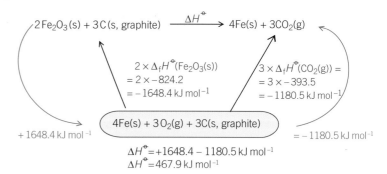

$$\Delta H^{\ominus} = +1648.4 - 1180.5 \text{ kJ mol}^{-1}$$
$$\Delta H^{\ominus} = 467.9 \text{ kJ mol}^{-1}$$

Calculating ΔS

You can calculate the entropy change of the reaction by finding the difference between the sum of the entropies of all the products and the sum of the entropies of all the reactants:

$$2Fe_2O_3(s) + 3C(s, \text{graphite}) \rightarrow 4Fe(s) + 3CO_2(g)$$

$$(2 \times 87.4) + (3 \times 5.7) \qquad (4 \times 27.3) + (3 \times 213.6)$$

$$191.9 \qquad\qquad\qquad 750.0$$

$$\Delta S = +558.1 \text{ J K}^{-1} \text{ mol}^{-1}$$

This value is large and positive, as you would expect from a reaction in which two solids produce a gas.

Putting these values into $\Delta G = \Delta H - T\Delta S$:

$$0 = +467.9 - T \times \frac{558.1}{1000}$$

$$T = 467.9 \times \frac{1000}{558.1}$$

$$= 838.4 \text{ K}$$

The reaction is not feasible below this temperature.

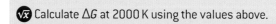 Calculate ΔG at 2000 K using the values above.

−648.3 kJ mol⁻¹

Study tip √x̄

Remember, to convert entropy in $J K^{-1} mol^{-2}$ to $kJ K^{-1} mol^{-1}$, divide by 1000.

Hint

In fact, in the blast furnace a higher temperature is used, above the melting point of iron (1808 K), so that the iron is formed as a liquid. Also, the carbon is not pure graphite, but coke.

Kinetic factors

Neither enthalpy changes nor entropy changes tell us *anything* about how quickly or slowly a reaction is likely to go. As such, you might predict that a certain reaction should occur spontaneously because of enthalpy and entropy changes, but the reaction might take place so slowly that for practical purposes it does not occur at all. In other words, there is a large activation energy barrier for the reaction.

Carbon gives an interesting example:

$$C(s, \text{graphite}) + O_2(g) \rightarrow CO_2(g) \qquad \Delta H^{\ominus} = -393.5 \text{ kJ mol}^{-1}$$

Synoptic link

Look back at Topic 5.1, Collision theory, to revise activation energy.

The reaction is exothermic and you can calculate the actual value of ΔS and so find ΔG.

Calculating ΔS

ΔS for the reaction is the sum of the entropies of the product minus the sum of the entropies of the reactants.

$$C(s, \text{graphite}) + O_2(g) \rightarrow CO_2(g) \qquad \Delta H^\ominus = -394\,\text{kJ}\,\text{mol}^{-1}$$
$$\phantom{C(s, \text{graphite})} 5.7 \qquad\quad 205.0 \qquad 213.6$$

So $\quad \Delta S = 213.6 - (5.7 + 205.0)$

$\quad\quad \Delta S = +2.9\,\text{J}\,\text{K}^{-1}\,\text{mol}^{-1}$, positive as predicted.

Calculating ΔG

$$\Delta G = \Delta H - T\Delta S$$

So under standard condition (approximately room temperature and pressure):

$$\Delta G = -394 - \left(298 \times \frac{2.9}{1000}\right)$$

Remember to divide the entropy value by 1000 to convert from $\text{J}\,\text{K}^{-1}\,\text{mol}^{-1}$ to $\text{kJ}\,\text{K}^{-1}\,\text{mol}^{-1}$.

$$\Delta G = -394 - 0.86$$

$$\Delta G = -394.86\,\text{kJ}\,\text{mol}^{-1}$$, negative so the reaction is feasible.

However, experience with graphite (the 'lead' in pencils) tells you that the reaction does not take place at room temperature – although it will take place at higher temperatures. At room temperature, the reaction is so slow that in practice it doesn't take place at all.

Since the branch of chemistry dealing with enthalpy and entropy changes is called **thermodynamics**, and that dealing with rates is called **kinetics**, graphite is said to be thermodynamically unstable but kinetically stable.

Study tip

Entropies of elements in their standard states are *not* zero.

▼ **Table 2** *Entropy values for some substances*

Substance	$S^\ominus$/$\text{J}\,\text{K}^{-1}\,\text{mol}^{-1}$
Mg(s)	32.7
MgO(s)	26.9
$MgCO_3$(s)	65.7
Zn(s)	41.6
ZnO(s)	43.6
$Pb(NO_3)_2$(s)	213.0
PbO(s)	68.7
NO_2(g)	240.0
O_2(g)	205.0
CO_2(g)	213.6
H_2O(l)	69.7
H_2O(g)	188.7

Summary questions

1 a Without doing a calculation, predict whether the entropy change for the following reactions will be significantly positive, significantly negative, or approximately zero and explain your reasoning.

   i $\quad Mg(s) + ZnO(s) \rightarrow MgO(s) + Zn(s)$

   ii $\quad 2Pb(NO_3)_2(s) \rightarrow 2PbO(s) + 4NO_2(g) + O_2(g)$

   iii $\quad MgO(s) + CO_2(g) \rightarrow MgCO_3(s)$

   iv $\quad H_2O(l) \rightarrow H_2O(g)$

  b Calculate $\Delta S^\ominus$ for each reaction using data in Table 2. Comment on your answers.

2 √x̄ For the reaction:

$$MgO(s) \rightarrow Mg(s) + \tfrac{1}{2}O_2(g)$$

$$\Delta H^\ominus = +602\,\text{kJ}\,\text{mol}^{-1}$$

$$\Delta S^\ominus = +109\,\text{J}\,\text{K}^{-1}\,\text{mol}^{-1}$$

a Using the equation $\Delta G = \Delta H - T\Delta S$, calculate ΔG at:

   i $\quad$ 1000 K

   ii $\quad$ 6000 K

   iii At which temperature is the reaction feasible?

b Calculate the temperature when $\Delta G = 0$.

3 √x̄ Calculate the entropy change for:

$$NH_3(g) + HCl(g) \rightarrow NH_4Cl(s)$$
ammonia   hydrogen   ammonium
         chloride   chloride

The entropy values are: $\quad S^\ominus\ NH_3 \quad 192\,\text{J}\,\text{K}^{-1}\,\text{mol}^{-1}$

$\quad\quad\quad\quad\quad\quad\quad\quad\quad\quad S^\ominus\ HCl \quad 187\,\text{J}\,\text{K}^{-1}\,\text{mol}^{-1}$

$\quad\quad\quad\quad\quad\quad\quad\quad\quad\quad S^\ominus\ NH_4Cl \quad 95\,\text{J}\,\text{K}^{-1}\,\text{mol}^{-1}$

Practice questions

1 (a) Figure 1 shows how the entropy of a molecular substance X varies with temperature.

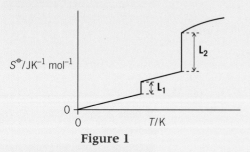

Figure 1

(i) Explain, in terms of molecules, why the entropy is zero when the temperature is zero Kelvin.

(2 marks)

(ii) Explain, in terms of molecules, why the first part of the graph in Figure 1 is a line that slopes up from the origin.

(2 marks)

(iii) On Figure 1, mark on the appropriate axis the boiling point T_b of sub<u>stance</u> X.

(1 mark)

(iv) In terms of the behaviour of molecules, explain why L_2 is longer than L_1 in Figure 1.

(2 marks)

(b) Figure 2 shows how the free-energy change for a particular gas-phase reaction varies with temperature.

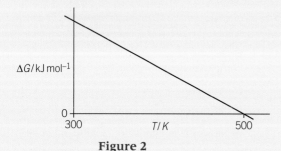

Figure 2

(i) Explain, with the aid of a thermodynamic equation, why this line obeys the mathematical equation for a straight line, $y = mx + c$.

(2 marks)

(ii) Explain why the magnitude of ΔG decreases as T increases in this reaction.

(1 mark)

(iii) State what you can deduce about the feasibility of this reaction at temperatures lower than 500 K.

(1 mark)

(c) The following reaction becomes feasible at temperatures above 5440 K.

$$H_2O(g) \rightarrow H_2(g) + \frac{1}{2}O_2(g)$$

The entropies of the species involved are shown in the following table.

	$H_2O(g)$	$H_2(g)$	$O_2(g)$
$S / J\,K^{-1}\,mol^{-1}$	189	131	205

(i) Calculate the entropy change ΔS for this reaction.

(1 mark)

(ii) Calculate a value, with units, for the enthalpy change for this reaction at 5440 K.

(3 marks)

AQA, 2013

2 The following equation shows the formation of ammonia.

$$\frac{1}{2}N_2(g) + \frac{3}{2}H_2(g) \rightarrow NH_3(g)$$

The graph shows how the free-energy change for this reaction varies with temperature above 240 K.

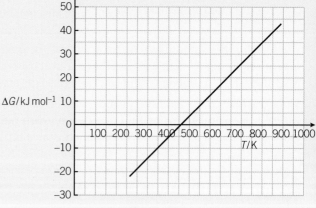

Figure 3

(a) Write an equation to show the relationship between ΔG, ΔH, and ΔS.

(1 mark)

(b) Use the graph to calculate a value for the slope (gradient) of the line. Give the units of this slope and the symbol for the thermodynamic quantity that this slope represents.

(3 marks)

(c) Explain the significance, for this reaction, of temperatures below the temperature value where the line crosses the temperature axis.

(2 marks)

(d) The line is not drawn below a temperature of 240 K because its slope (gradient) changes at this point.
Suggest what happens to the ammonia at 240 K that causes the slope of the line to change.

(1 mark)

AQA, 2012

3 This question is about magnesium oxide. Use data from the table below, where appropriate, to answer the following questions.

	$\Delta H^{\ominus}$/kJ mol^{-1}
First electron affinity of oxygen (formation of $O^-(g)$ from $O(g)$)	-142
Second electron affinity of oxygen (formation of $O^{2-}(g)$ from $O^-(g)$)	$+844$
Atomisation enthaply of oxygen	$+248$

(a) Define the term *enthalpy of lattice dissociation*.
(3 marks)

(b) In terms of the forces acting on particles, suggest one reason why the first electron affinity of oxygen is an exothermic process.
(1 mark)

(c) Complete the Born–Haber cycle for magnesium oxide by drawing the missing energy levels, symbols and arrows. The standard enthalpy change values are given in kJ mol^{-1}.

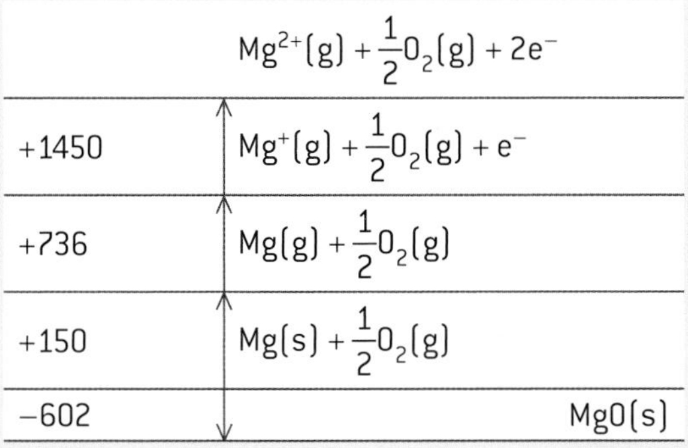

	$Mg^{2+}(g) + \frac{1}{2}O_2(g) + 2e^-$
+1450	$Mg^+(g) + \frac{1}{2}O_2(g) + e^-$
+736	$Mg(g) + \frac{1}{2}O_2(g)$
+150	$Mg(s) + \frac{1}{2}O_2(g)$
−602	$MgO(s)$

(4 marks)

(d) Use your Born–Haber cycle from part **c** to calculate a value for the enthalpy of lattice dissociation for magnesium oxide.
(2 marks)

(e) The standard free-energy change for the formation of magnesium oxide from magnesium and oxygen, $\Delta_f G^{\ominus} = -570$ kJ mol^{-1}. Suggest **one** reason why a sample of magnesium appears to be stable in air at room temperature, despite this negative value for $\Delta_f G^{\ominus}$.
(1 mark)

(f) Use the value of $\Delta_f G^{\ominus}$ given in part **e** and the value of $\Delta_f H^{\ominus}$ from part **c** to calculate a value for the entropy change $\Delta S^{\ominus}$ when one mole of magnesium oxide is formed from magnesium and oxygen at 298 K. Give the units of $\Delta S^{\ominus}$.
(3 marks)

(g) In terms of the reactants and products and their physical states, account for the sign of the entropy change that you calculated in part **f**.
(2 marks)

AQA, 2012

The main factors that affect the rate of chemical reactions are temperature, concentration, pressure, surface area, and catalysts. In this topic you will look at the measurement of reaction rates.

What is a reaction rate?

As a reaction, $A + 2B \rightarrow C$, takes place, the concentrations of the reactants A and B decrease with time and the concentration of product C increases with time. You could measure the concentration of A, B, or C with time and plot the results (Figure 1).

The **rate of the reaction** is defined as the change in concentration (of any of the reactants or products) with unit time, but notice how different the graphs are for A, B, and C. As [C] (the product) increases, [A] and [B] (the reactants) decrease. However, as the equation tells us, for every A that reacts there are two of B, so [B] decreases twice as fast as [A]. For this reason it is important to state whether you are following A, B, or C. Usually it is assumed that a rate is measured by following the concentration of a product(s), because the concentration of the product increases with time.

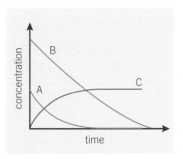

▲ **Figure 1** *Changes of concentration with time for A, B, and C*

The rate of reaction at any instant

You are often interested in the rate at a *particular* instant in time rather than over a period of time. To find the rate of change of [C] at a particular instant, draw a tangent to the curve at that time and then find its gradient (slope), as in Figure 2.

$$\text{rate} = \frac{a}{b} = \frac{\text{change in concentration}}{\text{time}}$$

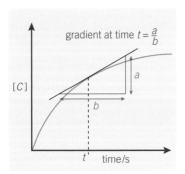

▲ **Figure 2** *The rate of change of [C] at time t, is the gradient of the concentration–time graph at t*

Worked example: Measuring a reaction rate

To measure a reaction rate, you need a method of measuring the concentration of one of the reactants or products over a period of time (keeping the temperature constant, because rate varies with temperature). The method chosen will depend on the substance whose concentration is being measured and also on the speed of the reaction.

Reaction rates are measured in $mol\ dm^{-3}\ s^{-1}$.

For example, in the reaction between bromine and methanoic acid, the solution starts off brown (from the presence of bromine) and ends up colourless:

$$Br_2(aq) + HCO_2H(aq) \rightarrow 2Br^-(aq) + 2H^+(aq) + CO_2(g)$$

So, a colorimeter can be used to measure the decreasing concentration of bromine. The reaction is slow enough to enable the colorimeter to be read every half a minute and the measurements recorded. A computer or data logger could also be used to measure the readings, and this may be essential for faster reactions. →

Table 1 shows some typical results.

In order to find the reaction rate at different times, the results are plotted on a graph and then you can measure the gradients of the tangents at the times required. For example, when $t = 0$, 300 s, and 600 s (Figure 3).

At $t = 0$ s, rate of reaction $= \dfrac{0.010}{240} = 0.000\,041\,6\,mol\,dm^{-3}\,s^{-1}$

At $t = 300$ s, rate of reaction $= \dfrac{0.0076}{540} = 0.000\,014\,mol\,dm^{-3}\,s^{-1}$

At $t = 600$ s, rate of reaction $= \dfrac{0.0046}{840} = 0.000\,005\,5\,mol\,dm^{-3}\,s^{-1}$

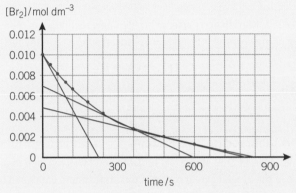

▲ **Figure 3** *Finding the rate of reaction at* t = 0, t = 300, *and* t = 600 s

▼ **Table 1** *[Br₂] measured over time*

Time / s	$[Br_2]$ / mol dm^{-3}
0	0.0100
30	0.0090
60	0.0081
90	0.0073
120	0.0066
180	0.0053
240	0.0044
360	0.0028
480	0.0020
600	0.0013
720	0.0007

Study tip

The rate at $t = 0$ is also called the initial rate – the rate at the start of the reaction.

Hint

The reaction can also be monitored by collecting the carbon dioxide gas.

Fast reactions

Measuring the rate of a chemical reaction requires the experimenter to measure the concentration of one of the reactants or products several times over the course of the reaction. This is fine for reactions that take a few hours or a few minutes, for example, the reaction mixture can be sampled every so often and a titration can be carried out to find the concentration of one of the components. However, some reactions can be over in a few seconds or less.

The British chemists George Porter and Ronald Norrish received the 1967 Nobel Prize for chemistry for devising a technique to follow reactions that are over in a microsecond (10^{-6} s). They shared the prize with Manfred Eigen. Their method is called flash photolysis and involves starting a reaction by firing a powerful pulse of light (the photolysis flash) into a reaction mixture. This breaks chemical bonds and produces highly reactive free radicals which react rapidly with each other and with other molecules. Shortly after the first flash, further flashes of light (the probe flashes) are shone through the reaction vessel at carefully timed intervals down to as little as a microsecond (the timing is done electronically). The probe flashes are used to record the amount of light absorbed by one of the species involved in the reaction and thereby measure its concentration.

Synoptic link

Free radicals are species with unpaired electrons. See Topic 12.3, Industrial cracking, and Topic 12.5, The formation of halogenoalkanes.

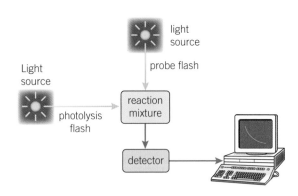

▲ **Figure 5** *Flash photolysis*

In the 1950s, Norrish and Porter used their new technique to measure the reaction of the chlorine monoxide free radical produced in the flash:

$$ClO_2 \rightarrow ClO^\bullet + O^\bullet$$

Pairs of these radicals reacted to give chlorine and oxygen:

$$2ClO^\bullet \rightarrow Cl_2 + O_2$$

At the time, this reaction (over in about $\frac{1}{1000}$ s) was thought to be of academic interest only. However, 30 years on, it was realised that it was involved in the breakdown of ozone in the atmosphere catalysed by chlorine resulting from CFC molecules in aerosol propellants and other items. So Norrish and Porter's work has come to have immense practical importance in enabling understanding this environmental problem.

In recent years, the use of lasers for the flashes has allowed chemists to measure even faster reactions, down to picoseconds and less (a picosecond is 10^{-12} s).

Suggest what might happen to the $O\bullet$ radical produced in the original reaction.

Summary questions

Answer the following questions about the reaction rate graph in Figure 4.

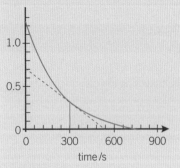

▲ **Figure 4**

1 Is the concentration being plotted that of a reactant or a product? Explain your answer.

2 🆅 The tangent to the curve at the time 300 seconds is drawn on the graph. Find the gradient of the tangent. Remember to include units.

3 What does this gradient represent?

4 Without drawing tangents, what can be said about the gradients of the tangents at time 0 seconds and time 600 seconds?

5 Explain your answer to question **4**.

The rate of a chemical reaction depends on the concentrations of some or all of the species in the reaction vessel – reactants and catalysts. But these do not necessarily all make the same contribution to how fast the reaction goes. The **rate expression** tells us about the contributions of the species that do affect the reaction rate.

For example, in the reaction $X + Y \rightarrow Z$, the concentration of X, [X], may have more effect than the concentration of Y, [Y]. Or, it may be that [X] has no effect on the rate and only [Y] matters. The detail of how each species contributes to the rate of the reaction can only be found out by experiment. A species that does not appear in the chemical equation may also affect the rate, for example, a catalyst.

The rate expression

The rate expression is the result of experimental investigation. It is an equation that describes how the rate of the reaction at a particular temperature depends on the concentration of species involved in the reaction. It is quite possible that one (or more) of the species that appear in the chemical equation will not appear in the rate expression. This means that they do not affect the rate. For example, the reaction:

$$X + Y \rightarrow Z$$

This reaction *might* have the rate expression:

$$\text{rate} \propto [X][Y]$$

The symbol $\propto$ means proportional to.

This would mean that both [X] and [Y] have an equal effect on the rate. Doubling either [X] or [Y] would double the rate of the reaction. Doubling the concentration of both would quadruple the rate.

But it might be that the rate expression for the reaction is:

$$\text{rate} \propto [X][Y]^2$$

This would mean that doubling [X] would double the rate of the reaction, but doubling [Y] would quadruple the rate.

A species that is not in the chemical equation may appear in the rate equation.

The rate constant *k*

By introducing a constant into the expression you can get rid of the proportionality sign. For example, suppose the rate expression were:

$$\text{rate} \propto [X][Y]^2$$

This can be written:

$$\text{rate} = k[X][Y]^2$$

k is called the **rate constant** for the reaction. *k* is different for every reaction and varies with temperature, so the temperature at which it was measured needs to be stated. If the concentrations of all the species in the rate equation are 1 mol dm^{-3}, then the rate of reaction is equal to the value of *k*.

Learning objectives:
→ Define the expressions order of reaction and overall order of reaction.
→ Define the expression rate equation.
→ State what a rate equation is.
→ Define the term rate constant of a rate equation.

Specification reference: 3.1.9

Study tip

Species in chemistry is a general term that includes molecules, ions, and atoms that might be involved in a chemical reaction.

Hint

When $X \propto Y$, if X is doubled then Y also doubles.

The order of a reaction

Suppose the rate expression for a reaction is:

$$\text{rate} = k[X][Y]^2$$

This means that [Y], which is raised to the power of 2, has double the effect on the rate than that of [X]. The **order of reaction**, with respect to one of the species, is the *power* to which the concentration of that species is raised in the rate expression. It tells us how the rate depends on the concentration of that species.

So, for rate $= k[X][Y]^2$ the order with respect to X is *one* ([X] and $[X]^1$ are the same thing), and the order with respect to Y is *two*.

The overall order of the reaction is the sum of the orders of all the species, which appear in the rate expression. In this case the overall order is *three*. So this reaction is said to be *first order* with respect to X, *second order* with respect to Y, and *third order* overall.

So if the rate expression for a reaction is rate $= k[A]^m[B]^n$, where m and n are the orders of the reaction with respect to A and B, the overall order of the reaction is $m + n$.

The chemical equation and the rate expression

The rate expression tells us about the species that affect the rate. Species that appear in the chemical equation do not necessarily appear in the rate equation. Also, the coefficient of a species in the chemical equation – the number in front of it – has no relevance to the rate expression. But catalysts, which do not appear in the chemical equation, *may* appear in the rate expression.

For example, in the reaction:

$$CH_3COCH_3(aq) + I_2(aq) \xrightarrow{H^+ \text{ catalyst}} CH_2ICOCH_3(aq) + HI(aq)$$

propanone   iodine   iodopropanone   hydrogen iodide

The rate expression has been found *by experiment* to be:

$$\text{rate} = k[CH_3COCH_3(aq)][H^+(aq)]$$

So the reaction is first order with respect to propanone, first order with respect to H^+ ions, and second order overall. The rate does not depend on $[I_2 (aq)]$, so you can say the reaction is *zero* order with respect to iodine. The H^+ ions act as a catalyst in this reaction.

Units of the rate constant

The units of the rate constant vary depending on the overall order of reaction.

For a **zero** order reaction:

$$\text{rate} = k$$

The units of rate are $mol\,dm^{-3}\,s^{-1}$.

For a **first** order reaction where:

$$\text{rate} = k[A]$$

$$k = \frac{\text{Rate}}{[A]}$$

The units of rate are $mol\,dm^{-3}\,s^{-1}$ and the units of [A] are $mol\,dm^{-3}$, so the units of k are s^{-1} obtained by cancelling:

$$k = \frac{\cancel{mol\,dm^{-3}}\,s^{-1}}{\cancel{mol\,dm^{-3}}}$$

Therefore, the units of k for a first order rate constant are s^{-1}.

For a **second** order reaction where:

$$rate = k[B]\,[C]$$

$$k = \frac{Rate}{[B]\,[C]}$$

The units of rate are $mol\,dm^{-3}\,s^{-1}$ and the units of both [B] and [C] are $mol\,dm^{-3}$, so the units of k are s^{-1} obtained by cancelling:

$$k = \frac{\cancel{mol\,dm^{-3}}\,s^{-1}}{\cancel{mol\,dm^{-3}}\,mol\,dm^{-3}}$$

Therefore the units of k for a second order rate constant are $mol^{-1}\,dm^3\,s^{-1}$.

For a **third** order reaction:

$$rate = k[D]\,[E]^2$$

$$k = \frac{Rate}{[D]\,[E]^2}$$

The unit of rate is $mol\,dm^{-3}\,s^{-1}$, the unit of [D] is $mol\,dm^{-3}$, and the unit of $[E]^2$ is $(mol\,dm^{-3})^2$.

$$k = \frac{\cancel{mol\,dm^{-3}}\,s^{-1}}{\cancel{mol\,dm^{-3}}\,(mol\,dm^{-3})^2}$$

Therefore, the units of k for a third order rate constant are $mol^{-2}\,dm^6\,s^{-1}$.

Hint $\sqrt{x}$

It is better to work out the units rather than try to remember them.

Study tip

It is important you understand the terms rate of reaction, order of reaction, and rate constant.

Summary questions

1 Write down the rate expression for a reaction that is first order with respect to [A], first order with respect to [B], and second order with respect to [C].

2 Consider the reaction:

$$BrO_3^-(aq) + 5Br^-(aq) + 6H^+(aq) \rightarrow 3Br_2(aq) + 3H_2O(l)$$

bromate ions   bromide ions   hydrogen ions   bromine   water

The rate expression is:

$$rate = k[BrO_3^-(aq)][Br^-(aq)][H^+(aq)]^2$$

a What is the order with respect to:

  i $BrO_3^-(aq)$ **ii** $Br^-(aq)$ **iii** $H^+(aq)$?

b What would happen to the rate if youe doubled the concentration of:

  i $BrO_3^-(aq)$ **ii** $Br^-(aq)$ **iii** $H^+(aq)$?

c What are the coefficients of the following in the chemical equation above.

  i $BrO_3^-(aq)$ **ii** $Br^-(aq)$ **iii** $H^+(aq)$

  iv $Br_2(aq)$ **v** $H_2O(l)$

d Work out the units for the rate constant.

3 $\sqrt{x}$ In the reaction $L + M \rightarrow N$ the rate expression is found to be:

$$rate = k[L]^2[H^+]$$

a What is k?

b What is the order of the reaction with respect to:

  i L   **ii** M   **iii** N   **iv** H^+?

c What is the overall order of the reaction?

d $\sqrt{x}$ The rate is measured in $mol\,dm^{-3}\,s^{-1}$. What are the units of k?

e Suggest the function of H^+ in the reaction.

4 In the reaction $G + 2H \rightarrow I + J$, which is the correct rate expression?

A $rate = k[G][H]^2$

B $rate = \dfrac{k[G][H]}{[I][J]}$

C $rate = k[G][H]$

D It is impossible to tell without experimental data.

Learning objectives:

→ Describe how the order of a reaction with respect to a reagent is found experimentally.

→ Describe how a change in concentration affects the value of the rate constant.

→ Describe how a change in temperature affects the value of the rate constant.

Specification reference: 3.1.9

The rate expression tells you how the rate of a reaction depends on the concentration of the species involved. It only includes the species that affect the rate of the reaction.

• If the rate is not affected by the concentration of a species, the reaction is *zero order* with respect to that species. The species is not included in the rate expression.

• If the rate is directly proportional to the concentration of the species, the reaction is *first order* with respect to that species.

• If the rate is proportional to the square of the concentration of the species, the reaction is *second order* with respect to that species, and so on.

Finding the order of a reaction by using rate–concentration graphs

One method of finding the order of a reaction with respect to a particular species, A, is by plotting a graph of rate against concentration.

Plot the original graph of [A] against time, and draw tangents at different values of [A]. The gradients of these tangents are the reaction rates (the changes in concentration over time) at different concentrations (Figure 1). The values for these rates can then be used to construct a second graph of rate against concentration (Figure 2).

• If the graph is a horizontal straight line (Figure 2a), this means that the rate is unaffected by [A] so the order is zero.

• If the graph is a sloping straight line through the origin (Figure 2b) then rate $\propto$ [A]^1 so the order is 1.

• If the graph is not a straight line (Figure 2c), the order cannot be found directly – it could be two. Try plotting rate against [A]^2. If this is a straight line, then the order is two.

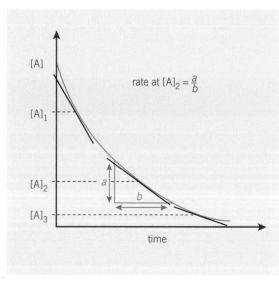

rate at $[A]_2 = \dfrac{a}{b}$

▲ **Figure 1** *Finding the rates of reaction for different values of [A]*

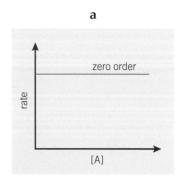

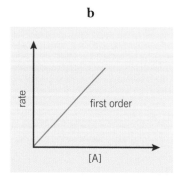

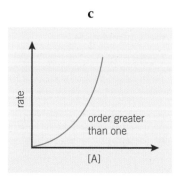

▲ **Figure 2** *Graphs of rate against concentration*

The initial rate method

With the initial rate method, a series of experiments is carried out at constant temperature. Each experiment starts with a different combination of initial concentrations of reactants, catalyst, and so on. The experiments are planned so that, between any pair of experiments, the concentration of only one species varies – the rest stay the same. Then, for each experiment, the concentration of one reactant is followed and a concentration–time graph plotted (Figure 3). The tangent to the graph at time = 0 is drawn. The gradient of this tangent is the **initial rate**. By measuring the *initial* rate, the concentrations of all substances in the reaction mixture are known *exactly* at this time.

Comparing the initial concentration and the initial rates for pairs of experiments allows the order with respect to each reactant to be found. For example, for the reaction:

$$2NO(g) \quad + \quad O_2(g) \quad \rightarrow \quad 2NO_2(g)$$

nitrogen monoxide $\qquad$ oxygen $\qquad$ nitrogen dioxide

The initial rates are shown in Table 1.

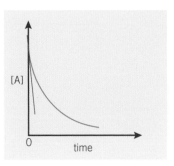

▲ **Figure 3** *Finding the initial rate of a reaction. The initial rate is the gradient at time = 0*

▼ **Table 1** *Results obtained for the reaction $2NO(g) + O_2(g) \rightarrow 2NO_2(g)$*

Experiment number	Initial [NO] / mol dm^{-3}	Initial [O$_2$] / mol dm^{-3}	Initial rate / mol dm^{-3} s^{-1}
1	1.0×10^{-3}	1.0×10^{-3}	7.0×10^{-4}
2	2.0×10^{-3}	1.0×10^{-3}	28.0×10^{-4}
3	3.0×10^{-3}	1.0×10^{-3}	63.0×10^{-4}
4	2.0×10^{-3}	2.0×10^{-3}	56.0×10^{-4}
5	3.0×10^{-3}	3.0×10^{-3}	189.0×10^{-4}

Comparing Experiment 1 with Experiment 2, [NO] is doubled whilst [O$_2$] stays the same. The rate quadruples (from 7.0×10^{-4} mol dm^{-3} s^{-1} to 28.0×10^{-4} mol dm^{-3} s^{-1}) which suggests rate $\propto$ [NO]^2. This is confirmed by considering Experiments 1 and 3 where [NO] is trebled whilst [O$_2$] stays the same. Here the rate is increased ninefold, as would be expected if rate $\propto$ [NO]^2. So the order with respect to nitrogen monoxide is two.

Now compare Experiment 2 with Experiment 4. Here [NO] is constant but [O$_2$] doubles. The rate doubles (from 28.0×10^{-4} mol dm^{-3} s^{-1} to 56.0×10^{-4} mol dm^{-3} s^{-1}) so it looks as if rate $\propto$ [O$_2$]. This is confirmed by considering Experiments 3 and 5. Again [NO] is constant, but [O$_2$] triples. The rate triples too, confirming that the order with respect to oxygen is one.

So $\qquad$ rate $\propto$ [NO]^2 and rate $\propto$ [O$_2$]^1

That is, $\qquad$ rate $\propto$ [NO]^2[O$_2$]^1

Provided that no other species affect the reaction rate, the overall order is three and the rate expression is:

$$\text{rate} = k[NO]^2[O_2]^1$$

Hint $\sqrt{x}$

It is easier to apply the technique to problems than to read about it. However, you can always work out the answer mathematically.

You know that in the example, rate is $= k[NO]^x[O_2]^y$, where x is the order with respect to NO and y is the order with respect to O$_2$.

So, $\dfrac{\text{rate of Experiment 2}}{\text{rate of Experiment 1}} =$

$$\frac{k[NO]_2{}^x[O_2]_2{}^y}{k[NO]_1{}^x[O_2]_1{}^y}$$

Putting in the numbers from the table:

$$\frac{28.0 \times 10^{-4}}{7.0 \times 10^{-4}} =$$

$$\frac{k[2.0 \times 10^{-3}]^x[1.0 \times 10^{-3}]^y}{k[1.0 \times 10^{-3}]^x[1.0 \times 10^{-3}]^y}$$

$$\frac{28.0 \times 10^{-4}}{7.0 \times 10^{-4}} =$$

$$\frac{k[2.0 \times 10^{-3}]^x[\cancel{1.0 \times 10^{-3}}]^y}{k[1.0 \times 10^{-3}]^x[\cancel{1.0 \times 10^{-3}}]^y}$$

$$4 = 2^x$$

$$x = 2$$

So the order with respect to NO is 2.

The iodine clock reaction

The iodine clock reaction is used to measure the rate of the reaction between hydrogen peroxide and potassium iodide in acidic conditions to form iodine (Reaction 1).

$$H_2O_2(aq) + 2I^-(aq) + 2H^+(aq) \rightarrow I_2(aq) + 2H_2O(l)$$ **Reaction 1**

This reaction can be timed by adding a known number of moles of sodium thiosulfate to the reaction mixture along with a little starch. In effect, you are measuring the initial rate of the reaction.

As soon as the iodine is produced by the reaction above, it reacts immediately with the thiosulfate ions in a 1 : 1 ratio by Reaction 2.

$$I_2(aq) + 2S_2O_3^{2-}(aq) \rightarrow S_4O_6^{2-}(aq) + 2I^-(aq)$$ **Reaction 2**

Reaction 2 acts solely as a timing device for Reaction 1. Once the same number of moles of iodine has been produced as the number of moles of thiosulfate added, the free iodine produced by Reaction 1 reacts immediately with the starch to give a dark blue/black colour. The appearance of this colour is sudden, and the time it takes to appear after the reactants have been mixed can be timed accurately. The shorter the time t for the iodine to appear, the faster the rate of Reaction 1. So the value $\frac{1}{t}$ is proportional to the reaction rate.

To carry out the reaction, a solution of hydrogen peroxide is added to one containing potassium iodide, sodium thiosulfate, and starch.

▲ **Figure 4** *Timing the iodine clock reaction*

Some results for an iodine clock reaction are shown below.

Concentration of hydrogen peroxide / mol dm^{-3}	Time for blue colour to appear / s	$\frac{1}{t / s^{-1}}$
1.0	20	0.05
0.5	40	0.025
0.25	80	0.0125

1 This is a redox reaction. What species has been oxidised?
2 What is the oxidising agent?
3 What happens to the rate of reaction when the concentration of hydrogen peroxide is doubled?
4 What is the order of the reaction with respect to hydrogen peroxide?
5 What can you say about the order of the reaction with respect to iodide ions *from these results*?

6 Suggest how the concentration of hydrogen peroxide can easily be varied.

7 State three factors that must be kept constant in order to find the order of the reaction with respect to hydrogen peroxide.

Finding the rate constant *k*

To find k in the reaction of NO and O_2 substitute any set of values of rate, [NO], and [O_2] in the equation.

Taking the values for Experiment 2:

$$28.0 \times 10^{-4} = k(2 \times 10^{-3})^2 \times 1 \times 10^{-3}$$

$$28.0 \times 10^{-4} = k \times 4 \times 10^{-9}$$

$$k = \frac{28.0}{4} \times 10^5$$

$$k = 7.0 \times 10^5$$

But you need to work out the units for k, as these are different for reactions of different overall order. Putting in the units gives:

$$28.0 \times 10^{-4} \, \text{mol dm}^{-3} \, \text{s}^{-1} = k \, (2 \times 10^{-3})^2 \, (\text{mol dm}^{-3}) \, (\text{mol dm}^{-3})$$
$$\times 1 \times 10^{-3} \, \text{mol dm}^{-3}$$

Units can be cancelled in the same way as numbers, so cancelling the units gives:

$$28.0 \times 10^{-4} \, \cancel{\text{mol dm}}^{-3} \, \text{s}^{-1} =$$

$$k \, (4 \times 10^{-6}) \, (\cancel{\text{mol dm}}^{-3}) \, (\text{mol dm}^{-3}) \times 1 \times 10^{-3} \, \text{mol dm}^{-3}$$

$$28.0 \times 10^{-4} = k \times 4 \times 10^{-9} \, \text{mol}^2 \, \text{dm}^{-6} \, \text{s}^1$$

$$k = \frac{28.0}{4} \times 10^5 \, \text{mol}^{-2} \, \text{dm}^6 \, \text{s}^{-1}$$

$$k = 7.0 \times 10^5 \, \text{dm}^6 \, \text{mol}^{-2} \, \text{s}^{-1}$$

Since the units of k vary for reactions of different orders, it is important to put the units for rate and the concentrations in and then cancel them to make sure you have the correct units for k.

The effect of temperature on *k*

Small changes in temperature produce large changes in reaction rates. A rough rule is that for every 10 K rise in temperature, the rate of a reaction doubles. Suppose the rate expression for a reaction is rate = k[A][B]. You know that [A] and [B] do not change with temperature, so the rate constant k must increase with temperature.

In fact, the rate constant k allows you to compare the speeds of different reactions at a given temperature. It is an inherent property of a particular reaction. It is the rate of the reaction at a particular temperature when the concentrations of all the species in the rate expression are 1 mol dm^{-3}. The larger the value of k, the faster the reaction. Look at Table 2. You can see that the value of k increases with temperature. This is true for all reactions.

Hint

You should get the same value of k using the figures for any of the experiments.

Study tip $\sqrt{x}$

Practise working out the units for a rate constant.

▼ **Table 2** *The values of the rate constant, k, at different temperatures for the reaction* $2HI(g) \rightarrow I_2(g) + H_2(g)$

Temperature / K	k / mol^{-1} dm^3 s^{-1}
633	0.0178×10^{-3}
666	0.107×10^{-3}
697	0.501×10^{-3}
715	1.05×10^{-3}
781	15.1×10^{-3}

Study tip

Remember, increasing temperature always increases the rate of reaction and the value of the rate constant k.

Synoptic link

Look back at Topic 5.1, Collision theory, to revise activation energy.

Reaction rate and temperature $\sqrt{x}$

Table 3 shows the results of an experiment to find how the initial rate of the reaction of sodium thiosulfate with hydrochloric acid varies with temperature.

$$Na_2S_2O_3(aq) + 2HCl(aq) \rightarrow S(s) + 2NaCl(aq) + H_2O(l) + SO_2(g)$$

▼ **Table 3**

Temperature / °C	Time / s
15	140
20	74
25	70
30	45
30	25
50	12
60	7

The time from mixing the reactants in a conical flask to a cross below the flask becoming obscured by the sulfur formed was measured over a range of temperatures. As a shorter time means a faster reaction, the value of $\frac{1}{t}$ gives us a measure of the **rate** of the reaction.

Use a spreadsheet program to calculate the values of $\frac{1}{t}$ and to plot a graph of Initial rate of reaction (i.e. $\frac{1}{t}$) vertically against temperature (horizontally). Or do this with a calculator and graph paper.

1 What does this graph tell you about the relationship between reaction rate and temperature?
2 Use the graph in Figure 5 to explain this relationship.

Why the rate constant depends on temperature

Temperature is a measure of the average kinetic energy of particles. Particles will only react together if their collisions have enough energy to start bond breaking. This energy is called the activation energy E_a. Figure 5 shows how the energies of the particles in a gas (or in a solution) are distributed at three different temperatures. Only molecules with energy greater than E_a can react.

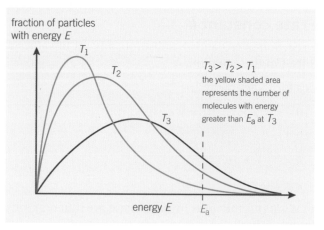

▲ **Figure 5** *The distribution of molecular energies at three temperatures*

The shape of the graph changes with temperature. As the temperature increases, a greater proportion of molecules have enough energy to react. This is the main reason for the increase in reaction rate with temperature.

Summary questions

1 For the reaction A + B → C, the following data were obtained:

Initial [A] / mol dm^{-3}	Initial [B] / mol dm^{-3}	Initial rate / mol dm^{-3} s^{-1}
1	1	3
1	2	12
2	2	24

a What is the order of reaction with respect to:
  i A   ii B?
b What is the overall order?
c What would be the initial rate if the initial [A] were 1 mol dm^{-3} and [B] were 3 mol dm^{-3}?
d What do these results suggest is the rate expression for this reaction?
e Can we be certain that this is the full rate expression? Explain your answer.

Introducing the Arrhenius equation

The rates of chemical reactions increase greatly for relatively small rises in temperature. As the temperature rises, the number of collisions between reactant molecules increases. However, the increase in the number of collisions is not great enough to account for the increase in the rate of the reaction.

Increasing the temperature of a reaction also increases the number of collisions that have energy greater than the activation energy. The shape of a Maxwell–Boltzmann distribution curve shows how the distribution of energies of the reactants is different at different temperatures.

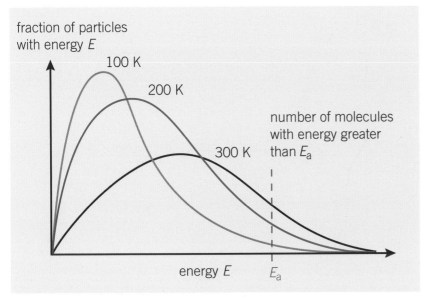

Figure 2 *Maxwell–Boltzmann distribution curve showing distribution of molecular energies at three temperatures*

The fraction of molecules with energy greater than the activation energy is given by:

$$e^{-E_a/RT}$$

E_a is the activation energy in $J\,mol^{-1}$, R is the gas constant ($8.3\,J\,K^{-1}\,mol^{-1}$), and T the temperature in K.

The activation energy can be linked to the rate constant by the **Arrhenius equation**:

$$k = A\, e^{-E_a/RT}$$

k is the rate constant, which is proportional to reaction rate.

A is the pre-exponential factor, which is related to the number of collisions between reactant molecules.

$e^{-E_a/RT}$ is the fraction of collisions with enough energy to react.

Logarithmic form of the Arrhenius equation

The Arrhenius equation is easier to use when you take logs of both sides to the base e, called natural logs.

$$\ln k = -\frac{E_a}{RT} + \ln A$$

or

$$\ln k = -\frac{E_a}{T} \times \frac{1}{T} + \ln A$$

This means that a graph of $\ln k$ against $\frac{1}{T}$ will be a straight line of gradient $-\frac{E_a}{R}$. You can use this to find a value for E_a experimentally by measuring the rate of a reaction at different temperatures.

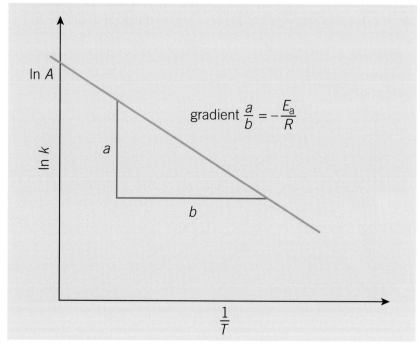

Figure 2 A graph of $\ln k$ against $\frac{1}{T}$ gives a gradient of $-\frac{E_a}{R}$. You can use this to calculate the activation energy

Using the Arrhenius equation

Look at the reaction of the decomposition of hydrogen iodide, HI:

$$2HI(g) \rightarrow I_2(g) + H_2(g)$$

The experimentally determined values of the rate constant k at different temperatures are shown in Table 1.

Table 1 Experimentally determined values for $\ln k$ and $\frac{1}{T}$ for the decomposition of hydrogen iodide

T/K	$\frac{1}{T}$ $/10^{-3}/K^{-1}$	$k/dm^3\,mol^{-1}\,s^{-1}$	$\ln k$
633	1.579	1.78×10^{-5}	−10.936
666	1.501	1.07×10^{-4}	−9.143
697	1.434	5.01×10^{-4}	−7.599
715	1.398	1.05×10^{-3}	−6.858
781	1.280	1.51×10^{-2}	−4.193

Hint

Check with your calculator that you get the values for $\ln k$ and $\frac{1}{T}$ shown in Table 1.

Using the data in Table 1, you can plot a graph of $\ln k$ against $\frac{1}{T}$ and measure the gradient (Figure 3). From this you can calculate E_a.

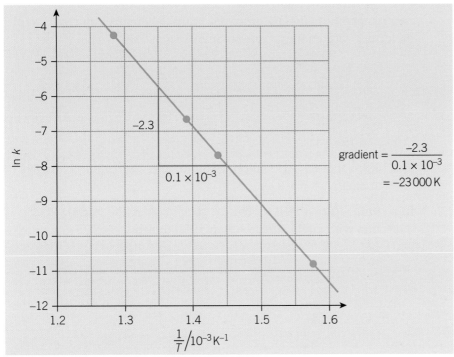

Figure 3 *Graph of $\ln k$ against $\frac{1}{T}$ / 10^{-3}*

Worked example: Calculating E_a from the gradient

The gradient of a graph of $\ln k$ against $\frac{1}{T}$ is $\frac{E_a}{R}$

So $E_a = \text{gradient} \times R$

$E_a = -23\,000\,\text{K} \times 8.3\,\text{J}\,\text{K}^{-1}\,\text{mol}^{-1}$

$E_a = 190\,900\,\text{J}\,\text{mol}^{-1}$

$\mathbf{E_a = 190\,kJ\,mol^{-1}}$

The activation energy for the decomposition of hydrogen iodide is $190\,\text{kJ}\,\text{mol}^{-1}$ to 2 s.f.

Hint

Remember that $\ln k$ has no units, it is just a number. The calculated value of E_a will be in $\text{J}\,\text{mol}^{-1}$ and must be converted to $\text{kJ}\,\text{mol}^{-1}$.

This value is realistic as it represents the energy required to break a covalent bond. Typical values for activation energy are between 40 and $400\,\text{kJ}\,\text{mol}^{-1}$.

The effect of increasing the temperature

Imagine a reaction with activation energy $50\,\text{kJ}\,\text{mol}^{-1}$ ($50\,000\,\text{J}\,\text{mol}^{-1}$), a fairly typical value.

At a temperature of $300\,\text{K}$ (that of a warm laboratory), the fraction of molecules with energy greater than $50\,\text{kJ}\,\text{mol}^{-1}$ is:

$$= e^{-\frac{50000}{8.3 \times 300}}$$

$$= e^{-20.08}$$

$$= 1.9 \times 10^{-9}$$

$$= 19 \times 10^{-10}$$

This means that only 19 molecules out of every 10^{10} molecules have enough energy to react.

At a temperature of $310\,\text{K}$ ($10\,\text{K}$ or $10\,°\text{C}$) higher, the fraction of molecules with energy greater than $50\,\text{kJ}\,\text{mol}^{-1}$:

$$= e^{-\frac{50000}{8.3 \times 310}}$$

$$= e^{-19.43}$$

$$= 3.6 \times 10^{-9}$$

$$= 36 \times 10^{-10}$$

Now 36 molecules out of every 10^{10} molecules have enough energy to react. The $10\,\text{K}$ rise in temperature has almost doubled the number of molecules that have energy greater than the activation energy, and therefore the rate will almost double. This is a general rule of thumb often used by chemists. However, it is only applicable to reactions where E_a is around $50\,\text{kJ}\,\text{mol}^{-1}$ and around room temperature.

Summary Questions

1 Calculate the proportion of molecules with energy greater than an activation energy of $100\,\text{kJ}\,\text{mol}^{-1}$ at:

   a 300 K

   b 310 K

2 The activation energy for the reaction $2HI(g) \rightarrow I_2(g) + H_2(g)$ without a catalyst is $190\,\text{kJ}\,\text{mol}^{-1}$. The reaction is catalysed by a number of metals. Using a metal the following data was obtained:

Temperature /K	Rate constant $k/\text{dm}^3\,\text{mol}^{-1}\,\text{s}^{-1}$
625	2.27
667	7.56
727	24.87
767	91.18
833	334.59

Plot a graph of $\ln k$ against $\frac{1}{T}$ and find the activation energy with this catalyst.

Most reactions take place in more than one step. The separate steps that lead from reactants to products are together called the **reaction mechanism**. For example, the reaction below involves 12 ions:

$$BrO_3^-(aq) + 6H^+(aq) + 5Br^-(aq) \rightarrow 3Br_2(aq) + 3H_2O(l)$$

This reaction *must* take place in several steps – it is very unlikely indeed that the 12 ions of the reactants will all collide at the same time. The steps in-between will involve very short-lived intermediates. These intermediate species, which give information about the mechanism of the reaction, are usually difficult or impossible to isolate and therefore identify. So, other ways of working out the mechanism of the reaction are used.

The rate-determining step

In a multi-step reaction, the steps nearly always follow *after* each other, so that the product of one step is the starting material for the next. Therefore the rate of the slowest step will govern the rate of the whole process. The slowest step may form a 'bottleneck', called the rate-determining step or **rate-limiting step**. Suppose you had everything you needed to make a cup of coffee, starting with cold water. The rate of getting your drink will be governed by the slowest step – waiting for the kettle to boil – no matter how quickly you get the cup out of the cupboard, the coffee out of the jar, or add the milk.

In a chemical reaction, any step that occurs *after* the rate-determining step will not affect the rate, provided that it is fast compared with the rate-determining step. So species that are involved in the mechanism after the rate-determining step do not appear in the rate expression. For example, the reaction:

$$A + B + C \rightarrow Y + Z$$

This reaction might occur in the following steps:

1   $A + B \xrightarrow{\text{fast}} D$ (first intermediate)

2   $D \xrightarrow{\text{slow}} E$ (second intermediate)

3   $E + C \xrightarrow{\text{fast}} Y + Z$

Step 2 is the slowest step and so determines the rate. Then, as soon as some E is produced, it rapidly reacts with C to produce Y and Z.

But the rate of Step 1 might affect the overall rate – the concentration of D depends on this. So, any species involved in or *before* the rate-determining step could affect the overall rate and therefore appear in the rate expression.

So, for the reaction $A + B + C \rightarrow Y + Z$, the rate equation will be:

$$\text{rate} \propto [A][B][D]$$

The reaction between iodine and propanone demonstrates this.

Learning objectives:

→ Define the expression rate-determining step of a reaction.

→ Describe the connection between the rate equation for a reaction and the reaction mechanism.

Specification reference: 3.1.9

The overall reaction is:

The rate expression is found to be rate = $k[CH_3COCH_3][H^+]$
The mechanism is:

The rate-determining step is the first one, which explains why $[I_2]$ does *not* appear in the rate expression.

Using the order of a reaction to find the rate-determining step

Here is a simple example. The three structural isomers with formula C_4H_9Br all react with alkali. The overall reaction is represented by the following equation:

$$C_4H_9Br + OH^- \rightarrow C_4H_9OH + Br^-$$

Two mechanisms are possible.

a   A two-step mechanism:

Step 1: $C_4H_9Br \xrightarrow{\text{slow}} C_4H_9^+ + Br^-$

Step 2: $C_4H_9^+ + OH^- \xrightarrow{\text{fast}} C_4H_9OH$

The slow step involves breaking the C—Br bond whilst the second (fast) step is a reaction between oppositely charged ions.

b A one-step mechanism:

$$C_4H_9Br + OH^- \xrightarrow{\text{slow}} C_4H_9OH + Br^-$$

The C—Br bond breaks at the same time as the C—OH bond is forming.

The three isomers of formula C_4H_9Br are:

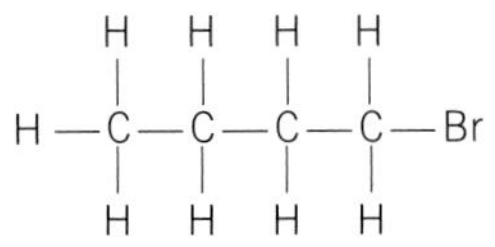

1-bromobutane 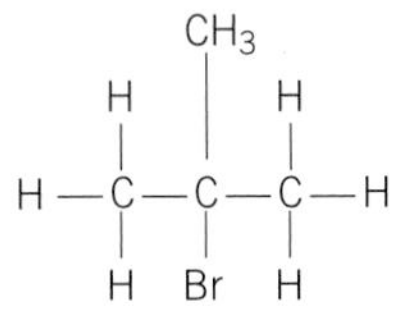   2-bromo-2-methylpropane 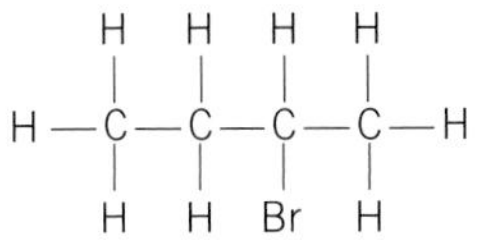   2-bromobutane

Experiments show that 1-bromobutane reacts by a second order mechanism:

$$\text{rate} = k[C_4H_9Br][OH^-]$$

The rate depends on the concentration of *both* the bromobutane *and* the OH^- ions, suggesting mechanism **b**, a one-step reaction.

Experiments show that 2-bromo-2-methylpropane reacts by a first order mechanism:

$$\text{rate} = k[C_4H_9Br]$$

This suggests mechanism **a** in which a slow step, breaking the C—Br bond, is followed by a rapid step in which two oppositely charged ions react together. So, the breaking of the C—Br bond is the rate-determining step.

The compound 2-bromobutane reacts by a mixture of both mechanisms and has a more complex rate expression.

Study tip

The species in the rate equation are the reactants involved in reactions occurring before the rate-determining step.

Summary questions

1 The following reaction schemes show possible mechanisms for the overall reaction:

$$A + E \xrightarrow{\text{catalyst}} G$$

Scheme 1	Scheme 2	Scheme 3
(i) $A + B \xrightarrow{\text{slow}} C$	(i) $A + B \xrightarrow{\text{fast}} C$	(i) $A + B \xrightarrow{\text{fast}} C$
(ii) $C \xrightarrow{\text{fast}} D + B$	(ii) $C \xrightarrow{\text{fast}} D + B$	(ii) $C \xrightarrow{\text{slow}} D + B$
(iii) $D + E \xrightarrow{\text{fast}} F$	(iii) $D + E \xrightarrow{\text{slow}} F$	(iii) $D + E \xrightarrow{\text{fast}} F$
(iv) $F \xrightarrow{\text{fast}} G$	(iv) $F \xrightarrow{\text{slow}} G$	(iv) $F \xrightarrow{\text{fast}} G$

a In Scheme 2, which species is the catalyst?

b Which species *cannot* appear in the rate expression for Scheme 1?

c Which is the rate-determining step in Scheme 3?

Practice questions

1. This question involves the use of kinetic data to calculate the order of a reaction and also a value for a rate constant.

   (a) The data in this table were obtained in a series of experiments on the rate of the reaction between compounds **E** and **F** at a constant temperature.

Experiment	Initial concentration of E / mol dm^{-3}	Initial concentration of F / mol dm^{-3}	Initial rate of reaction / mol dm^{-3} s^{-1}
1	0.15	0.24	0.42×10^{-3}
2	0.45	0.24	3.78×10^{-3}
3	0.90	0.12	7.56×10^{-3}

   (i) Deduce the order of reaction with respect to **E**. *(1 mark)*
   (ii) Deduce the order of reaction with respect to **F**. *(1 mark)*

   (b) √x̄ The data in the following table were obtained in two experiments on the rate of the reaction between compounds **G** and **H** at a constant temperature.

Experiment	Initial concentration of G / mol dm^{-3}	Initial concentration of H / mol dm^{-3}	Initial rate of reaction / mol dm^{-3} s^{-1}
4	3.8×10^{-2}	2.6×10^{-2}	8.6×10^{-4}
5	6.3×10^{-2}	7.5×10^{-2}	To be calculated

   The rate equation for this reaction is

   $$\text{rate} = k[\mathbf{G}]^2[\mathbf{H}]$$

   (i) Use the data from Experiment **4** to calculate a value for the rate constant k at this temperature. Deduce the units of k.

   (3 marks)

   (ii) Calculate a value for the initial rate of reaction in Experiment **5**.

   (1 mark)
   AQA, 2013

2. Gases **P** and **Q** react as shown in the following equation.

   $$2P(g) + 2Q(g) \rightarrow R(g) + S(g)$$

   The initial rate of the reaction was measured in a series of experiments at a constant temperature. The following rate equation was determined.

   $$\text{rate} = k[\mathbf{P}]^2[\mathbf{Q}]$$

   (a) Complete the table of data for the reaction between **P** and **Q**.

Experiment	Initial [G] / mol dm^{-3}	Initial [Q] / mol dm^{-3}	Initial rate / mol dm^{-3} s^{-1}
1	2.5×10^{-2}	1.8×10^{-2}	5.0×10^{-5}
2	7.5×10^{-2}	1.8×10^{-2}	
3	5.0×10^{-2}		5.0×10^{-5}
4		5.4×10^{-2}	4.5×10^{-4}

   (3 marks)

   (b) ● √x̄ Use the data from Experiment 1 to calculate a value for the rate constant k at this temperature. Deduce the units of k.

   (3 marks)
   AQA, 2012

3 Propanone and iodine react in acidic conditions according to the following equation.

$$CH_3COCH_3 + I_2 \longrightarrow ICH_2COCH_3 + HI$$

A student studied the kinetics of this reaction using hydrochloric acid and a solution containing propanone and iodine. From the results the following rate equation was deduced.

$$\text{rate} = k[CH_3COCH_3][H^+]$$

(a) Give the overall order for this reaction.

(1 mark)

(b) $\sqrt{x}$ When the initial concentrations of the reactants were as shown in the table below, the initial rate of reaction was found to be 1.24×10^{-4} mol dm^{-3} s^{-1}.

	Initial concentration / mol dm^{-3}
CH_3COCH_3	4.40
I_2	5.00×10^{-3}
H^+	0.820

Use these data to calculate a value for the rate constant, k for the reaction and give its units.

(3 marks)

(c) Deduce how the initial rate of reaction changes when the concentration of iodine is doubled but the concentrations of propanone and of hydrochloric acid are unchanged.

(1 mark)

(d) The following mechanism for the overall reaction has been proposed.

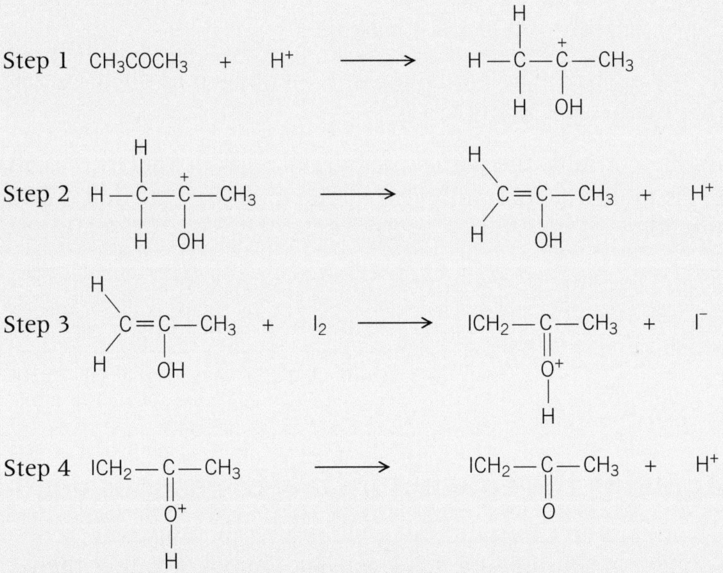

Step 1 CH_3COCH_3 + H^+ $\longrightarrow$ (structure)

Step 2 $H-C-C-CH_3$ $\longrightarrow$ $C=C-CH_3$ + H^+

Step 3 $C=C-CH_3$ + I_2 $\longrightarrow$ ICH_2-C-CH_3 + I^-

Step 4 ICH_2-C-CH_3 $\longrightarrow$ ICH_2-C-CH_3 + H^+

Use the rate equation to suggest which of the four steps could be the rate-determining step. Explain your answer.

(2 marks)

(e) Use your understanding of reaction mechanisms to predict a mechanism for Step 2 by adding one or more curly arrows as necessary to the structure of the carbocation below.

Step 2 $H-C-C-CH_3$ $\longrightarrow$ $C=C-CH_3$ + H^+

(1 mark)

AQA, 2010

19 Equilibrium constant K_p
19.1 Equilibrium constant K_p for homogeneous systems

Learning objectives:

→ State what is meant by partial pressure.

→ Apply the equilibrium law to gaseous equilibria.

→ Predict the effect of changing pressure and temperature on a gaseous equilibrium.

Specification reference 3.1.10

Synoptic link

Revise the material on equilibrium in Chapter 6, Equilibria, in particular the equilibrium law.

Hint $\sqrt{x}$

The SI unit of pressure is the pascal (Pa). 1 pascal is a pressure of 1 newton per square metre ($N\,m^{-2}$). This is roughly the weight of an apple spread over the area of an opened newspaper.

Gaseous equilibria

You have seen how you can apply the equilibrium law to reversible reactions that occur in solution and how to derive an expression for an equilibrium constant K_c for such reactions. Many reversible reactions take place in the gas phase. These include many important industrial reactions such as the synthesis of ammonia and a key stage of the contact process for making sulfuric acid. Gaseous equilibria also obey the equilibrium law but it is usual to express their concentrations in a different way using the idea of partial pressure.

Partial pressure

In a mixture of gases, each gas contributes to the total pressure. This contribution is called its partial pressure p and is the pressure that the gas would exert if it occupied the container on its own. The sum of the partial pressures of all the gases in a mixture is the total pressure. For example, air is a mixture of approximately 20% oxygen molecules and 80% nitrogen molecules and has a pressure (at sea level) of approximately 100 kPa (kilopascals).

So the approximate partial pressure of oxygen in the air is 20 kPa and that of nitrogen is 80 kPa.

Mathematically, the partial pressure of a gas in a mixture is given by its mole fraction multiplied by the total pressure and is given the symbol p.

partial pressure p of A = mole fraction of A × total pressure

$$\text{The mole fraction of a gas A} = \frac{\text{number of moles of gas A in the mixture}}{\text{total number of moles of gas in the mixture}}$$

Applying the equilibrium law to gaseous equilibria

An equilibrium constant can be found in the same way as for a reaction in solution but it is given the symbol K_p rather than K_c.

For a reaction $aA(g) + bB(g) \rightleftharpoons yY(g) + zZ(g)$

$$K_p = \frac{p^y Y(g)_{eqm}\ p^z Z(g)_{eqm}}{p^a A(g)_{eqm}\ p^b B(g)_{eqm}}$$

Note how this corresponds to the equilibrium law expressed in terms of concentration that you have seen earlier.

Worked example: Equilibrium law in gaseous equilibria

1 For the equilibrium

$$H_2(g) + I_2(g) \rightleftharpoons 2HI(g)$$

$$K_p = \frac{(pHI(g)_{eqm})^2}{pH_2(g)_{eqm}\, pI_2(g)_{eqm}}$$

This particular K_p has no units as they cancel.

2 For the equilibrium

$$3H_2(g) + N_2(g) \rightleftharpoons 2NH_3(g)$$
(the key step in the Haber process)

$$K_p = \frac{p^2NH_3(g)_{eqm}}{p^3H_2(g)_{eqm}\, pN_2(g)_{eqm}}$$

Here K_p would have units of Pa^{-2}

Worked example: Calculating partial pressure

This example shows how you can use the expression for K_p to calculate the composition of an equilibrium mixture

K_p is 0.020 for the reaction

$$2HI(g) \rightleftharpoons H_2(g) + I_2(g)$$

If the reaction started with pure HI, and the initial pressure of HI was 100 kPa, what would be the partial pressure of hydrogen when equilibrium is reached?

Set out the problem in the same way as when using K_c.

	2HI(g)	$\rightleftharpoons$	H$_2$(g)	+	I$_2$(g)
Start:	100 kPa		0 kPa		0 kPa
At eqm:	(100 − 2x) kPa		x kPa		x kPa

The chemical equation tells us:

- that there will be the same number of moles of H$_2$ and I$_2$ at equilibrium, therefore $pH_{2eqm} = pI_{2eqm} = x$
- that for each mole of hydrogen (and of iodine) that is produced, *two* moles of hydrogen iodide are used up so that if $pH_{2eqm} = x$, $pHI_{eqm} = (100 - 2x)$

$$K_p = \frac{pH_2(g)_{eqm}\, pI_2(g)_{eqm}}{(pHI(g)_{eqm})^2}$$

Putting in the figures gives: $0.02 = \dfrac{x^2}{(100 - 2x)^2}$

Taking the square root of each side gives: $0.141 = \dfrac{x}{(1200 - 2x)}$

$$0.141 \times (100 - 2x) = x$$

$$14.1 - 0.282x = x$$

$$14.1 = 1.282x$$

$$x = \frac{14.1}{1.282}\ 10.99$$

$pH_2(g) = 11$ kPa (to 2 s.f.)

The chemical equation tells us that pI$_2$(g) must be the same as pH_2(g) and that pHI(g) must be $100 - (2 \times 11) = 78$ kPa.

The effect of changing temperature and pressure on a gaseous equilibrium

Le Chatelier's principle applies to gaseous equilibria in the same way as to equilibria in solution. The only difference is that the partial pressure of reactants and products replace concentration.

So, for a reaction that is exothermic going left to right, increasing the temperature forces the equilibrium to the left, so that the reaction absorbs heat. In other words increasing the temperature decreases K_p. So for the Haber process reaction

$$3H_2(g) + N_2(g) \rightleftharpoons 2NH_3(g) \quad \Delta H = -92 \text{ kJ mol}^{-1}$$

So increasing the temperature *decreases* the yield of ammonia at equilibrium.

Increasing the pressure forces the equilibrium in such a way as to reduce the total pressure, that is, to the side with fewer molecules. So increasing the total pressure *increases* the yield of ammonia.

Changing the total pressure only affects the equilibrium position when there is a change in the total number of molecules on either side of the reaction. So for the equilibrium

$$2HI(g) \rightleftharpoons H_2(g) + I_2(g)$$

Pressure will have no effect on the equilibrium position.

Increasing the pressure on a gas phase reaction will increase the rate at which equilibrium is reached as there will be more collisions between molecules. Increasing temperature will also increase the rate at which equilibrium is attained as will the uses of a catalyst.

Summary questions

1   Using Le Châtelier's principle, predict the effect of increasing: (i) the pressure and (ii) the temperature on the following reactions:

   a $2SO_2(g) + O_2(g) \rightleftharpoons 2SO_3(g)$

   $\Delta H = -197 \text{ kJ mol}^{-1}$

   b $N_2O_4(g) \rightleftharpoons 2NO_2(g) \quad \Delta H = +58 \text{ kJ mol}^{-1}$

   c $H_2(g) + CO_2(g) \rightleftharpoons H_2O(g) + CO(g)$

   $\Delta H = +40 \text{ kJ mol}^{-1}$

2   $A(g) + B(g) \rightleftharpoons C(g) + D(g)$ represents an exothermic reaction and $K_p = pC(g)\, pD(g)\, /\, pA(g)\, pB(g)$.

   In the above expression, what would happen to K_p:

   a   if the temperature were decreased

   b   if more A were added to the mixture

   c   if a catalyst were added?

3   State how the position of each of the following gaseous equilibria will be affected by

   i   Increasing temperature

   ii   Decreasing total pressure

   iii Using a catalyst

   Give the units of K_p in each case

   a $2SO_2(g) + O_2(g) \rightleftharpoons 2SO_3(g)$

   $\Delta H = -192 \text{ kJ mol}^{-1}$

   b $H_2(g) + CO_2(g) \rightleftharpoons H_2O(g) + CO(g)$

   $\Delta H = -40 \text{ kJ mol}^{-1}$

   c $N_2O_4(g) \rightleftharpoons 2NO_2(g) \quad \Delta H = +57 \text{ kJ mol}^{-1}$

Practice questions

1 √x̄ Consider the equilibrium system below.
$2SO_2(g) + O_2(g) \rightleftharpoons 2SO_3(g)$
The partial pressures for the gases in the equilibrium, mixture are
$pSO_2 = 0.080$ atm
$pO_2 = 0.90$ atm
$pSO_3 = 5.0$ atm
Calculate k_p for this system. Give your answer to an appropriate number of significant figures. Include the unit.

(3 marks)

2 √x̄ Calculate the value of K_p for the system shown below.
$2NO_2(g) \rightleftharpoons N_2O_4(g)$
At 65 °C the partial pressures of the gases at equilibrium are
$pNO_2 = 0.80$ atm
$pN_2O_4 = 0.25$ atm
Give your answer to three significant figures and include the unit.

(3 marks)

3 √x̄ A chemist analysed the equilibrium system below
$3H_2(g) + N_2(g) \rightleftharpoons 2NH_3(g)$
And found that there was 26.0 moles of NH_3, 13.0 moles of H_2, and 65.0 moles of N_2 present in the equilibrium mixture.
The total pressure of the system was 12.0 atm.
(i) Calculate the mole fraction of each gas at equilibrium.

(1 mark)

(ii) Calculate the partial pressure of each gas at equilibrium.

(1 mark)

(iii) Calculate K_p for this system. Give your answer to three decimal points and include any units.

(3 marks)

4 √x̄ A chemist investigated the equilibrium system below.
$H_2(g) + I_2(g) \rightleftharpoons 2HI(g)$
At 450 °C and a pressure of 3.00 atm.
At equilibrium there was 0.30 mol of H_2, 0.40 mol of I_2, and 1.40 mol of HI.
Calculate the K_p. Give your answer to two significant figures and include any units.

(5 marks)

5 √x̄ Phosphorus pentachloride, PCl_5 decomposes on heating to form phosphorus trichloride PCl_3 and chlorine, Cl_2 according to the equation below.
$PCl_5 \rightleftharpoons PCl_3 + Cl_2$
At a temperature of 350 °C and a pressure of 12.0 atm the amount of gas present at equilibrium was 0.40 mol of PCl_5, 0.75 mol of PCl_3, and 0.90 mol of Cl_2.
Calculate the value of K_p. Give your answer to two significant figures and include any units.

(5 marks)

6 √x̄ Calculate the value of K_p for the system shown below.
$3H_2(g) + N_2(g) \rightleftharpoons 2NH_3(g)$
At 80 °C the partial pressures of the gases at equilibrium are:
$pH_2 = 0.80$ atm
$pN_2 = 0.25$ atm
$pNH_3 = 0.35$ atm
Give your answer to two significant figures and include any units.

(3 marks)

20 Electrode potentials and electrochemical cells
20.1 Electrode potentials and the electrochemical series

Learning objectives:

→ Illustrate how half equations are written for the reactions at an electrode.

→ Explain the term standard electrode potential.

→ Describe how standard electrode potentials are measured.

→ Describe the conventional representation of a cell.

Specification reference: 3.1.11

Synoptic link

You will need to know redox equations studied in Chapter 7, Oxidation, reduction, and redox reactions.

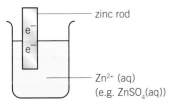

▲ **Figure 2** *A zinc electrode*

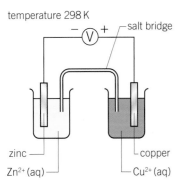

▲ **Figure 3** *Two electrodes connected together with a voltmeter to measure the potential difference*

If you place two different metals in a salt solution and connect them together (Figure 1) an electric current flows so that electrons pass from the more reactive metal to the less reactive. This is the basis of batteries that power everything from MP3 players to milk floats.

This topic and the following two look at how electricity is produced by electrochemical cells and how this can be used to explain and predict redox reactions (which are all about electron transfer).

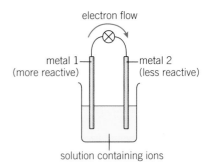

▲ **Figure 1** *Flow of electric current*

Half cells

When a rod of metal is dipped into a solution of its own ions, an equilibrium is set up.

For example, dipping zinc into zinc sulfate solution sets up the following equilibrium:

$$Zn(s) \rightleftharpoons Zn^{2+}(aq) + 2e^-$$

This arrangement is called an electrode, or a half cell, as two half cells can be joined together to make an electrical cell (Figure 2).

If you could measure this potential, it would tell us how readily electrons are released by the metal, that is, how good a reducing agent the metal is. (Remember that reducing agents release electrons.)

However, electrical potential cannot be measured directly, only potential *difference* (often called voltage). What you *can* do is to connect together two different electrodes and measure the potential difference between them with a voltmeter (Figure 3) for copper and zinc electrodes.

The electrical circuit is completed by a **salt bridge**, the simplest form of which is a piece of filter paper soaked in a solution of a salt (usually saturated potassium nitrate). A salt bridge is used rather than a piece of wire, to avoid further metal/ion potentials in the circuit.

If you connect the two electrodes to the voltmeter (Figure 3) you get a potential difference (voltage) of $1.10\,V$ (if the solutions are $1.00\,mol\,dm^{-3}$ and the temperature $298\,K$). The voltmeter shows that the zinc electrode is the more negative.

The fact that the zinc electrode is negative tells you that zinc loses its electrons more readily than does copper – zinc is a better reducing agent. If the voltmeter were removed and electrons allowed to flow, they would do so from zinc to copper. The following changes would take place:

1 Zinc would dissolve to form $Zn^{2+}(aq)$, increasing the concentration of $Zn^{2+}(aq)$.
2 The electrons would flow through the wire to the copper rod where they would combine with $Cu^{2+}(aq)$ ions (from the copper sulfate solution) so depositing fresh copper on the rod and decreasing the concentration of $Cu^{2+}(aq)$.

The following two **half reactions** would take place:

$$Zn(s) \rightarrow Zn^{2+}(aq) + 2e^-$$

and

$$Cu^{2+}(aq) + 2e^- \rightarrow Cu(s)$$

adding: $Zn(s) + Cu^{2+}(aq) + \cancel{2e^-} \rightarrow Zn^{2+}(aq) + Cu(s) + \cancel{2e^-}$

When the two half reactions are added together, the electrons cancel out and you get the overall reaction:

$$Zn(s) + Cu^{2+}(aq) \rightarrow Zn^{2+}(aq) + Cu(s)$$

This is the reaction you get on putting zinc directly into a solution of copper ions. It is a redox reaction with zinc being oxidised and copper ions reduced. If the two half cells are connected they generate electricity. This forms an electrical cell called the Daniell cell (Figure 5).

The hydrogen electrode

To compare the tendency of different metals to release electrons, a standard electrode is needed to which any other half cell can be connected for comparison. The half cell chosen is called the standard hydrogen electrode (Figure 6).

Hydrogen gas is bubbled into a solution of $H^+(aq)$ ions. Since hydrogen doesn't conduct, electrical contact is made via a piece of unreactive platinum metal (coated with finely divided platinum to increase the surface area and allow any reaction to proceed rapidly). The electrode is used under standard conditions of $[H^+(aq)] = 1.00$ mol dm^{-3}, pressure 100 kPa, and temperature 298 K.

> **Hint**
>
> A perfect voltmeter does not allow any current to flow – it merely measures the electrical 'push' or pressure (the potential difference) which tends to make current flow.

> **Study tip**
>
> The salt chosen for the salt bridge must not react with either of the solutions in the half cells.

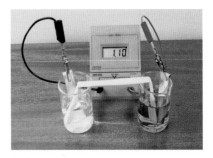

▲ **Figure 4** *Measuring the potential difference of zinc and copper*

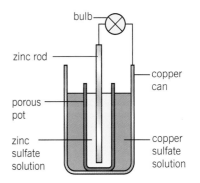

▲ **Figure 5** *A Daniell cell lighting a bulb. The porous pot acts like a salt bridge*

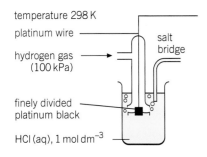

▲ **Figure 6** *The standard hydrogen electrode*

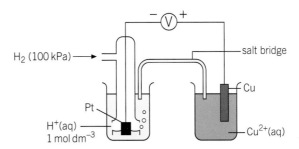

▲ **Figure 7** *Measuring $E^{\ominus}$ for a copper electrode*

The potential of the standard hydrogen electrode is defined as zero, so if it is connected to another electrode (Figure 7), the measured voltage, called the electromotive force E (emf), is the electrode potential of that cell. If the second cell is at standard conditions ([metal ions] = 1.00 mol dm^{-3}, temperature = 298 K), then the emf is given the symbol $E^{\ominus}$. Electrodes with negative values of $E^{\ominus}$ are better at releasing electrons (better reducing agents) than hydrogen.

Changing the conditions, such as the concentration of ions or temperature, of an electrode will change its electrical potential.

The electrochemical series

A list of some $E^{\ominus}$ values for metal/metal ion standard electrodes is given in Table 1.

In Table 1, the equilibria are written as reduction reactions (with the electrons on the left of the arrow). These are called electrode potentials, sometimes known as reduction potentials. (Remember from OIL RIG – Reduction Is Gain.)

Arranged in this order with the most negative values at the top, this list is called the electrochemical series. The number of electrons involved in the reaction has no effect on the value of $E^{\ominus}$.

The voltage obtained by connecting two standard electrodes together is found by the difference between the two $E^{\ominus}$ values. So connecting an Al^{3+}(aq)/Al(s) standard electrode to a Cu^{2+}(aq)/Cu(s) standard electrode would give a voltage of 2.00 V (Figure 8).

▼ **Table 1** *Some E$^{\ominus}$ values. Good reducing agents have negative values for E$^{\ominus}$*

Half reaction	$E^{\ominus}$/V
Li$^+$(aq) + e$^-$ → Li(s)	−3.03
Ca^{2+}(aq) + 2e$^-$ → Ca(s)	−2.87
Al^{3+}(aq) + 3e$^-$ → Al(s)	−1.66
Zn^{2+}(aq) + 2e$^-$ → Zn(s)	−0.76
Pb^{2+}(aq) + 2e$^-$ → Pb(s)	−0.13
2H$^+$(aq) + 2e$^-$ → H$_2$(g)	0.00
Cu^{2+}(aq) + 2e$^-$ → Cu(s)	+0.34
Ag$^+$(aq) + e$^-$ → Ag(s)	+0.80

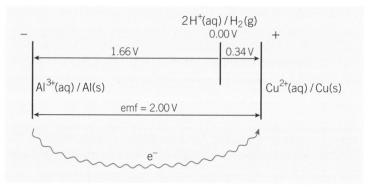

▲ **Figure 8** *Calculating the value of the voltage when two electrodes are connected*

If you connect an Al^{3+}(aq)/Al(s) standard electrode to a Cu^{2+}(aq)/Cu(s) standard electrode, the emf will be 2.00 V and the Al^{3+}(aq)/Al(s) electrode will be negative.

If you connect an Al^{3+}(aq)/Al(s) standard electrode to a Pb^{2+}(aq)/Pb(s) standard electrode the emf will be 1.53 V and the Al^{3+}(aq)/Al(s) electrode will be the negative electrode of the cell (Figure 9).

It is worth sketching diagrams like the ones shown in Figures 8 and Figure 9. It will prevent you getting confused with signs. Remember that more negative values are drawn to the left on the diagrams.

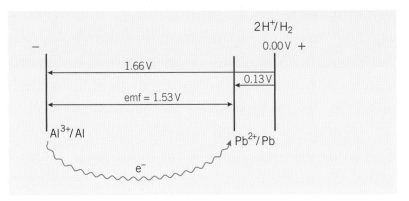

▲ **Figure 9** *Calculating the value of the emf for an Al^{3+}/Al electrode connected to a Pb^{2+}/Pb electrode*

For example, if you connect an $Al^{3+}(aq)/Al(s)$ standard electrode to a $Zn^{2+}(aq)/Zn(s)$ standard electrode (Figure 10) the voltmeter will read 0.90 V and the $Al^{3+}(aq)/Al(s)$ electrode will be negative.

Representing cells

There is shorthand for writing down the cell formed by connecting two electrodes. The conventions are those recommended by IUPAC (International Union of Pure and Applied Chemistry). The usual apparatus diagram is shown in Figure 10 and the cell diagram is written using the following conventions:

- A vertical solid line indicates a **phase boundary**, for example, between a solid and a solution.

- A double vertical line shows a salt bridge.

- The species with the highest oxidation state is written next to the salt bridge.

- When giving the value of the emf $E^{\ominus}$ state the polarity (i.e., whether it is positive or negative) of the right-hand electrode, as the cell representation is written. In the case of the aluminium and copper cells in Figure 8 the copper half cell is more positive (it is connected to the positive terminal of the voltmeter) and, if allowed to flow, electrons would go from aluminium to copper.

$Al(s)|Al^{3+}(aq)||Cu^{2+}(aq)|Cu(s)$ $\qquad$ $E^{\ominus}_{cell} = +2.00\,V$

You could also have written the cell:

$Cu(s)|Cu^{2+}(aq)||Al^{3+}(aq)|Al(s)$ $\qquad$ $E^{\ominus}_{cell} = -2.00\,V$

This still tells us that electrons flow from aluminium to copper as the polarity of the right-hand electrode is always given.

So emf $= E^{\ominus}(R) - E^{\ominus}(L)$

Where $E^{\ominus}(R)$ represents the emf of the right-hand electrode and $E^{\ominus}(L)$ that of the left-hand electrode.

The cell representation for a silver electrode connected to a $Pb^{2+}(aq)/Pb(s)$ half cell would be:

$Pb(s)|Pb^{2+}(aq)||Ag^{+}(aq)|Ag(s)$ $\qquad$ $E^{\ominus}_{cell} = +0.93\,V$

Study tip

The phase indicates a state of matter such as solid, liquid, or gas. It also refers to aqueous solutions.

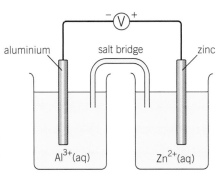

▲ **Figure 10** *Connecting a pair of electrodes*

Summary questions

1 a Represent the following on a conventional cell diagram:

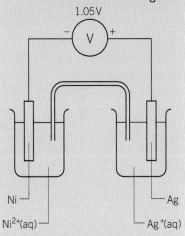

  b If the voltmeter was replaced by a wire, in which direction would the electrons flow? Write equations for the reactions occurring in each beaker and write an equation for the overall cell reaction.

2 $\sqrt{x}$ Calculate $E^{\ominus}_{cell}$ for:

  a $Zn(s)|Zn^{2+}(aq)\ ||$
    $Pb^{2+}(aq)|Pb(s)$

  b $Pb(s)|Pb^{2+}(aq)\ ||\ Zn^{2+}(aq)$
    $Zn(s)$

It is possible to use standard electrode potentials to decide on the feasibility of a redox (i.e., electron transfer) reaction. When you connect a pair of electrodes, the electrons will flow from the more negative to the more positive and not in the opposite direction. So the signs of the electrodes tell us the direction of a redox reaction.

Think of the following electrodes (Figure 1):

$$Zn^{2+}(aq) + 2e^- \rightleftharpoons Zn(s) \quad \text{written in short as } Zn^{2+}(aq)/Zn(s)$$

$$E^\ominus = -0.76 \text{ V and}$$

$$Cu^{2+}(aq) + 2e^- \rightleftharpoons Cu(s) \quad \text{written in short as } Cu^{2+}(aq)/Cu(s)$$

$$E^\ominus = +0.34 \text{ V}$$

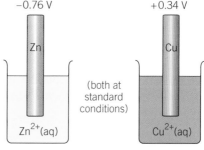

▲ **Figure 1** *The two electrodes. Their potentials are measured with respect to a standard hydrogen electrode*

Figure 2 shows these two electrodes connected together. Electrons will tend to flow from zinc (the more negative) to copper (the more positive).

So you know which way the two half reactions must go. You can use a diagram to represent the cell and if you then include $E^\ominus$ values for the two electrodes, you can find the emf for the cell. Figure 3 is the diagram for the zinc/copper cell. You can see how it is related to the apparatus.

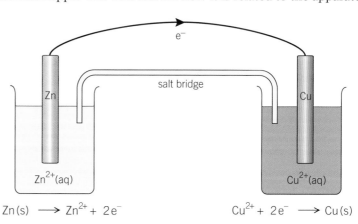

$$Zn(s) \longrightarrow Zn^{2+} + 2e^- \qquad Cu^{2+} + 2e^- \longrightarrow Cu(s)$$

▲ **Figure 2** *Connecting $Zn^{2+}(aq)/Zn(s)$ and $Cu^{2+}(aq)/Cu(s)$ electrodes*

So the two equations are:

$$Zn(s) \rightarrow Zn^{2+}(aq) + 2e^-$$

$$Cu^{2+}(aq) + 2e^- \rightarrow Cu(s)$$

The overall effect is:

$$Zn(s) \rightarrow Zn^{2+}(aq) + 2e^-$$

$$Cu^{2+}(aq) + 2e^- \rightarrow Cu(s)$$
$$\overline{Cu^{2+}(aq) + Zn(s) + \cancel{2e^-} \rightarrow Cu(s) + Zn^{2+}(aq) + \cancel{2e^-}}$$

So this reaction is feasible and is the reaction that actually happens, either by connecting the two electrodes or more directly by adding Zn to $Cu^{2+}(aq)$ ions in a test tube. The reverse reaction is *not* feasible and does not occur.

$$Cu(s) + Zn^{2+}(aq) \rightarrow Zn(s) + Cu^{2+}(aq)$$

This reaction would require electrons to flow from positive to negative.

You can go through this process whenever you want to predict the outcome of a redox reaction.

With redox systems that only involve metal ions but no metal (e.g., Fe^{3+} / Fe^{2+}), a beaker containing all the relevant ions and a platinum electrode to make electrical contact is used in order to measure $E^\ominus$ by connecting to a hydrogen electrode (Figure 4).

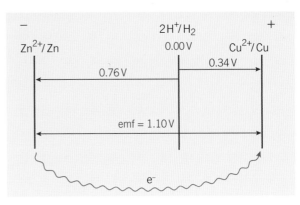

▲ **Figure 3** *Predicting the direction of electron flow when a Zn^{2+}/Zn electrode is connected to a Cu^{2+}/Cu electrode*

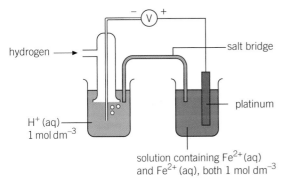

▲ **Figure 4**

The cell diagram would be written as follows:

$$Pt \mid H_2(g) \mid 2H^+(aq) \parallel Fe^{3+}(aq), Fe^{2+}(aq) \mid Pt$$

Remember the rule that the most oxidised species (in this case H^+ and Fe^{3+}) go next to the salt bridge.

Further examples of predicting the direction of redox reactions

You can extend the electrochemical series to systems other than simple metal / metal ion ones (Table 1).

▼ **Table 1** $E^\ominus$ *values for more reduction half equations*

Reduction half equation	$E^\ominus / V$
$Li^+(aq) + e^- \rightarrow Li(s)$	−3.03
$Ca^{2+}(aq) + 2e^- \rightarrow Ca(s)$	−2.87
$Al^{3+}(aq) + 3e^- \rightarrow Al(s)$	−1.66
$Zn^{2+}(aq) + 2e^- \rightarrow Zn(s)$	−0.76
$Cr^{3+}(aq) + e^- \rightarrow Cr^{2+}(aq)$	−0.41
$Pb^{2+}(aq) + 2e^- \rightarrow Pb(s)$	−0.13
$2H^+(aq) + 2e^- \rightarrow H_2(g)$	0.00
$Cu^{2+}(aq) + e^- \rightarrow Cu+(aq)$	+0.15
$Cu^{2+}(aq) + 2e^- \rightarrow Cu(s)$	+0.34
$I_2(aq) + 2e^- \rightarrow 2I^-(aq)$	+0.54
$Fe^{3+}(aq) + e^- \rightarrow Fe^{2+}(aq)$	+0.77
$Ag^+(aq) + e^- \rightarrow Ag(s)$	+0.79
$Br_2(aq) + 2e^- \rightarrow 2Br^-(aq)$	+1.07
$Cl_2(aq) + 2e^- \rightarrow 2Cl^-(aq)$	+1.36
$MnO_4^- + 8H^+(aq) + 5e^- \rightarrow Mn^{2+}(aq) + 4H_2O(l)$	+1.51
$Ce^{4+}(aq) + e^- \rightarrow Ce^{3+}(aq)$	+1.70

Worked example: emf for an iron-chlorine electrochemical cell

Will the following reaction occur or not?

$$Fe^{3+}(aq) + Cl^-(aq) \rightarrow Fe^{2+}(aq) + \frac{1}{2} Cl_2(aq)$$

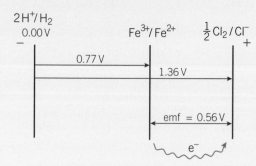

▲ **Figure 4** *Working out the emf for $Fe^{3+}(aq) + Cl^-(aq) \rightarrow Fe^{2+}(aq) + \frac{1}{2} Cl_2(aq)$*

Figure 4 shows that the emf is 0.53 V with iron the more negative. So, electrons will flow from the Fe^{3+}/Fe^{2+} standard electrode to the $\frac{1}{2} Cl_2/Cl^-$ standard electrode.

$$Fe^{2+} \rightarrow Fe^{3+} + e^-$$
$$\frac{1}{2}Cl_2 + e^- \rightarrow Cl^-$$

$$Fe^{2+} + \frac{1}{2}Cl_2 + \cancel{e^-} \rightarrow Fe^{3+} + Cl^- + \cancel{e^-}$$

This is the reaction that will occur. The following reaction is not feasible.

$$Fe^{3+}(aq) + Cl^-(aq) \not\rightarrow Fe^{2+}(aq) + \frac{1}{2}Cl_2(aq)$$

So, chlorine will oxidise iron(II) ions to iron(III) ions.

Hint

$E^{\ominus}$ values tell you whether a reaction is feasible or not. It doesn't give any information about the speed of the reaction. A feasible reaction may be so slow that in practice it does not take place at all at room temperature.

Worked example: emf for an iron-iodine electrochemical cell

Will the following reaction occur or not?

$$Fe^{3+}(aq) + I^-(aq) \rightarrow Fe^{2+}(aq) + \frac{1}{2} I_2(aq)$$

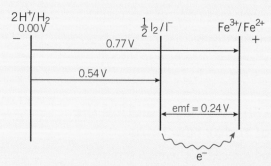

▲ **Figure 5** *Working out the emf for $Fe^{3+}(aq) + I^-(aq) \rightarrow Fe^{2+}(aq) + \frac{1}{2}I_2(aq)$*

Figure 5 shows that the emf is 0.23 V with iodine the more negative.

So, electrons will flow from the $\frac{1}{2}$ I_2/I^- electrode to the Fe^{3+}/Fe^{2+} electrode.

$$I^- \rightarrow \frac{1}{2} I_2 + e^-$$
$$Fe^{3+} + e^- \rightarrow Fe^{2+}$$

$$I^- + Fe^{3+} + \cancel{e^-} \rightarrow \frac{1}{2} I_2 + Fe^{2+} + \cancel{e^-}$$

This is the reaction that will occura. The following reaction is not feasible.

$$Fe^{2+}(aq) + \frac{1}{2} I_2(aq) \not\rightarrow Fe^{3+}(aq) + I^-(aq)$$

So, iron(III) ions will oxidise iodide ions to iodine.

Hint

$$E(R) - E(L) = emf$$

If the emf is positive then the reaction is feasible.

Summary questions

1 $\sqrt{x}$ What will be the value of the emf for an $Al^{3+}(aq)/Al(s)$ standard electrode connected to a $Zn^{2+}(aq)/Zn(s)$ standard electrode? Draw a diagram like Figure 3 to illustrate your answer.

2 $\sqrt{x}$ Use the values of $E^{\ominus}$ in Table 1 to calculate the emf for the following:

   a $Ce^{4+}(aq) + Fe^{2+}(aq) \rightarrow Ce^{3+}(aq) + Fe^{3+}(aq)$

   b $I_2(aq) + 2Br^-(aq) \rightarrow Br_2(aq) + 2I^-(aq)$

   c $MnO_4^-(aq) + 8H^+(aq) + 5I^-(aq) \rightarrow Mn^{2+}(aq) + 4H_2O(l) + 2\frac{1}{2}I_2(aq)$

   d $2H^+(aq) + Pb(s) \rightarrow Pb^{2+}(aq) + H_2(g)$

3 Which of the halogens could possibly oxidise $Ag(s)$ to $Ag^+(aq)$ ions?

4 a Is the reaction $Br_2(aq) + 2Cl^-(aq) \rightarrow Cl_2(aq) + 2Br^-(aq)$ feasible?

   b Is the reaction $Fe^{3+}(aq) + Br^-(aq) \rightarrow Fe^{2+}(aq) + \frac{1}{2}Br_2(aq)$ feasible?

Learning objectives

→ Describe the differences between non-rechargeable, rechargeable, and fuel cells.

→ Describe the electrode reactions in a hydrogen–oxygen fuel cell.

→ Describe the benefits and risks to society associated with each type of cell.

Specification reference: 3.1.11

▲ **Figure 1**

▲ **Figure 2** *In the Victorian era, Morse code telegraphic communications used Daniell cells*

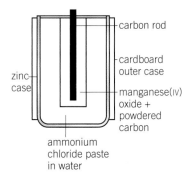

▲ **Figure 3** *A Leclanché cell*

Modern life would not be the same without batteries – both rechargeable and single use for tablets, phones, MP3 players, and so on. Nowadays there is a huge variety of types and brands advertised with slogans like long life and high power. Batteries are based on the principles of electrochemical cells. Strictly, a battery refers to a number of cells connected together, but in everyday speech the word has come to mean almost any portable source of stored electricity. You will need to be able to apply the principles of electrochemical cells, but you will not be expected to learn the details of the construction of the cells described below.

Non-rechargeable cells

Zinc/copper cells

The Daniell cell provides an emf of 1.1 V. It was developed by the British chemist John Daniell in the 1830s and was used to provide the electricity for old fashioned telegraphs which sent messages by Morse code (Figure 2). However, it was not practical for portable devices, because of the liquids that it contained. It works on the general principle of electrons being transferred from a more reactive metal to a less reactive one. The voltage can be worked out from the difference between the electrode potentials in the electrochemical series.

Zinc/carbon cells

The electrodes can be made from materials other than metals. For example, in the Leclanché cell (which is named after the Frenchman George Leclanché), the positive electrode is carbon which acts like the inert platinum electrode in the hydrogen electrode (Figure 3). The Leclanché cell is the basis of most ordinary disposable batteries. The electrolyte is a paste rather than a liquid.

The commercial form of this type of cell consists of a zinc canister filled with a paste of ammonium chloride, NH_4Cl, and water – the electrolyte. In the centre is a carbon rod. It is surrounded by a mixture of manganese(IV) oxide and powdered carbon. The half equations are:

$$Zn(s) \rightleftharpoons Zn^{2+}(aq) + 2e^- \qquad E \approx -0.8\,V$$
$$2NH_4^+(aq) + 2e^- \rightleftharpoons 2NH_3(g) + H_2(g) \qquad E \approx +0.7\,V$$

These are not $E^\ominus$ values as the conditions are far from standard.

The reactions that take place are:

- at the zinc:

$$Zn(s) \rightarrow Zn^{2+}(aq) + 2e^-$$

- at the carbon rod:

$$2NH_4^+(aq) + 2e^- \rightarrow 2NH_3(g) + H_2(g)$$

So the overall reaction as the cell discharges is:

$$2NH_4^+(aq) + Zn(s) \rightarrow 2NH_3(g) + H_2(g) + Zn^{2+}(aq)$$

emf ≈ 1.5 V with the zinc as the negative terminal

The labels in Figure 3: carbon rod; cardboard outer case; zinc case; manganese(IV) oxide + powdered carbon; ammonium chloride paste in water

The hydrogen gas is oxidised to water by the manganese(IV) oxide (preventing a build up of pressure), whilst the ammonia dissolves in the water of the paste.

As the cell discharges, the zinc is used up and the walls of the zinc canister become thin and prone to leakage. The ammonium chloride electrolyte is acidic and can be corrosive. That is why you should remove spent batteries from equipment. This cell is ideal for doorbells, for example which need a small current intermittently.

A variant of this cell is the zinc chloride cell. It is similar to the Leclanché but uses zinc chloride as the electrolyte. Such cells are better at supplying high currents than the Leclanché and are marketed as extra life batteries for radios, torches, and shavers.

Long life alkaline batteries are also based on the same system, but with an electrolyte of potassium hydroxide. Powdered zinc is used, whose greater surface area allows the battery to supply high currents. The cell is enclosed in a steel container to prevent leakage. These cells are suitable for equipment taking continuous high currents such as personal stereos. In this situation they can last up to 16 times as long as ordinary zinc/carbon batteries, but they are more expensive.

Many other electrode systems are in use, especially for miniature batteries such as those used in watches, hearing aids, cameras, and electronic equipment. These include zinc/air, mercury(II) oxide/zinc, silver oxide/zinc, and lithium/manganese(IV) oxide. Which is used for which application depends on the precise requirements of voltage, current, size, and cost.

Rechargeable batteries

These can be recharged by reversing the cell reactions. This is done by applying an external voltage greater than the voltage of the cell to drive the electrons in the opposite direction.

Lead–acid batteries

Lead–acid batteries are rechargeable batteries used to operate the starter motors of cars. They consist of six 2 V cells connected in series to give 12 V. Each cell consists of two plates dipped into a solution of sulfuric acid. The positive plate is made of lead coated with lead(IV) oxide, PbO_2, and the negative plate is made of lead (Figure 4).

On discharging, the following reactions occur as the battery drives electrons from the lead plate to the lead(IV) oxide coated one.

At the lead plate:

$$Pb(s) + SO_4^{2-}(aq) \rightarrow PbSO_4(s) + 2e^-$$

At the lead-dioxide-coated plate:

$$PbO_2(s) + 4H^+(aq) + SO_4^{2-}(aq) + 2e^- \rightarrow PbSO_4(s) + 2H_2O(l)$$

The overall reaction as the cell discharges is:

$$PbO_2(s) + 4H^+(aq) + 2SO_4^{2-}(aq) + Pb(s) \rightarrow 2PbSO_4(s) + 2H_2O(l)$$

$$emf \approx 2 V$$

These reactions are reversed as the battery is charged up and electrons flow in the reverse direction, driven by the car's generator.

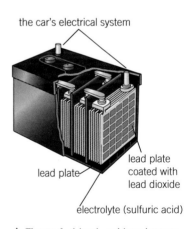

the car's electrical system

lead plate

lead plate coated with lead dioxide

electrolyte (sulfuric acid)

▲ **Figure 4** *A lead–acid car battery*

Hint

Technically, a battery consists of two or more simple cells connected together. So a 1.5 V zinc–carbon battery is really a cell, whilst a car battery is a true battery.

▲ **Figure 5** *Some of the enormous variety of batteries available today*

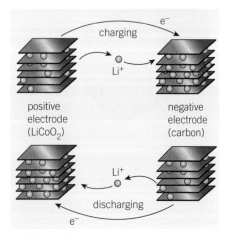

▲ **Figure 6**

Portable batteries

There are now rechargeable batteries that come in all shapes and sizes.

Nickel/cadmium

These are now available in standard sizes to replace traditional zinc–carbon batteries. Although more expensive to buy, they can be recharged up to 500 times, reducing the effective cost significantly. These cells are called nickel/cadmium and have an alkaline electrolyte. The two half equations are:

$$Cd(OH)_2(s) + 2e^- \rightleftharpoons Cd(s) + 2OH^-(aq)$$
$$NiO(OH)(s) + H_2O(l) + e^- \rightleftharpoons Ni(OH)_2(s) + OH^-(aq)$$

Overall:

$$2NiO(OH)(s) + Cd(s) + 2H_2O(l) \rightleftharpoons 2Ni(OH)_2(s) + Cd(OH)_2(s)$$
$$\text{emf} \approx +1.2\,V$$

The reaction goes from left to right on discharge (electrons flowing from Cd to Ni) and right to left on charging.

Lithium ion

The rechargeable lithium ion cell is used in laptops, tablets, smartphones, and other mobile gadgets. It is light because lithium is the least dense metal. The electrolyte is a solid polymer rather than a liquid or paste so it cannot leak, and its charge can be topped up at any time without the memory effect of some other rechargeable batteries, which can only be recharged efficiently when they have been fully discharged. The cell can even be bent or folded without leaking.

The positive electrode is made of lithium cobalt oxide, $LiCoO_2$, and the negative electrode is carbon. These are arranged in layers with a sandwich of solid electrolyte in between. On charging, electrons are forced through the external circuit from positive to negative electrode and at the same time lithium ions move through the electrolyte towards the positive electrode to maintain the balance of charge.

The reactions that occur on charging are:

negative electrode: $Li^+ + e^- \rightarrow Li$ $\quad E^\ominus = -3\,V$

positive electrode: $Li^+ + CoO_2 + e^- \rightarrow Li^+(CoO_2)^-$ $\quad E^\ominus = +1\,V$

On discharging, the processes are reversed so electrons flow from negative to positive. A single cell gives a voltage of between 3.5 V and 4.0 V, compared with around 1.5 V for most other cells.

Alkaline hydrogen–oxygen fuel cell

The reactions that occur in an alkaline hydrogen–oxygen fuel cell are:

$$2H_2(g) + 4OH^-(aq) \rightarrow 4H_2O(l) + 4e^- \quad E^\ominus = -0.83\,V$$
$$O_2(g) + 2H_2O(l) + 4e^- \rightarrow 4OH^-(aq) \quad E^\ominus = +0.40\,V$$

Overall:

$$2H_2(g) + O_2(g) \rightarrow 2H_2O(l) \quad E = 1.23\,V$$

The cell has two electrodes of a porous platinum-based material. They are separated by a semi-permeable membrane and the electrolyte is sodium hydroxide solution. Hydrogen enters at the negative electrode and the following half reaction takes place:

$$2H_2(g) + 4OH^-(aq) \rightarrow 4H_2O(l) + 4e^- \qquad E^\ominus = -0.83\,V$$

This releases electrons, which flow through the circuit to the other electrode where oxygen enters and the following reaction takes place:

$$O_2(g) + 2H_2O(l) + 4e^- \rightarrow 4OH^-(aq) \qquad E^\ominus = +0.4\,V$$

This accepts electrons from the other electrode and releases OH^- ions which travel through the semi-permeable membrane to that electrode.

The overall effect is for hydrogen and oxygen to react together to produce water and generate an emf of 1.23 V.

$$O_2(g) + 4H_2(g) \rightarrow 2H_2O(l) \qquad emf = 1.23\,V$$

This is the same reaction as burning hydrogen and oxygen but it takes place at a low temperature so there is no production of nitrogen oxides, which form if hydrogen is burnt directly.

This type of fuel cell is used to generate electricity on spacecraft because the only by-product is pure water, which can be used as drinking water by the astronauts. In terrestrial use it is important because unlike many other sources of electrical energy, it produces no carbon dioxide.

The hydrogen economy

At first sight, this type of fuel cell appears to be very 'green' because the only product is water. However you have to consider the source of the hydrogen. At present, most hydrogen is made from crude oil – a non-renewable resource. It could be made by electrolysis of water but most electricity is made by burning fossil fuels, which emit carbon dioxide. Also, hydrogen-powered vehicles will need an infrastructure of hydrogen filling stations to be built, which also raises the issues of storing and transporting a highly flammable gas. For example, in the 1930s two airships which used hydrogen as their lifting gas – the Hindenberg and the R101 – exploded with serious loss of life. A major problem is storing hydrogen because it is a gas. For example, burning 1 g of hydrogen gives out around three times as much energy as burning 1 g of petrol but the hydrogen takes up around 8000 times as much space.

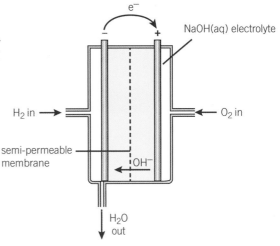

▲ Figure 7

Study tip

The hydrogen/oxygen fuel cell does not need to be electrically recharged.

▲ Figure 8 *A fuel-cell powered vehicle*

Hydrogen storage

One way of storing hydrogen safely is by absorbing it into solid compounds called metal hydrides. The hydrogen is absorbed under pressure and released by gentle heating.

Summary question

1 List the advantages and disadvantages of each cell, described with regard to cost, practicality, safety, and the environment.

Practice questions

1 Hydrogen–oxygen fuel cells can operate in acidic or in alkaline conditions but commercial cells use porous platinum electrodes in contact with concentrated aqueous potassium hydroxide. The table below shows some standard electrode potentials measured in acidic and in alkaline conditions.

Half-equation	$E^{\ominus}$/V
$O_2(g) + 4H^+(aq) + 4e^- \rightarrow 2H_2O(l)$	+1.23
$O_2(g) + 2H_2O(l) + 4e^- \rightarrow 4OH^-(aq)$	+0.40
$2H^+(aq) + 2e^- \rightarrow H_2(g)$	0.00
$2H_2O(l) + 2e^- \rightarrow 2OH^-(aq) + H_2(g)$	−0.83

(a) State why the electrode potential for the standard hydrogen electrode is equal to 0.00 V. *(1 mark)*

(b) ✓𝓍 Use data from the table to calculate the e.m.f. of a hydrogen–oxygen fuel cell operating in alkaline conditions. *(1 mark)*

(c) Write the conventional representation for an alkaline hydrogen–oxygen fuel cell. *(2 marks)*

(d) ✓𝓍 Use the appropriate half-equations to construct an overall equation for the reaction that occurs when an alkaline hydrogen–oxygen fuel cell operates. Show your working. *(2 marks)*

(e) Give **one** reason, other than cost, why the platinum electrodes are made by coating a porous ceramic material with platinum rather than by using platinum rods. *(1 mark)*

(f) Suggest why the e.m.f. of a hydrogen–oxygen fuel cell, operating in acidic conditions, is exactly the same as that of an alkaline fuel cell. *(1 mark)*

(g) Other than its lack of pollution, state briefly the main advantage of a fuel cell over a rechargeable cell such as the nickel–cadmium cell when used to provide power for an electric motor that propels a vehicle. *(1 mark)*

(h) Hydrogen–oxygen fuel cells are sometimes regarded as a source of energy that is carbon neutral. Give **one** reason why this may **not** be true.

(1 mark)

AQA, 2010

2 Where appropriate, use the standard electrode potential data in the table below to answer the questions which follow.

Standard electrode potential	$E^{\ominus}$/V
$Zn^{2+}(aq) + 2e^- \rightarrow Zn(s)$	−0.76
$V^{3+}(aq) + e^- \rightarrow V^{2+}(aq)$	−0.26
$SO_4^{2-}(aq) + 2H^+(aq) + 2e^- \rightarrow SO_3^{2-}(aq) + H_2O(l)$	+0.17
$VO^{2+}(aq) + 2H^+(aq) + e^- \rightarrow V^{3+}(aq) + H_2O(l)$	+0.34
$Fe^{3+}(aq) + e^- \rightarrow Fe^{2+}(aq)$	+0.77
$VO_2^+(aq) + 2H^+(aq) + e^- \rightarrow VO^{2+}(aq) + H_2O(l)$	+1.00
$Cl_2(aq) + 2e^- \rightarrow 2Cl^-(aq)$	+1.36

(a) From the table above select the species which is the most powerful reducing agent. *(1 mark)*

(b) From the table above select
  (i) a species which, in acidic solution, will reduce $VO_2^+(aq)$ to $VO^{2+}(aq)$ but will **not** reduce $VO^{2+}(aq)$ to $V^{3+}(aq)$,
  (ii) a species which, in acidic solution, will oxidise $VO^{2+}(aq)$ to $VO_2^+(aq)$.

(2 marks)

(c) The cell represented below was set up under standard conditions.

Pt|Fe^{2+}(aq), Fe^{3+}(aq)||Tl^{3+}(aq), Tl^{+}(aq)|Pt   Cell emf = + 0.48 V

(i) Deduce the standard electrode potential for the following half reaction.

Tl^{3+}(aq) + 2e$^-$ → Tl^{+}(aq)

(ii) Write an equation for the spontaneous cell reaction.

(3 marks)

AQA, 2005

3 **Table 3** shows some standard electrode potential data.

	$E^{\ominus}$/V
ZnO(s) + H$_2$O(l) + 2e$^-$ → Zn(s) + 2OH$^-$(aq)	−1.25
Fe^{2+}(aq) + 2e$^-$ → Fe(s)	−0.44
O$_2$(g) + 2H$_2$O(l) + 4e$^-$ → 4OH$^-$(aq)	+0.40
2HOCl(aq) + 2H$^+$(aq) + 2e$^-$ → Cl$_2$(g) + 2H$_2$O(l)	+1.64

(a) Give the conventional representation of the cell that is used to measure the standard electrode potential of iron as shown in **Table 3**.

(2 marks)

(b) With reference to electrons, give the meaning of the term reducing agent.

(1 mark)

(c) Identify the weakest reducing agent from the species in **Table 3**. Explain how you deduced your answer.

(2 marks)

(d) When HOCl acts as an oxidising agent, one of the atoms in the molecule is reduced.

(i) Place a tick next to the atom that is reduced.

Atom that is reduced	Tick (√)
H	
O	
Cl	

(ii) Explain your answer to Question 3(d)(i) in terms of the change in the oxidation state of this atom.

(1 mark)

(e) Using the information given in **Table 3**, deduce an equation for the redox reaction that would occur when hydroxide ions are added to HOCl.

(2 marks)

(f) √x̄ The half-equations from **Table 3** that involve zinc and oxygen are simplified versions of those that occur in hearing aid cells. A simplified diagram of a hearing aid cell is shown in **Figure 1**.

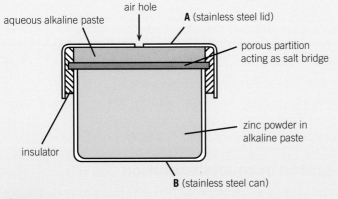

▲ **Figure 1**

Use data from **Table 3** to calculate the e.m.f. of this cell.

(1 mark)

AQA, 2014

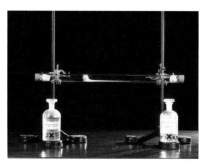

▲ **Figure 1** The white ring of ammonium chloride is formed when hydrogen chloride (left) and ammonia (right) react

The Brønsted–Lowry description of acidity (developed in 1923 by Thomas Lowry and Johannes Brønsted independently) is the most generally useful current theory of acids and bases.

An acid is a substance that can donate a proton (H^+ ion) and a base is a substance that can accept a proton.

Proton transfer

Hydrogen chloride gas and ammonia gas react together to form ammonium chloride – a white ionic solid:

$$HCl(g) \ + \ NH_3(g) \ \rightarrow \ NH_4Cl(s)$$
hydrogen   ammonia   ammonium chloride
chloride

Here, hydrogen chloride is acting as an acid by donating a proton to ammonia. Ammonia is acting as a base by accepting a proton. Acids and bases can only react in pairs – one acid and one base.

So, you could think of the reaction in these terms:

$$HCl(g) \ + \ NH_3(g) \ \rightarrow \ [NH_4^+Cl^-](s)$$
acid   base

Another example is a mixture of concentrated sulfuric acid, H_2SO_4, and concentrated nitric acid, HNO_3. They behave as an acid–base pair:

$$H_2SO_4 \ + \ HNO_3 \ \rightarrow \ H_2NO_3^+ \ + \ HSO_4^-$$

Sulfuric acid donates a proton to nitric acid, so is acting as the acid, whilst in this example nitric acid is acting as a base. In fact, whether a species is acting as an acid or a base depends on the reactants. Water is a good example of this.

Water as an acid and a base

Hydrogen chloride can donate a proton to water, so that water acts as a base:

$$HCl \ + \ H_2O \ \rightarrow \ H_3O^+ \ + \ Cl^-$$

H_3O^+ is called the **oxonium ion**, but the names hydronium ion and hydroxonium ion are also used.

Water may also act as an acid. For example:

$$H_2O \ + \ NH_3 \ \rightarrow \ OH^- \ + \ NH_4^+$$

Here water is donating a proton to ammonia.

The proton in aqueous solution

It is important to realise that the H^+ ion is just a proton. The hydrogen atom has only one electron and if this is lost all that remains is a proton (the hydrogen nucleus). This is about 10^{-15} m in diameter, compared to 10^{-10} m or more for any other chemical entity. This extremely small size and consequent intense electric field cause it to have unusual properties compared with other positive ions. It is never found isolated. In aqueous solutions it is always bonded to at least one water molecule to form the ion H_3O^+. For simplicity, protons are represented in an aqueous solution by $H^+(aq)$ rather than $H_3O^+(aq)$.

Since the H^+ ion has no electrons of its own, it can only form a bond with another species that has a lone pair of electrons.

The ionisation of water

Water is slightly ionised:

$$H_2O(l) \rightleftharpoons H^+(aq) + OH^-(aq)$$

This may be written:

$$H_2O(l) + H_2O(l) \rightleftharpoons H_3O^+(aq) + OH^-(aq)$$

This emphasises that this is an acid–base reaction in which one water molecule donates a proton to another.

This equilibrium is established in water and all aqueous solutions:

$$H_2O(l) \rightleftharpoons H^+(aq) + OH^-(aq)$$

You can write an equilibrium expression:

$$K_c = \frac{[H^+(aq)][OH^-(aq)]}{[H_2O(l)]}$$

The concentration of water $[H_2O(l)]$ is constant and is incorporated into a modified equilibrium constant K_w, where $K_w = K_c \times [H_2O(l)]$.

So,
$$K_w = [H^+(aq)] [OH^-(aq)]$$

K_w is called the **ionic product** of water and at 298 K it is equal to 1.0×10^{-14} mol^2 dm^{-6}. Each H_2O that dissociates (splits up) gives rise to one H^+ and one OH^- so, in pure water, at 298 K:

$$[OH^-(aq)] = [H^+(aq)]$$

So,
$$1.0 \times 10^{-14} = [H^+(aq)]^2$$

$$[H^+(aq)] = 1.0 \times 10^{-7} \text{ mol dm}^{-3} = [OH^-(aq)] \text{ at 298 K (25 °C)}$$

The concentration of water $\sqrt{x}$

What is the concentration of water? This question often catches out even experienced chemists. 1 dm^3 of water weighs 1000 g. The M_r of water is 18.0

1   How many moles of water is 1000 g
2   What does this make the concentration of water in mol dm^{-3}?

2 55.5 mol dm^{-3}
1 55.5 moles

Synoptic link

Look back at Topic 6.4, The Equilibrium constant K_c, to revise equilibrium constants.

Study tip

The important point to remember at this stage is that the product of $[H^+(aq)]_{eqm}$ and $[OH^-(aq)]_{eqm}$ is constant at any given temperature so that if the concentration of one of these ions increases, the other must decrease proportionately.

Summary questions

1   Identify which reactant is an acid and which a base in the following:

    a $HNO_3 + OH^- \rightarrow NO_3^- + H_2O$

    b $CH_3COOH + H_2O \rightarrow CH_3COO^- + H_3O^+$

2   At 298 K in an acidic solution, $[H^+]$ is 1×10^{-4} mol dm^{-3}. What is $[OH^-(aq)]$?

3   What species are formed when the following bases accept a proton?

    a OH^- b NH_3 c H_2O d Cl^-

21.2 The pH scale

Learning objectives:

→ Define pH.

→ Explain why a logarithmic scale is used.

→ Describe how pH is measured.

→ Describe how the pH of a solution is used to find the concentration of $H^+(aq)$ and $OH^-(aq)$ ions.

→ Describe how the pH of a solution is calculated from the concentration of $H^+(aq)$ ions.

Specification reference: 3.1.12

Study tip

Always give pH values to two decimal places.

Maths link 🖩

You must be able to use your calculator to look up logs to the base 10 ($\log_{10}$) and antilogs. See Section 5, Mathematical skills, if you are not sure about these.

The acidity of a solution depends on the concentration of $H^+(aq)$ and is measured on the pH scale.

$$pH = -\log_{10}[H^+(aq)]$$

Remember that square brackets, [ ], mean the concentration in $mol\,dm^{-3}$.

How the pH scale was invented

Did you know that the pH scale was first introduced by a brewer? In 1909, the Danish biochemist Søren Sørenson was working for the Carlsberg company studying the brewing of beer. Brewing requires careful control of acidity to produce conditions in which yeast (which aids the fermentation process) will grow but unwanted bacteria will not. The concentrations of acid with which Sørenson was working were very small, such as one ten-thousandth of a mole per litre, and so he looked for a way to avoid using numbers such as 0.0001 (1×10^{-4}). Taking the $\log_{10}$ of this number gave -4, and for further convenience he took the negative of it giving 4. So the pH scale was born.

This expression is more complicated than simply stating the concentration of $H^+(aq)$. However, using the logarithm of the concentration does away with awkward numbers like 10^{-13}, etc., which occur because the concentration of $H^+(aq)$ in most aqueous solutions is so small. The minus sign makes almost all pH values positive (because the logs of numbers less than 1 are negative).

On the pH scale:

- The *smaller* the pH, the *greater* the concentration of $H^+(aq)$.
- A difference of *one* pH number means a *tenfold* difference in $[H^+]$ so that, for example, pH 2 has ten times the H^+ concentration of pH 3.

Remember that, at 298 K, $K_w = [H^+(aq)][OH^-(aq)]$ $= 1.00 \times 10^{-14}\ mol^2\,dm^{-6}$. This means that in neutral aqueous solutions:

$$[H^+(aq)] = [OH^-(aq)] = 1.0 \times 10^{-7}\ mol\,dm^{-3}$$

$$pH = -\log_{10}[H^+(aq)] = -\log_{10}[1.0 \times 10^{-7}] = 7.00$$

so the pH is 7.00.

Mixing bathroom cleaners

▲ **Figure 1** *Household bleach products*

Bathroom cleaners come in essentially two types – bleach-based for removing coloured stains, and acid-based used, for example, for removing limescale in the toilet bowl.

> 1 Limescale is made up of calcium carbonate, $CaCO_3$. Write the equation for the reaction of hydrochloric acid with calcium carbonate.

Most bathroom cleaners have a warning on the label not to mix them with other types of cleaner. Why is this?

The active ingredient in household bleach is chloric(I) acid (HClO), whilst acid-based cleaners contain hydrochloric acid (HCl). These react together to form chlorine gas.

$$HClO(aq) + HCl(aq) \rightleftharpoons Cl_2(g) + H_2O(aq)$$

Imagine you have put a large amount of bleach in the toilet bowl (so that chloric(I) acid is in excess) and then you add a squirt (say 50 cm^3) of acid-based cleaner of concentration 1 mol dm^{-3}. Assume the equilibrium is forced completely to the right.

2   How many moles of HCl have you added?
3   How many moles of chlorine would be produced?
4   What volume of chlorine is this?
5   Why would there be less chlorine gas in the bathroom than you have calculated in 3?

This is a significant amount of chlorine and, considering that it was used as a poisonous gas in the First World War, something to be avoided.

pH and temperature

The equilibrium reaction below is endothermic in the forward direction.

$$H_2O(l) \rightleftharpoons H^+(aq) + OH^-(aq) \qquad \Delta H = +57.3 \text{ kJ mol}^{-1}$$

Therefore the value of K_w increases with temperature and the pH of water is different at different temperatures (Table 1). So, for example, at 373 K (boiling point), the pH of water is about 6.

This does not mean that water is acidic (water is always neutral because it always has an equal number of H^+ ions and OH^- ions) but merely that the neutral value for the pH is 6 at this temperature, rather than pH 7 at room temperature. So in boiling water $[H^+]$ (and $[OH^-]$) are both about 1×10^{-6} mol dm^{-3}.

Table 1 shows that the pH of sea water at the poles and at the Equator will be different.

1   Use Table 1 to calculate the concentration of H^+ at 313 K.
2   What is the concentration of OH^- at the same pH?

▼ **Table 1** *The effect of temperature on the pH of water*

T / K	K_w / mol^2 dm^{-6}	pH (neutral)
273	0.114×10^{-14}	7.47
283	0.293×10^{-14}	7.27
293	0.681×10^{-14}	7.08
298	1.008×10^{-14}	7.00
303	1.471×10^{-14}	6.92
313	2.916×10^{-14}	6.77
323	5.476×10^{-14}	6.63
373	51.300×10^{-14}	6.14

2 OH$^-$ concentration will be the same 1.7×10^{-7} mol dm^{-3}
1 1.7×10^{-7} mol dm^{-3}

Measuring pH

pH can be measured using an indicator paper or a solution, such as universal indicator. This is made from a mixture of dyes that change colour at different $[H^+(aq)]$. This is fine for measurements to the nearest whole number, but for more precision a pH meter is used. A pH meter has an electrode which dips into a solution and produces a voltage related to $[H^+(aq)]$. The pH readings can then be read directly on the meter or fed into a computer or data logger, for continuous monitoring of a chemical process or medical procedure, for example.

◀ **Figure 2** *Using a pH meter*

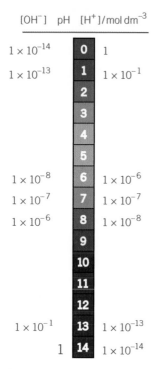

▲ **Figure 3** *The pH scale*

pH measures alkalinity as well

pH measures alkalinity as well as acidity, because as $[H^+(aq)]$ goes up, $[OH^-(aq)]$ goes down. At 298 K, if a solution contains more $H^+(aq)$ than $OH^-(aq)$, its pH will be less than 7 and it is called acidic. If a solution contains more $OH^-(aq)$ than $H^+(aq)$, its pH will be greater than 7 and it is called alkaline (Figure 3).

Working with the pH scale

Finding [H⁺(aq)] from pH

You can work out the concentration of hydrogen ions $[H^+]$ in an aqueous solution if you know the pH. It is the antilogarithm of the pH value.

For example, an acid has a pH of 3.00:

$$pH = -\log_{10}[H^+(aq)]$$

$$3.00 = -\log_{10}[H^+(aq)]$$

$$-3.00 = \log_{10}[H^+(aq)]$$

Take the antilog of both sides:

$$[H^+(aq)] = 1.0 \times 10^{-3} \text{ mol dm}^{-3}$$

Finding [OH⁻(aq)] from pH

With bases, you need two steps. Suppose the pH of a solution is 10.00:

$$pH = -\log_{10}[H^+(aq)]$$

$$10.00 = -\log_{10}[H^+(aq)]$$

$$-10.00 = \log_{10}[H^+(aq)]$$

Take the antilog of both sides:

$$[H^+(aq)] = 1.0 \times 10^{-10}$$

You know $[H^+(aq)] [OH^-(aq)] = 1.0 \times 10^{-14} \text{ mol}^2 \text{dm}^{-6}$

Substituting your value for $[H^+(aq)] = 1.0 \times 10^{-10}$ into the equation:

$$[1.0 \times 10^{-10}] [OH^-(aq)] = 1.0 \times 10^{-14} \text{ mol}^2 \text{dm}^{-6}$$

$$[OH^-(aq)] = \frac{1.0 \times 10^{-14}}{1.0 \times 10^{-10}} = 1.0 \times 10^{-4} \text{ mol dm}^{-3}$$

The pH of strong acid solutions – worked example

HCl dissociates completely in dilute aqueous solution to $H^+(aq)$ ions and $Cl^-(aq)$ ions, that is, the reaction goes to completion:

$$HCl(aq) \rightarrow H^+(aq) + Cl^-(aq)$$

Acids that dissociate completely like this are called **strong acids**.

So in 1.00 mol dm^{-3} HCl:

$$[H^+(aq)] = 1.00 \text{ mol dm}^{-3}$$

$$\log[H^+(aq)] = \log 1.00 = 0.00$$

$$-\log [H^+(aq)] = 0.00$$

So the pH of 1 mol dm^{-3} HCl = 0.00

In a 0.16 mol dm^{-3} solution of HCl:

$$[H^+(aq)] = 0.16\,mol\,dm^{-3}$$

$$\log [H^+(aq)] = \log 0.160 = -0.796$$

$$-\log[H^+(aq)] = 0.796$$

So the pH of 0.16 mol dm^{-3} HCl = 0.80 to 2 d.p.

Worked example: The pH of alkaline solutions

In alkaline solutions, it takes two steps to calculate [H$^+$(aq)].

To find the [H$^+$(aq)] of an alkaline solution at 298 K:

1 Calculate [OH$^-$(aq)].

2 Then use: [H$^+$(aq)] [OH$^-$(aq)] = 1.00 × 10^{-14} mol^2 dm^{-6} to calculate [H$^+$(aq)].

The pH can then be calculated.

For example, to find the pH of 1.00 mol dm^{-3} sodium hydroxide solution:

Sodium hydroxide is fully dissociated in aqueous solution – it is called a strong alkali.

$$NaOH(aq) \rightarrow Na^+(aq) + OH^-(aq)$$

$$[OH^-(aq)] = 1.00\,mol\,dm^{-3}$$

but $$[OH^-(aq)]\,[H^+(aq)] = 1.00 \times 10^{-14}\,mol\,dm^{-3}$$

$$[H^+(aq)] = 1.00 \times 10^{-14}\,mol\,dm^{-3}$$

and $$[H^+(aq)] = -\log(1.00 \times 10^{-14})$$

$$pH = 14.00$$

In a 0.100 mol dm^{-3} sodium hydroxide solution:

$$[OH^-(aq)] = 1.00 \times 10^{-1}\,mol\,dm^{-3}$$

$$[OH^-(aq)]\,[H^+(aq)] = 1.00 \times 10^{-14}\,mol\,dm^{-3}$$

$$[H^+(aq)] \times 10^{-1} = 1.00 \times 10^{-14}\,mol\,dm^{-3}$$

$$[H^+(aq)] = 1.00 \times 10^{-13}\,mol\,dm^{-3}$$

$$\log [H^+(aq)] = -13.00$$

$$pH = 13.00$$

Study tip

In very concentrated solution even strong acids are not fully ionised.

Study tip

The pH of solutions of strong acids which have a concentration greater than 1 mol dm^{-3} is negative. The pH of solutions of strong bases which have concentration greater than 1 mol dm^{-3} is larger than 14.00.

Study tip

When writing out calculations on pH it is acceptable to omit the 'aq' in expressions such as [H$^+$(aq)], so we could write simply pH = $-\log_{10}$ [H$^+$].

Summary questions

1 What is the pH of a solution in which [H$^+$] is 1.00 × 10^{-2} mol dm^{-3}?

2 What is [H$^+$] in a solution of pH = 6.00?

3 At 298 K what is [OH$^-$] in a solution of pH = 9.00?

4 Calculate the pH of a 0.020 mol dm^{-3} solution of HCl.

5 Calculate the pH of 0.200 mol dm^{-3} sodium hydroxide.

In Topic 21.2 you looked at the pH of acids such as hydrochloric acid which dissociate completely into ions when dissolved in water. Acids that completely dissociate into ions in aqueous solutions are called **strong acids**. The word strong refers *only* to the extent of dissociation and not in any way to the concentration. So it is perfectly possible to have a very dilute solution of a strong acid.

The same arguments apply to bases. Strong bases are completely dissociated into ions in aqueous solutions. For example, sodium hydroxide is a strong base:

$$NaOH(aq) \rightarrow Na^+(aq) + OH^-(aq)$$

Weak acids and bases

Many acids and bases are only slightly ionised (not fully dissociated) when dissolved in water. Ethanoic acid (the acid in vinegar, also known as acetic acid) is a typical example. In a $1 \, mol \, dm^{-3}$ solution of ethanoic acid, only about four in every thousand ethanoic acid molecules are dissociated into ions (so the degree of dissociation is $\frac{4}{1000}$)– the rest remain dissolved as wholly covalently bonded molecules. In fact an equilibrium is set up:

$CH_3COOH(aq)$	$\rightleftharpoons$	$H^+(aq)$	$+$	$CH_3COO^-(aq)$
ethanoic acid		hydrogen ions		ethanoate ions

	ethanoic acid	hydrogen ions	ethanoate ions
Before dissociation:	1000	0	0
At equilibrium:	996	4	4

Acids like this are called **weak acids**. Weak refers *only* to the degree of dissociation. In a $5 \, mol \, dm^{-3}$ solution, ethanoic acid is still a weak acid, while in a $10^{-4} \, mol \, dm^{-3}$ solution, hydrochloric acid is still a strong acid.

When ammonia dissolves in water, it forms an alkaline solution. The equilibrium lies well to the left and ammonia is weakly basic:

$$NH_3(aq) + H_2O(l) \rightleftharpoons NH_4^+(aq) + OH^-(aq)$$

The dissociation of weak acids

Weak acids

Imagine a weak acid HA which dissociates:

$$HA(aq) \rightleftharpoons H^+(aq) + A^-(aq)$$

The equilibrium constant is given by:

$$K_c = \frac{[H^+(aq)]_{eqm}[A^-(aq)]_{eqm}}{[HA(aq)]_{eqm}}$$

For a weak acid, this is usually given the symbol K_a and called the **acid dissociation constant**.

$$K_a = \frac{[H^+(aq)]_{eqm}[A^-(aq)]_{eqm}}{[HA(aq)]_{eqm}}$$

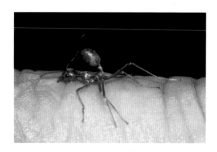

▲ **Figure 1** *Formic acid (methanoic acid) is quite concentrated when used as a weapon by the stinging ant and, although it is a weak acid, being sprayed with it can be a painful experience*

The larger the value of K_a, the further the equilibrium is to the right, the more the acid is dissociated, and the stronger it is. Acid dissociation constants for some acids are given in Table 1.

K_a has units and it is important to state these. They are found by multiplying and cancelling the units in the expression for K_a.

$$K_a = \frac{\cancel{mol\,dm^{-3}} \times mol\,dm^{-3}}{\cancel{mol\,dm^{-3}}} = mol\,dm^{-3}$$

Calculating the pH of weak acids
In Topic 20.2 you calculated the pH of solutions of strong acids, by assuming that they are fully dissociated. For example, in a $1.00\ mol\,dm^{-3}$ solution of nitric acid, $[H^+] = 1.00\ mol\,dm^{-3}$. In weak acids this is no longer true, and you must use the acid dissociation expression to calculate $[H^+]$.

Calculating the pH of 1.00 mol dm⁻³ ethanoic acid
The concentrations in $mol\,dm^{-3}$ are:

$$CH_3COOH(aq) \rightleftharpoons CH_3COO^-(aq) + H^+(aq)$$

Before dissociation: 1.00   0   0

At equilibrium: $1.00 - [CH_3COO^-(aq)]$ $[CH_3COO^-(aq)]$ $[H^+(aq)]$

$$K_a = \frac{[CH_3COO^-(aq)][H^+(aq)]}{[CH_3COOH(aq)]}$$

But as each CH_3COOH molecule that dissociates produces one CH_3COO^- ion and one H^+ ion:

$$[CH_3COO^-(aq)] = [H^+(aq)]$$

Since the degree of dissociation of ethanoic acid is so small (it is a weak acid), $[H^+(aq)]_{eqm}$ is very small and, to a good approximation, $1.00 - [H^+(aq)] \approx 1.00$.

$$K_a = \frac{[H^+(aq)]^2}{1.00}$$

From Table 1, $K_a = 1.70 \times 10^{-5}\ mol\,dm^{-3}$

$$1.70 \times 10^{-5} = [H^+(aq)]^2$$

$$[H^+(aq)] = \sqrt{1.75 \times 10^{-5}}$$

$$[H^+(aq)] = 4.12 \times 10^{-3}\ mol\,dm^{-3}$$

Taking logs: $\log[H^+(aq)] = -2.385$

$$pH = 2.385 = 2.39 \text{ to 2 d.p.}$$

Calculating the pH of 0.100 mol dm⁻³ ethanoic acid
Using the same method, you get:

$$K_a = \frac{[H^+(aq)]^2}{0.10 - [H^+(aq)]}$$

Again, $0.100 - [H^+(aq)] \approx 0.10$

so: $1.70 \times 10^{-5} = \dfrac{[H^+(aq)]^2}{0.10}$

$$1.70 \times 10^{-6} = [H^+(aq)]^2$$

$$[H^+(aq)] = 1.30 \times 10^{-3}\ mol\,dm^{-3}$$

$$pH = 2.89 \text{ to 2 d.p.}$$

▼ **Table 1** *Values of K_a for some weak acids*

Acid	K_a / mol dm⁻³
chloroethanoic	1.30×10^{-3}
benzoic	6.30×10^{-5}
ethanoic	1.70×10^{-5}
hydrocyanic	4.90×10^{-10}

Study tip

It is acceptable to omit the eqm subscripts unless they are specifically asked for.

Hint $\sqrt{x}$

The symbol $\approx$ means approximately equal to.

pK_a

For a weak acid pK_a is often referred to. This is defined as:

$$pK_a = -\log_{10} K_a$$

Think of p as meaning $-\log_{10}$ of.

pK_a can be useful in calculations (Table 2). It gives a measure of how strong a weak acid is – the smaller the value of pK_a, the stronger the acid.

▼ **Table 2** *Values of K_a and pK_a for some weak acids*

Acid	K_a/mol dm^{-3}	pK_a
chloroethanoic	1.30×10^{-3}	2.88
benzoic	6.30×10^{-5}	4.20
ethanoic	1.70×10^{-5}	4.77
hydrocyanic	4.90×10^{-10}	9.31

Summary questions

1 Which is the strongest acid in Table 2?

2 What can you say about the concentration of H$^+$ ions compared with the concentration of ethanoate ions in all solutions of pure ethanoic acid?

3 $\sqrt{x}$ Calculate the pH of the following solutions:

  a 0.100 mol dm^{-3} chloroethanoic acid

  b 0.0100 mol dm^{-3} benzoic acid

21.4 Acid–base titrations

A titration is used to find the concentration of a solution by gradually adding to it a second solution with which it reacts. One of the solutions is of known concentration. To use a titration, you must know the equation for the reaction.

pH changes during acid–base titrations

In an acid–base titration, an acid of known concentration is added from a burette to a measured amount of a solution of a base (an alkali) until an indicator shows that the base has been neutralised. Alternatively, the base is added to the acid until the acid is neutralised. You can then calculate the concentration of the alkali from the volume of acid used.

You can also follow a neutralisation reaction by measuring the pH with a pH meter (Figure 2) in which case you do not need an indicator. The pH meter is calibrated by placing the probe in a buffer solution of a known pH (Topic 21.6).

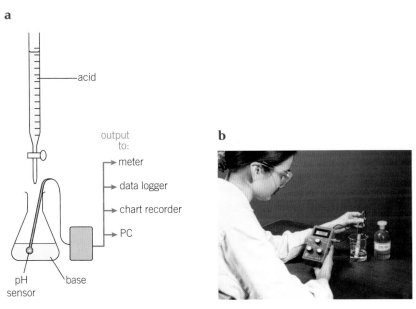

a

▲ **Figure 2** **a** *Apparatus to investigate pH changes during a titration* **b** *A pH meter*

▲ **Figure 1** *An acid–base titration, to find the concentration of a base. A volumetric pipette is used to deliver an accurately measured volume of base of unknown concentration into the flask. The acid of known concentration is in the burette*

Titration procedure

The titration curves in Figure 3 were determined using the apparatus shown in Figure 2. Using a pipette, 25 cm^3 acid was placed in the conical flask. The base was added, 1 cm^3 at a time, from a burette and the pH recorded. At areas on the curve where the pH was changing rapidly, the experiment was repeated adding the base 0.1 cm^3 at a time to find the shape of the curve more precisely.

Titration curves

Figure 3 shows the results obtained for four cases using monoprotic acids. In these cases the *base* is added from the burette and the acid has been accurately measured into a flask. The shape of each titration curve is typical for the type of acid–base titration.

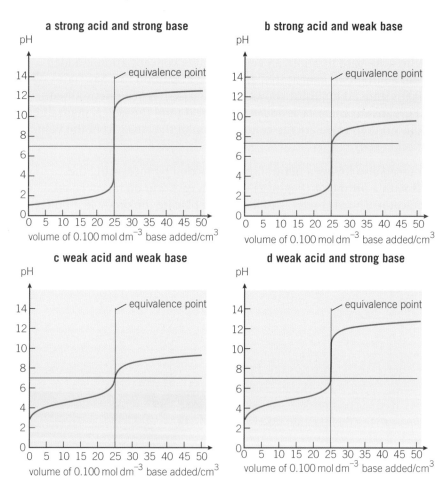

▲ **Figure 3** *Graphs of pH changes for titrations of different acids and bases*

The first thing to notice about these curves is that the pH does not change in a linear manner as the base is added. Each curve has almost horizontal sections where a lot of base can be added without changing the pH much. There is also a very steep portion of each curve, except weak acid–weak base, where a single drop of base changes the pH by several units.

In a titration, the **equivalence point** is the point at which sufficient base has been added to just neutralise the acid (or vice-versa). In each of the titrations in Figure 3 the equivalence point is reached after 25.0 cm³ of base has been added. However, the pH at the equivalence point is not always exactly 7.

In each case, except the weak acid–weak base titration, there is a large and rapid change of pH at the equivalence point (i.e., the curve is almost vertical) even though this is may not be centred on pH 7.

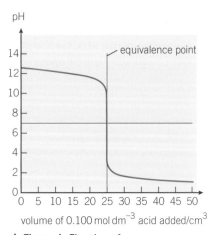

▲ **Figure 4** *Titration of a strong base–strong acid, adding 0.100 mol dm⁻³ HCl(aq) to 25.0 cm³ of 0.100 mol dm⁻³ NaOH(aq)*

This is relevant to the choice of indicator for a particular titration (see Topic 20.5).

You can add the acid to the base for these pH curves and the shape will be flipped around the pH 7 line. For example, Figure 4 shows a strong acid–strong base curve.

Working out concentrations

The equivalence point can be used to work out the concentration of the unknown acid (or base).

Worked example: A monoprotic acid

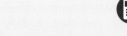

In a titration, the equivalence point is reached when $25.00\,cm^3$ of $0.0150\,mol\,dm^{-3}$ sodium hydroxide is neutralised by $15.00\,cm^3$ hydrochloric acid. What is the concentration of the acid?

$$HCl(aq) + NaOH(aq) \rightarrow NaCl(aq) + H_2O(l)$$

The equivalence point shows that $15.0\,cm^3$ hydrochloric acid of concentration A contains the same number of moles as $25.00\,cm^3$ of $0.0150\,mol\,dm^{-3}$ sodium hydroxide.

$$\text{number of moles in solution} = c \times \frac{V}{1000}$$

Where c is concentration in $mol\,dm^{-3}$ and V is volume in cm^3

From the equation, number of moles HCl = number of moles NaOH

$$25.00 \times \frac{0.0150}{1000} = 15.00 \times \frac{A}{1000}$$

$$A = 0.025$$

So, the concentration of the acid is $0.0250\,mol\,dm^{-3}$.

Worked example: A diprotic acid

In a titration, the equivalence point is reached when $20.00\,cm^3$ of $0.0100\,mol\,dm^{-3}$ sodium hydroxide is neutralised by $15.00\,cm^3$ sulfuric acid. What is the concentration of the acid?

$$H_2SO_4(aq) + 2NaOH(aq) \rightarrow Na_2SO_4(aq) + 2H_2O(l)$$

The equivalence point shows that $15.00\,cm^3$ sulfuric acid of concentration B contains the same number of moles as $20.00\,cm^3$ of $0.0100\,mol\,dm^{-3}$ sodium hydroxide.

$$\text{number of moles in solution} = c \times \frac{V}{1000}$$

Where c is concentration in $mol\,dm^{-3}$ and V is volume in cm^3.

$$\text{Number of moles of NaOH} = 20.00 \times \frac{0.0100}{1000} = \frac{0.2}{1000}$$

From the equation, number of moles $H_2SO_4 = \frac{1}{2}$ number of moles NaOH

So number of moles of $H_2SO_4 = \frac{0.1}{1000}$

$$\text{Number of moles of } H_2SO_4 = 15.00 \times \frac{B}{1000} = \frac{0.1}{1000}$$

So, the concentration, B, of the acid is $0.0067\,mol\,dm^{-3}$.

Summary questions

1 ⎷ $25.0\,cm^3$ sodium hydroxide is neutralised by $15.0\,cm^3$ sulfuric acid, H_2SO_4, of concentration $0.100\,mol\,dm^{-3}$.

  a Write the equation for this reaction.

  b From the equation, how many moles of sulfuric acid will neutralise 1.00 mol of sodium hydroxide?

  c How many moles of sulfuric acid are used in the neutralisation?

  d What is the concentration of the sodium hydroxide?

2 The graph below shows two titration curves of two acids labelled A and B with a base.

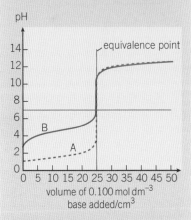

  a Which curve represents
    i a strong acid
    ii a weak acid?

  b Which one could represent:
    i ethanoic acid, CH_3COOH
    ii hydrochloric acid, HCl?

  c Was the base strong or weak? Explain your answer.

An acid–base titration uses an indicator to find the concentration of a solution of an acid or alkali. The equivalence point is the volume at which exactly the same number of moles of hydrogen ions (or hydroxide ions) has been added as there are moles of hydroxide ions (or hydrogen ions). The **end point** is the volume of alkali or acid added when the indicator just changes colour. Unless you choose the right indicator, the equivalence point and the end point may not always give the same answer.

A suitable indicator for a particular titration needs the following properties:

• The colour change must be sharp rather than gradual at the end point, that is, no more than one drop of acid (or alkali) is needed to give a complete colour change. An indicator that changes colour gradually over several cubic centimetres would be unsuitable and would not give a sharp end point.

• The end point of the titration given by the indicator must be the same as the equivalence point, otherwise the titration will give the wrong answer.

• The indicator should give a distinct colour change. For example, the colourless to pink change of phenolphthalein is easier to see than the red to yellow of methyl orange.

Some common indicators are given in Table 1 with their approximate colour changes. Notice that the colour change of most indicators takes place over a pH range of around two units, centred around the value of pK_a for the indicator. For this reason, not all indicators are suitable for all titrations. Universal indicator is not suitable for any titration because of its gradual colour changes.

▼ **Table 1** *Some common indicators. Universal indicator is a mixture of indicators that change colour at different pHs*

Indicator	pK_a value	Colour change
universal indicator		
methyl orange	3.7	red — change — yellow
bromophenol blue	4.0	yellow — change — blue
methyl red	5.1	red — change — yellow
bromothymol blue	7.0	yellow — change — blue
phenolphthalein	9.3	colourless — change — red

0 1 2 3 4 5 6 7 8 9 10 11 12 13 14 pH

very acidic — neutral — very alkaline

The following examples compare the suitability of two common indicators – phenolphthalein and methyl orange – for four different types of acid–base titration. In each case, the base is being added to the acid.

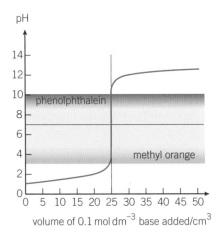

▲ **Figure 1** *Titration of a strong acid–strong base, adding 0.1 mol dm⁻³ NaOH(aq) to 25 cm³ of 0.1 mol dm⁻³ HCl(aq)*

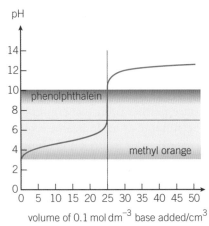

▲ **Figure 2** *Titration of a weak acid–strong base, adding 0.1 mol dm⁻³ NaOH(aq) to 25 cm³ of 0.1 mol dm⁻³ CH₃COOH(aq)*

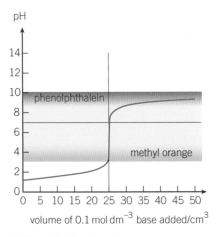

▲ **Figure 3** *Titration of a strong acid–weak base, adding 0.1 mol dm⁻³ NH₃(aq) to 25 cm³ of 0.1 mol dm⁻³ HCl(aq)*

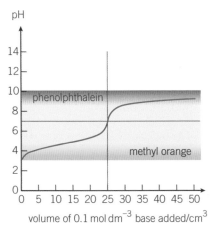

▲ **Figure 4** *Titration of a weak acid–weak base, adding 0.1 mol dm⁻³ NH₃(aq) to 25 cm³ of 0.1 mol dm⁻³ CH₃COOH(aq)*

1 Strong acid–strong base, for example, hydrochloric acid and sodium hydroxide

Figure 1 is the graph of pH against volume of base added. The pH ranges over which two indicators change colour are shown. To fulfil the first two criteria above, the indicator must change within the vertical portion of the pH curve. Here either indicator would be suitable, but phenolphthalein is usually preferred because of its more easily seen colour change.

2 Weak acid–strong base titration, for example, ethanoic acid and sodium hydroxide

Methyl orange is not suitable (Figure 2). It does not change in the vertical portion of the curve and will change colour in the 'wrong' place and over the addition of many cubic centimetres of base. Phenolphthalein will change sharply at exactly 25 cm³, the equivalence point, and would therefore be a good choice.

3 Strong acid–weak base titration, for example, hydrochloric acid and ammonia

Here methyl orange will change sharply at the equivalence point but phenolphthalein would be of no use (Figure 3).

4 Weak acid–weak base, for example, ethanoic acid and ammonia

Here neither indicator is suitable (Figure 4). In fact, no indicator could be suitable as an indicator requires a vertical portion of the curve over two pH units at the equivalence point to give a sharp change.

The half-neutralisation point

If you look at the titration curves you can see that there is always a very gently sloping, almost horizontal, part to the curve before you reach the steep line on which the equivalence point lies. As you add acid (or base), there is very little change to the pH, almost up to the volume of the equivalence point. The point half-way between the zero and the equivalence point is the **half-neutralisation point**. This is significant because the knowledge that you can add acid (or base) up to this point with the certainty that the pH will change very little is relevant to the theory of buffers. It also allows you to find the pK_a of the weak acid.

$$HA + OH^- \rightarrow H_2O + A^-$$

At the half-neutralisation pointhalf the HA has been converted into A^- and half remains, so:

$$[HA] = [A^-]$$

Therefore

$$K_a = \frac{[H^+]\cancel{[A^-]}}{\cancel{[HA]}}$$

$$K_a = [H^+]$$

And

$$-\log_{10} K_a = -\log_{10} [H^+]$$

$$pK_a = pH$$

Study tip

This provides a method of determining pK_a values experimentally.

Summary questions

1 The indicator bromocresol purple changes colour between pH 5.2 and 6.8. For which of the following titration types would it be suitable:

- **a** weak acid–weak base
- **b** strong acid–weak base
- **c** weak acid–strong base
- **d** strong acid–strong base

2 For which of the above titrations would bromophenol blue be suitable?

21.6 Buffer solutions

Buffers are solutions that can resist changes of pH. When small amounts of acid or alkali are added to them.

How buffers work

Buffers are designed to keep the concentration of hydrogen ions and hydroxide ions in a solution almost unchanged. They are based on an equilibrium reaction which will move in the direction to remove either additional hydrogen ions or hydroxide ions if these are added.

Acidic buffers

Acidic buffers are made from weak acids. They work because the dissociation of a weak acid is an equilibrium reaction.

Consider a weak acid, HA. It will dissociate in solution:

$$HA(aq) \rightleftharpoons H^+(aq) + A^-(aq)$$

From the equation, $[H^+(aq)] = [A^-(aq)]$, and as it is a weak acid, $[H^+(aq)]$ and $[A^-(aq)]$ are both very small because most of the HA is undissociated.

Adding alkali

If a little alkali is added, the OH^- ions from the alkali will react with HA to produce water molecules and A^-:

$$HA(aq) + OH^-(aq) \rightarrow H_2O(aq) + A^-(aq)$$

This removes the added OH^- so the pH tends to remain almost the same.

Adding acid

If H^+ is added, the equilibrium shifts to the left– H^+ ions combining with A^- ions to produce undissociated HA. But, since $[A^-]$ is small, the supply of A^- soon runs out and there is no A^- left to combine the added H^+. So the solution is not a buffer.

However, you can add to the solution a supply of extra A^- by adding a soluble salt of HA, which fully ionises, such as Na^+A^-. This increases the supply of A^- so that more H^+ can be used up. So, there is a way in which both added H^+ and OH^- can be removed.

An acidic buffer is made from a mixture of a weak acid and a soluble salt of that acid. It will maintain a pH of below 7 (acidic).

The function of the weak acid component of a buffer is to act as a source of HA which can remove any added OH^-:

$$HA(aq) + OH^-(aq) \rightarrow A^-(aq) + H_2O(l)$$

The function of the salt component of a buffer is to act as a source of A^- ions which can remove any added H^+ ions:

$$A^-(aq) + H^+(aq) \rightarrow HA(aq)$$

Buffers don't ensure that *no* change in pH occurs. The addition of acid or alkali will still change the pH, but only slightly and by far less than the change that adding the same amount to a non-buffer would cause.

Learning objectives:
→ State the definition of a buffer.
→ Describe how buffers work.
→ Describe how the pH of an acidic buffer solution can be calculated.
→ Describe what buffers are used for.

Specification reference: 3.1.12

Study tip

Practise writing equations to explain how a buffer solution reacts when small amounts of acid or base are added.

Hint

Buffers are important in brewing – the enzymes that control many of the reactions involved work best at specific pHs.

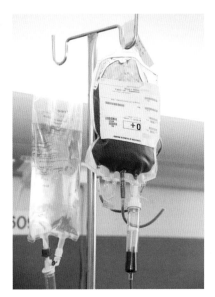

▲ **Figure 1** *Blood is buffered to a pH of 7.40*

▲ **Figure 2** *Most shampoos are buffered so that they are slightly alkaline*

It is also possible to saturate a buffer – add so much acid or alkali that all of the available HA or A⁻ is used up.

Another way of achieving a mixture of weak acid and its salt is by neutralising some of the weak acid with an alkali such as sodium hydroxide. If you neutralise half the acid (Topic 20.5) you end up with a buffer whose pH is equal to the pK_a of the acid as it has an equal supply of HA and A⁻.

$$\text{At half-neutralisation: pH} = \text{p}K_a$$

This is a very useful buffer because it is equally efficient at resisting a change in pH whether acid or alkali is added.

Basic buffers

Basic buffers also resist change but maintain a pH at above 7. They are made from a mixture of a weak base and a salt of that base.

A mixture of aqueous ammonia and ammonium chloride, $NH_4^+Cl^-$, acts as a basic buffer. In this case:

- The aqueous ammonia removes added H⁺:

$$NH_3(aq) + H^+(aq) \rightarrow NH_4^+(aq)$$

- the ammonium ion, NH_4^+, removes added OH⁻:

$$NH_4^+(aq) + OH^-(aq) \rightarrow NH_3(aq) + H_2O(l)$$

Examples of buffers

An important example of a system involving a buffer is blood, the pH of which is maintained at approximately 7.4. A change of as little as 0.5 of a pH unit may be fatal.

Blood is buffered to a pH of 7.4 by a number of mechanisms. The most important is:

$$H^+(aq) + HCO_3^-(aq) \rightleftharpoons CO_2(aq) + H_2O(l)$$

Addition of extra H⁺ ions moves this equilibrium to the right, removing the added H⁺. Addition of extra OH⁻ ions removes H⁺ by reacting to form water. The equilibrium above moves to the left releasing more H⁺ ions. (The same equilibrium reaction acts to buffer the acidity of soils.)

There are many examples of buffers in everyday products, such as detergents and shampoos. If either of these substances become too acidic or too alkaline, they could damage fabric or skin and hair.

Calculations on buffers

Different buffers can be made which will maintain different pHs. When a weak acid dissociates:

$$HA(aq) \rightleftharpoons H^+(aq) + A^-(aq)$$

You can write the expression:

$$K_a = \frac{[H^+(aq)][A^-(aq)]}{[HA(aq)]}$$

You can use this expression to calculate the pH of buffers.

Worked example: Calculating pH of a buffer solution 1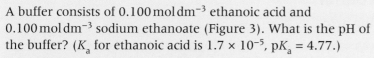

A buffer consists of $0.100 \, \text{mol dm}^{-3}$ ethanoic acid and $0.100 \, \text{mol dm}^{-3}$ sodium ethanoate (Figure 3). What is the pH of the buffer? (K_a for ethanoic acid is 1.7×10^{-5}, $pK_a = 4.77$.)

Calculate $[\text{H}^+(\text{aq})]$ from the equation.

$$K_a = \frac{[\text{H}^+(\text{aq})][\text{A}^-(\text{aq})]}{[\text{HA}(\text{aq})]}$$

Sodium ethanoate is fully dissociated, so $[\text{A}^-(\text{aq})] = 0.100 \, \text{mol dm}^{-3}$

Ethanoic acid is almost undissociated, so $[\text{HA}(\text{aq})] \approx 0.100 \, \text{mol dm}^{-3}$

$$1.7 \times 10^{-5} = [\text{H}^+(\text{aq})] \times \frac{0.100}{0.100}$$

$$1.7 \times 10^{-5} = [\text{H}^+(\text{aq})] \text{ and pH} = -\log_{10}[\text{H}^+(\text{aq})]$$

$$\text{pH} = 4.77$$

When you have equal concentrations of acid and salt, pH of the buffer is equal to pK_a of acid used, and this is exactly the same situation as the half-neutralisation point.

Changing the concentration of HA or A^- will affect the pH of the buffer. If you use $0.200 \, \text{mol dm}^{-3}$ ethanoic acid and $0.100 \, \text{mol dm}^{-3}$ sodium ethanoate, the pH will be 4.50. Check that you can do this by doing a calculation like the one above.

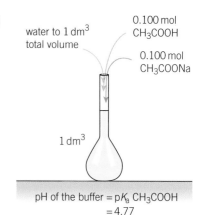

pH of the buffer = pK_a CH$_3$COOH
= 4.77

▲ **Figure 3** *Making a buffer*

Study tip √x̄

In a buffer solution $[\text{H}^+] \neq [\text{A}^-]$ so do *not* use the simplified expression

$$K_a = \frac{[\text{H}^+]^2}{[\text{HA}]}$$

Worked example: Calculating pH of a buffer solution 2

Calculate the pH of the buffer formed when $500 \, \text{cm}^3$ of $0.400 \, \text{mol dm}^{-3}$ NaOH is added to $500 \, \text{cm}^3$ $1.00 \, \text{mol dm}^{-3}$ HA. $K_a = 6.25 \times 10^{-5}$.

Some of the weak acid is neutralised by the sodium hydroxide leaving a solution containing A^- and HA , which will act as a buffer.

$$\text{moles HA} = c \times \frac{V}{1000} = 1.00 \times \frac{500}{1000} = 0.500 \, \text{mol}$$

$$\text{moles NaOH} = \text{moles OH}^- = c \times \frac{V}{1000} = 0.400 \times \frac{500}{1000} = 0.200 \, \text{mol}$$

Equation:	HA	+	NaOH	→	H₂O	+	NaA
Initially:	0.500 mol		0.200 mol				0
Finally:	0.300 mol		0 mol				0.200 mol

This leaves $1000 \, \text{cm}^3$ of a solution containing $0.300 \, \text{mol}$ HA and $0.200 \, \text{mol}$ A^- since all the NaA is dissociated to give A^-.

The concentrations are $[\text{HA}] = 0.300 \, \text{mol dm}^{-3}$ and $[\text{A}-] = 0.200 \, \text{mol dm}^{-3}$

$$K_a = \frac{[\text{H}^+(\text{aq})][\text{A}^-(\text{aq})]}{[\text{HA}(\text{aq})]}$$

$$6.25 \times 10^{-5} = [\text{H}^+(\text{aq})] \times \frac{0.200}{0.300}$$

$$[\text{H}^+(\text{aq})] = 6.25 \times 10^{-5} \times \frac{3}{2} = 9.375 \times 10^{-5}$$

$$\text{So pH} = -\log[\text{H}^+(\text{aq})] = 4.03$$

The pH change when an acid or a base is added to a buffer

Adding acid

You can calculate how the pH changes when acid is added to a buffer. Suppose you start with $1.00 \, dm^3$ of a buffer solution of ethanoic acid at concentration $0.10 \, mol \, dm^{-3}$ and sodium ethanoate at concentration $0.10 \, mol \, dm^{-3}$. K_a is 1.7×10^{-5}. This has a pH of 4.77, as shown in the calculation previously.

Now add $10.0 \, cm^3$ of hydrochloric acid of concentration $1.00 \, mol \, dm^{-3}$ to this buffer. Virtually all the added H^+ ions will react with the ethanoate ions, $[A^-]$, to form molecules of ethanoic acid, $[HA]$.

Before adding the acid:

* Number of moles of ethanoic acid = 0.10
* Number of moles of sodium ethanoate = 0.10

number of moles of hydrochloric acid added is $c \times \dfrac{V}{1000}$

$$1.00 \times \frac{10.0}{1000} = 0.010$$

After adding the acid, this means:

* The amount of acid is increased by 0.010 mol to 0.110 mol.
* The amount of salt is decreased by 0.010 mol to 0.090 mol.

So, the concentration of acid $[HA]$ is now $0.110 \, mol \, dm^{-3}$.

And, the concentration of salt $[A^-]$ is now $0.090 \, mol \, dm^{-3}$.

In calculating these concentrations you have ignored the volume of the added hydrochloric acid, $10 \, cm^3$ in $1000 \, cm^3$, only a 1% change.

$$K_a = \frac{[H^+(aq)][A^-(aq)]}{[HA(aq)]}$$

So $\qquad 1.7 \times 10^{-5} = [H^+(aq)] \dfrac{0.090}{0.110}$

$$[H^+(aq)] = 1.7 \times 10^{-5} \times \frac{0.110}{0.090} = 2.08 \times 10^{-5}$$

$$pH = 4.68$$

Note how small the pH change is (from 4.77 to 4.68).

 Making a buffer solution

You may need to make up buffer solutions of specified pHs, to calibrate a pH meter, for example.

To find suitable concentrations of weak acid and its salt, you use the equation

$$K_a = \frac{[H^+] \, [A^-]}{[HA]}$$

You can rearrange this equation to make $[H^+]$ the subject and then taking logs of both sides. This results in an equation that is easier to use for calculations on buffers. It is called the **Henderson–Hasselbalch equation**.

$$pH = pKa - \log\left(\frac{[HA]}{[A^-]}\right)$$

It tells you that the pH of a buffer solution depends on the pK_a of the weak acid on which it is based and on the **ratio** of the concentration of the acid to that of its salt.

For example, to make a buffer of pH = 4.50 you first select a weak acid whose pK_a is close to the required pH. Benzoic acid has a pK_a of 4.20, so you could make a buffer from benzoic acid and sodium benzoate.

Then substituting into the Henderson equation:

$$4.50 = 4.20 - \log\left(\frac{[HA]}{[A^-]}\right)$$

$$\log\left(\frac{[HA]}{[A^-]}\right) = -0.30$$

Taking antilogs:

$$\left(\frac{[HA]}{[A^-]}\right) = 0.50$$

This means that the concentration of acid is half the concentration of the salt.

So, for example, a solution that is 0.05 mol dm^{-3} in benzoic acid and 0.10 mol dm^{-3} in sodium benzoate would be a suitable buffer.

The formula of benzoic acid is C_6H_5COOH and that of sodium benzoate is $C_6H_5COO^-Na^+$.

> ### Study tip
>
> Always quote pH values to two decimal places.

1 Calculate the relative molecular mass of **a** benzoic acid and **b** sodium benzoate.
2 Calculate the mass of **a** benzoic acid and **b** sodium benzoate required to make up 250 cm^3 of a buffer solution that is 0.05 mol dm^{-3} in benzoic acid and 0 .10 mol dm^{-3} in sodium benzoate.
3 Describe the procedure for making up such a solution using a 250 cm^3 graduated flask.

Adding base

If you add 10 cm^3 of 0.10 mol dm^{-3} sodium hydroxide to the original buffer, it will react with the H$^+$ ions and more HA will ionise, so this time you decrease the concentration of the acid [HA] by 0.010 and increase the concentration of ethanoate ions by 0.010.

Using similar steps to those above gives the new pH as 4.89. Check that you agree with this answer.

Note how small the pH change is (from 4.77 to 4.89).

Summary question

1 Find the pH of the following buffers.
   a Using [ethanoic acid] = 0.10 mol dm^{-3}, [sodium ethanoate] = 0.20 mol dm^{-3} (K_a of ethanoic acid = 1.7×10^{-5}).
   b Using [benzoic acid] = 0.10 mol dm^{-3}, [sodium benzoate] = 0.10 mol dm^{-3} (K_a of benzoic acid = 6.3×10^{-5}).

Practice questions

1   $\sqrt{x}$ In this question give all values of pH to 2 decimal places.
The acid dissociation constant K_a for propanoic acid has the value 1.35×10^{-5} mol dm^{-3} at 25°C.

$$K_a = \frac{[H^+][CH_3CH_2COO^-]}{[CH_3CH_2COOH]}$$

(a)   Calculate the pH of a 0.169 mol dm^{-3} solution of propanoic acid.

(3 marks)

(b)   A buffer solution contains 0.250 mol of propanoic acid and 0.190 mol of sodium propanoate in 1000 cm^3 of solution.
A 0.015 mol sample of solid sodium hydroxide is then added to this buffer solution.
(i)   Write an equation for the reaction of propanoic acid with sodium hydroxide.
(ii)   Calculate the number of moles of propanoic acid and of propanoate ions present in the buffer solution after the addition of the sodium hydroxide.
(iii)   Hence, calculate the pH of the buffer solution after the addition of the sodium hydroxide.

(6 marks)
AQA, 2008

2   $\sqrt{x}$ In this question, give all values of pH to 2 decimal places.
(a)   The ionic product of water has the symbol K_w
(i)   Write an expression for the ionic product of water.

(1 mark)

(ii)   At 42 °C, the value of K_w is 3.46×10^{-14} mol^2 dm^{-6}.
Calculate the pH of pure water at this temperature.

(2 marks)

(iii)   At 75 °C, a 0.0470 mol dm^{-3} solution of sodium hydroxide has a pH of 11.36
Calculate a value for K_w at this temperature.

(2 marks)

(b)   Methanoic acid (HCOOH) dissociates slightly in aqueous solution.
(i)   Write an equation for this dissociation.

(1 mark)

(ii)   Write an expression for the acid dissociation constant K_a for methanoic acid.

(1 mark)

(iii)   The value of K_a for methanoic acid is 1.78×10^{-4} mol dm^{-3} at 25 °C. Calculate the pH of a 0.0560 mol dm^{-3} solution of methanoic acid.

(3 marks)

(iv)   The dissociation of methanoic acid in aqueous solution is endothermic.
Deduce whether the pH of a solution of methanoic acid will increase, decrease or stay the same if the solution is heated. Explain your answer.

(3 marks)

(c)   The value of K_a for methanoic acid is 1.78×10^{-4} mol dm^{-3} at 25 °C.
A buffer solution is prepared containing 2.35×10^{-2} mol of methanoic acid and 1.84×10^{-2} mol of sodium methanoate in 1.00 dm^3 of solution.
(i)   Calculate the pH of this buffer solution at 25 °C.

(3 marks)

(ii)   A 5.00 cm^3 sample of 0.100 mol dm^{-3} hydrochloric acid is added to the buffer solution in part **(c) (i)**.
Calculate the pH of the buffer solution after this addition.

(4 marks)
AQA, 2013

3   $\sqrt{x}$ In this question give all values of pH to 2 decimal places.
(a)   The dissociation of water can be represented by the following equilibrium.

$$H_2O(l) \rightleftharpoons H^+(aq) + OH^-(aq)$$

(i)   Write an expression for the ionic product of water, K_w.
(ii)   The pH of a sample of pure water is 6.63 at 50°C.
Calculate the concentration in mol dm^{-3} of H$^+$ ions in this sample of pure water.

(iii) Deduce the concentration in $mol\,dm^{-3}$ of OH^- ions in this sample of pure water.

(iv) Calculate the value of K_w at this temperature.

(4 marks)

(b) At 25 °C the value of K_w is $1.00 \times 10^{-14}\,mol^2\,dm^{-6}$.
Calculate the pH of a $0.136\,mol\,dm^{-3}$ solution of KOH at 25°C.

(2 marks)

AQA, 2008

4   Titration curves labelled A, B, C, and D for combinations of different aqueous solutions of acids and bases are shown below.
All solutions have a concentration of $0.1\,mol\,dm^{-3}$.

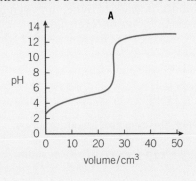

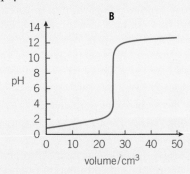

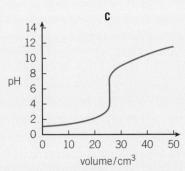

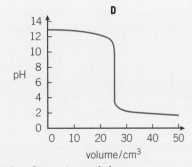

(a) In this part of the question write the appropriate letter in each box.
From the curves **A**, **B**, **C**, and **D**, choose the curve produced by the addition of:

(i)   ammonia to 25 cm³ of hydrochloric acid

(ii) sodium hydroxide to 25 cm³ of ethanoic acid

(iii) nitric acid to 25 cm³ of potassium hydroxide.

(3 marks)

(b) A table of acid–base indicators is shown below.
The pH ranges over which the indicators change colour and their colours in acid and alkali are also shown.

Indicator	pH range	Colour in acid	Colour in alkali
Trapaeolin	1.3 – 3.0	red	yellow
Bromocresol green	3.8 – 5.4	yellow	blue
Cresol purple	7.6 – 9.2	yellow	purple
Alizarin yellow	10.1 – 12.0	yellow	orange

(i)   Select from the table an indicator that could be used in the titration that produces curve **B** but **not** in the titration that produces curve **A**.

(1 mark)

(ii) Give the colour change at the end point of the titration that produces curve **D** when cresol purple is used as the indicator.

(1 mark)

AQA, 2011

83

Section 1 practice questions

1 The oxides nitrogen monoxide, NO, and nitrogen dioxide, NO_2, both contribute to atmospheric pollution.
The table gives some data for these oxides and for oxygen.

	$S^{\ominus}$/JK mol^{-1}	$\Delta_f H$/kJ mol^{-1}
$O_2(g)$	211	0
$NO(g)$	205	+90
$NO_2(g)$	240	+34

Nitrogen monoxide is formed in internal combustion engines. When nitrogen monoxide comes into contact with air, it reacts with oxygen to form nitrogen dioxide.

$$NO(g) + \frac{1}{2}O_2(g) \rightarrow NO_2(g)$$

(a) √x̄ Calculate the enthalpy change for this reaction.

(2 marks)

(b) √x̄ Calculate the entropy change for this reaction.

(2 marks)

(c) √x̄ Calculate the temperature below which this reaction is spontaneous.

(2 marks)

(d) Suggest **one** reason why nitrogen dioxide is **not** formed by this reaction in an internal combustion engine.

(1 mark)

(e) Write an equation to show how nitrogen monoxide is formed in an internal combustion engine.

(1 mark)

(f) Use your equation from part **(e)** to explain why the free-energy change for the reaction to form nitrogen monoxide stays approximately constant at different temperatures.

(2 marks)
AQA, 2012

2 The balance between enthalpy change and entropy change determines the feasibility of a reaction. The table below contains enthalpy of formation and entropy data for some elements and compounds.

	$N_2(g)$	$O_2(g)$	$NO(g)$	C(graphic)	C(diamond)
$\Delta_f H^{\ominus}$/kJ mol^{-1}	0	0	+90.4	0	+1.9
$S^{\ominus}$/JK^{-1} mol^{-1}	192.2	205.3	211.1	5.7	2.4

(a) Explain why the entropy value for the element nitrogen is much greater than the entropy value for the element carbon (graphite).

(2 marks)

(b) Suggest the condition under which the element carbon (diamond) would have an entropy value of zero.

(1 mark)

(c) Write the equation that shows the relationship between ΔG, ΔH and ΔS for a reaction.

(1 mark)

(d) State the requirement for a reaction to be feasible.

(1 mark)

(e) √x̄ Consider the following reaction that can lead to the release of the pollutant NO into the atmosphere.

$$\frac{1}{2}N_2(g) + \frac{1}{2}O_2(g) \rightarrow NO(g)$$

Use data from the table to calculate the minimum temperature above which this reaction is feasible.

(5 marks)

(f) At temperatures below the value calculated in part **2 (e)**, decomposition of NO into its elements should be spontaneous. However, in car exhausts this decomposition reaction does **not** take place in the absence of a catalyst. Suggest why this spontaneous decomposition does **not** take place.

(1 mark)

(g) ⓥ̄ A student had an idea to earn money by carrying out the following reaction.

$$C(graphite) \rightarrow C(diamond)$$

Use the data to calculate values for ΔH and ΔS for this reaction. Use these values to explain why this reaction is **not** feasible under standard pressure at any temperature.

(3 marks)
AQA, 2011

3 Use the table below, where appropriate, to answer the questions which follow.

Standard electrode potentials	$E^{\ominus}/V$
$2H^+(aq) + 2e^- \rightarrow H_2(g)$	0.00
$Br_2(aq) + 2e^- \rightarrow 2Br^-(aq)$	+1.09
$2BrO_3^-(aq) + 12H^+(aq) + 10e^- \rightarrow Br_2(aq) + 6H_2O(l)$	+1.52

Each of the above reactions can be reversed under suitable conditions.

(a) State the hydrogen ion concentration and the hydrogen gas pressure when, at 298 K, the potential of the hydrogen electrode is 0.00 V.

(2 marks)

(b) A diagram of a cell using platinum electrodes **X** and **Y** is shown below.

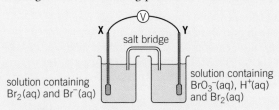

(i) ⓥ̄ Use the data in the table above to calculate the emf of the above cell under standard conditions.

(ii) Write a half equation for the reaction occurring at electrode **X** and an overall equation for the cell reaction which occurs when electrodes **X** and **Y** are connected.

(4 marks)
AQA, 2004

4 In this question, give all values of pH to 2 decimal places.

(a) ⓥ̄ The ionic product of water has the symbol K_w.

(i) Write an expression for the ionic product of water.

(1 mark)

(ii) At 42°C, the value of K_w is 3.46×10^{-14} mol^2 dm^{-6}.
Calculate the pH of pure water at this temperature.

(2 marks)

(iii) At 75°C, a 0.0470 mol dm^{-3} solution of sodium hydroxide has a pH of 11.36
Calculate a value for K_w at this temperature.

(2 marks)

(b) Methanoic acid, HCOOH, dissociates slightly in aqueous solution.

(i) Write an equation for this dissociation.

(1 mark)

(ii) Write an expression for the acid dissociation constant K_a for methanoic acid.

(1 mark)

(iii) The value of K_a for methanoic acid is 1.78×10^{-4} mol dm^{-3} at 25°C.
Calculate the pH of a 0.0560 mol dm^{-3} solution of methanoic acid.

(3 marks)

(iv) The dissociation of methanoic acid in aqueous solution is endothermic.
Deduce whether the pH of a solution of methanoic acid will increase, decrease or stay the same if the solution is heated. Explain your answer.

(3 marks)

(c) ✓x The value of K_a for methanoic acid is 1.78×10^{-4} mol dm^{-3} at 25°C.
A buffer solution is prepared containing 2.35×10^{-2} mol of methanoic acid and 1.84×10^{-2} mol of sodium methanoate in 1.00 dm3 of solution.

(i) Calculate the pH of this buffer solution at 25°C.

(3 marks)

(ii) A 5.00 cm^3 sample of 0.100 mol dm^{-3} hydrochloric acid is added to the buffer solution in part **(c)** (i).
Calculate the pH of the buffer solution after this addition.

(4 marks)

AQA, 2012

5 Formic acid is a weak acid sometimes used in products to remove limescale from toilet bowls, sinks and kettles. The structure of formic acid is:

$$H-C(=O)-O-H$$

(a) What is the systematic name of formic acid? *(1 mark)*
(b) What is meant by the term *weak acid*? *(1 mark)*
(c) The equation for the dissociation of formic acid is:

$$H-C(=O)-O-H \ (aq) \rightleftharpoons H-C(=O)-O^- + H^+$$

Explain why the hydrogen atom bonded to the oxygen is the one that dissociates rather than that bonded to the carbon atom. *(1 mark)*

(d) Write an expression for K_a, the dissociation constant of formic acid.
You may use the symbol HA for formic acid and A$^-$ for the formate ion. *(2 marks)*

(e) The values of pK_a for some other weak acids are in the table.

Acid	pK_a
sulfamic	0.99
formic	3.8
ethanoic	4.8

Which is the strongest acid? Explain your answer. *(2 marks)*

(f) ✓x Use the expression for K_a to calculate [H$^+$] in a 1 mol dm^{-3} solution of formic acid and hence the pH of this solution. *(4 marks)*

(g) Limescale contains calcium carbonate, $CaCO_3$. Write an equation for the reaction of calcium carbonate with formic acid. *(2 marks)*

(h) What mass of calcium carbonate would react with 4.5 g of formic acid? *(2 marks)*

(i) Bleach bases toilet cleaners contain chloric(I) acid in which the following equilibrium is set up:
$Cl_2(g) + OH^-(aq) \rightarrow HClO(aq) + Cl^-(aq)$
Suggest what would happen if formic acid is added to such a cleaner.
Why would this be potentially dangerous. *(3 marks)*

6 (a) Explain why the atomic radii of the elements decrease across Period 3 from sodium to chlorine.

(2 marks)

(b) Explain why the melting point of sulfur, S_8, is greater than that of phosphorus, P_4.

(2 marks)

(c) Explain why sodium oxide forms an alkaline solution when it reacts with water.

(2 marks)

(d) Write an ionic equation for the reaction of phosphorus(V) oxide with an excess of sodium hydroxide solution.

(1 mark)

7 Butadiene dimerises according to the equation

$$2C_4H_6 \rightarrow C_8H_{12}$$

The kinetics of the dimerisation are studied and the graph of the concentration of a sample of butadiene is plotted against time. The graph is shown in **Figure 1**.

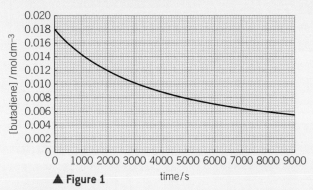

▲ **Figure 1**

(a) Draw a tangent to the curve when the concentration of butadiene is 0.0120 mol dm^{-3}.

(1 mark)

(b) The initial rate of reaction in this experiment has the value

$$4.57 \times 10^{-6} \, mol \, dm^{-3} \, s^{-1}.$$

Use this value, together with a rate obtained from your tangent, to justify that the order of the reaction is 2 with respect to butadiene.

(5 marks)

8 The human stomach typically contains 1 dm^3 of hydrochloric acid (HCl) which is used to aid digestion of food. The concentration of this acid is approximately 0.01 mol dm^{-3}. Hydrochloric acid is a strong acid.

(a) What is meant by the term *strong acid*? *(1 mark)*

(b) Write an equation for the dissociation of HCl in water. *(1 mark)*

(c) Calculate:

(i) the number of moles of hydrochloric acid in the stomach and

(ii) the pH of the acid in the stomach. *(2 marks)*

Too much acid in the stomach can cause heartburn and many medicines contain antacids – compounds that react to neutralise excess acid. One such medicine contains, in the recommended dose 0.267 g of sodium hydrogencarbonate (NaHCO$_3$) and 0.160 g of calcium carbonate (CaCO$_3$).

(d) Write equation for the reactions of:

(i) sodium hydrogencarbonate and

(ii) calcium carbonate with hydrochloric acid. *(4 marks)*

(e) Using these equations, calculate how many moles of hydrochloric acid are neutralised by:

(i) the sodium hydrogencarbonate

(ii) the calcium carbonate

(iii) the total number of moles of hydrochloric acid that is neutralised by the recommended dose of the antacid. *(4 marks)*

(f) (i) Calculate the number of moles of hydrochloric acid that remain in the stomach

(ii) Calculate the pH of the stomach acid remaining. (Remember that the acid has a volume of 1 dm^3). *(2 marks)*

Section 1 summary

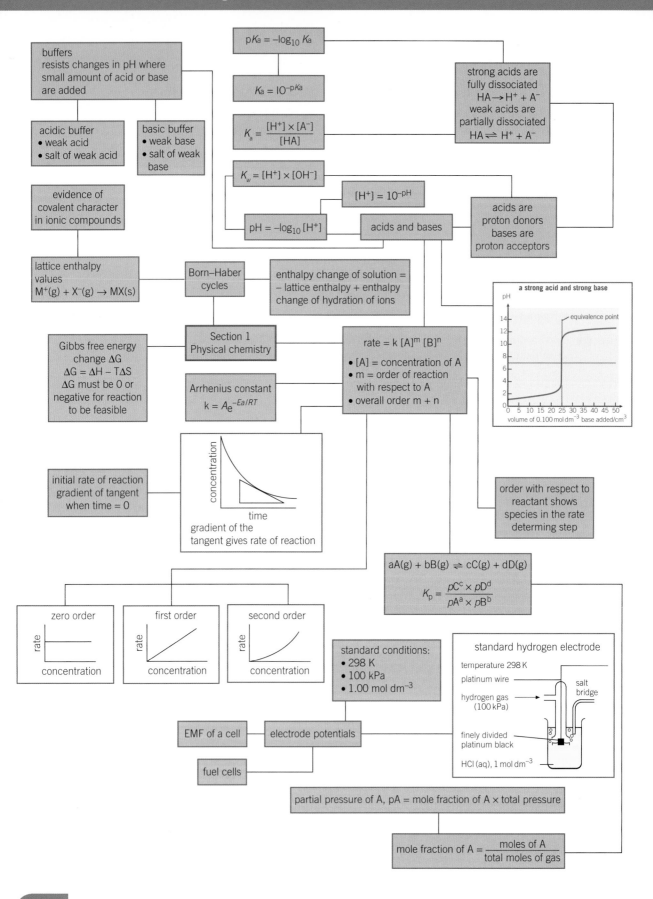

buffers
resists changes in pH where small amount of acid or base are added

$pK_a = -\log_{10} K_a$

strong acids are fully dissociated
$HA \rightarrow H^+ + A^-$
weak acids are partially dissociated
$HA \rightleftharpoons H^+ + A^-$

acidic buffer
• weak acid
• salt of weak acid

basic buffer
• weak base
• salt of weak base

$K_a = 10^{-pK_a}$

$K_a = \dfrac{[H^+] \times [A^-]}{[HA]}$

$K_w = [H^+] \times [OH^-]$

$[H^+] = 10^{-pH}$

acids are proton donors
bases are proton acceptors

evidence of covalent character in ionic compounds

$pH = -\log_{10} [H^+]$

acids and bases

lattice enthalpy values
$M^+(g) + X^-(g) \rightarrow MX(s)$

Born–Haber cycles

enthalpy change of solution = – lattice enthalpy + enthalpy change of hydration of ions

a strong acid and strong base

Gibbs free energy change ΔG
$\Delta G = \Delta H - T\Delta S$
ΔG must be 0 or negative for reaction to be feasible

Section 1 Physical chemistry

rate = k $[A]^m [B]^n$
• [A] = concentration of A
• m = order of reaction with respect to A
• overall order m + n

Arrhenius constant
$k = Ae^{-Ea/RT}$

initial rate of reaction gradient of tangent when time = 0

gradient of the tangent gives rate of reaction

order with respect to reactant shows species in the rate determing step

$aA(g) + bB(g) \rightleftharpoons cC(g) + dD(g)$

$K_p = \dfrac{pC^c \times pD^d}{pA^a \times pB^b}$

zero order

first order

second order

standard conditions:
• 298 K
• 100 kPa
• 1.00 mol dm^{-3}

standard hydrogen electrode

temperature 298 K
platinum wire
hydrogen gas (100 kPa)
salt bridge
finely divided platinum black
HCl (aq), 1 mol dm^{-3}

EMF of a cell

electrode potentials

fuel cells

partial pressure of A, pA = mole fraction of A × total pressure

mole fraction of A = $\dfrac{\text{moles of A}}{\text{total moles of gas}}$

Practical skills

In this section you have met the following ideas:

- Finding the entropy change for the vaporisation of water.
- Finding the order of reaction for a reactant.
- Making simple cells and using them to measure an unknown electrode potential.
- Finding how changing conditions effects the EMF of a cell.
- Using $E^{\ominus}$ values to predict the direction of simple redox reaction.
- Finding out the value of K_a for a weak acid.
- Exploring how the pH changes during neutralisation reactions.
- Making and testing a buffer solution.

Maths skills

In this section you have met the following maths skills:

- Using the expression $\Delta G = \Delta H - T\Delta S$ to solve problems.
- Determining the value of ΔS and ΔH from a graph of ΔG versus T.
- Working out the rate equation from given data.
- Using a concentration – time graph to calculate the rate of reaction.
- Calculating the value of the rate constant from the gradient of a zero order concentration – time graph.
- Calculating the partial pressures of reactants and products at equilibrium.
- Finding out the value of K_p
- Calculating the pH of a solution from $[H^+]$ and the value of $[H^+]$ from the pH.
- Using standard form to solve equations
- Finding the pH of strong bases.
- Calculating the pH of a buffer solution.

Extension

Produce a report exploring how ideas about entropy and enthalpy can be applied to everyday life.

Suggested resources:

Atkins, P[2010], *The Laws of Thermodynamics: A Very Short Introduction*. Oxford University Press, UK. ISBN 978-0-19-957219-9

Price, G[1998], *Thermodynamics of Chemical Processes: Oxford Chemistry Primers*. Oxford University Press, UK. ISBN 978-0-19-855963-4

Section 2
Inorganic Chemistry 2

Chapters in this section

22 Periodicity

23 The transition metals

24 Reactions of inorganic compounds in aqueous solution

Periodicity introduces the chemical properties of the Period 3 elements, and some of their compounds are described in order to establish patterns and trends in chemical behaviour across a period.

The transition metals have unique chemical structures which give their compounds characteristic (and useful) properties. These include coloured compounds, complex formation, catalytic activity, and variable oxidation states.

Reactions of inorganic compounds in aqueous solution looks at some reactions of some metal ions in solution – including acid-base reactions and ligand substitution reactions – forming metal-aqua ions.

What you already know:

The material in this unit builds on knowledge and understanding that you have built up at AS level. In particular the following:

- ◯ Atoms may be held together by covalent, ionic, or metallic bonds.
- ◯ Elements in the Periodic Table show patterns in their properties, which are related to their electronic structures.
- ◯ The transition metals form a block in the periodic table between the s-block elements and the p-block elements. In general they are hard and strong and have typical metallic properties.
- ◯ Acid-base reactions involve the transfer of H^+ ions.

Learning objectives:

→ State how, and under what conditions, sodium and magnesium react with water.

→ State how the elements from sodium to sulfur react with oxygen.

Specification reference: 3.2.4

As you move across a period in the Periodic Table from left to right, there are a number of trends in the properties of the elements.

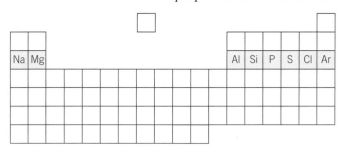

You have already looked at the physical properties of the elements from sodium to argon in Period 3. Here you will examine some of the chemical reactions of these elements.

The elements

The most obvious physical trend in the elements is from metals on the left to non-metals on the right.

- Sodium, magnesium, and aluminium are metallic – they are shiny (when freshly exposed to air), conduct electricity, and react with dilute acids to give hydrogen and salts.

- Silicon is a semi-metal (or metalloid) – it conducts electricity to some extent, a property that is useful in making semiconductor devices.

- Phosphorus, sulfur, and chlorine are typical non-metals – in particular, they do not conduct electricity and have low melting and boiling points.

- Argon is a noble gas. It is chemically unreactive and exists as separate atoms.

The redox reactions of the elements

The reactions of the elements in Period 3 are all redox reactions, since every element starts with an oxidation state of zero, and, after it has reacted, ends up with a positive or a negative oxidation state.

Reactions with water

Sodium and magnesium are the only metal elements in Period 3 that react with cold water. (Chlorine is the only non-metal that reacts with water.)

Sodium

The reaction of sodium with water is vigorous – the sodium floats on the surface of the water and fizzes rapidly, melting because of the heat energy released by the reaction. A strongly alkaline solution of sodium hydroxide is formed (pH 13–14). The oxidation state changes are shown as small numbers above the following symbol equations:

$$\overset{0}{2Na(s)} + \overset{+1\ -2}{2H_2O(l)} \rightarrow \overset{+1\ -2+1}{2NaOH(aq)} + \overset{0}{H_2(g)}$$

Magnesium

The reaction of magnesium is very slow at room temperature, only a few bubbles of hydrogen are formed after some days. The resulting solution is less alkaline than in the case of sodium because magnesium hydroxide is only sparingly soluble (pH around 10).

$$\underset{0}{Mg(s)} \; + \; \underset{+1 \; -2}{2H_2O(l)} \; \rightarrow \; \underset{+2 \; -2 \; +1}{Mg(OH)_2(aq)} \; + \; \underset{0}{H_2(g)}$$

The reaction is much faster with heated magnesium and steam and gives magnesium oxide and hydrogen.

$$\underset{0}{Mg(s)} + \underset{+1 \; -2}{H_2O(g)} \rightarrow \underset{+2 \; -2}{MgO(s)} + \underset{0}{H_2(g)}$$

All these reactions are redox ones, in which the oxidation state of the metal increases and that of some of the hydrogen atoms decreases.

Reaction with oxygen

All the elements in Period 3 (except for argon) are relatively reactive. Their oxides can all be prepared by direct reaction of the element with oxygen. The reactions are exothermic.

Sodium burns brightly in air (with a characteristic yellow flame) to form white sodium oxide:

$$\underset{0}{2Na(s)} + \underset{0}{\tfrac{1}{2}O_2(g)} \rightarrow \underset{+1 \; -2}{Na_2O(s)}$$

Magnesium

A strip of magnesium ribbon burns in air with a bright white flame. The white powder that is produced is magnesium oxide. If burning magnesium is lowered into a gas jar of oxygen the flame is even more intense (Figure 1).

$$magnesium + oxygen \rightarrow magnesium\ oxide$$
$$\underset{0}{2Mg(s)} + \underset{0}{O_2(g)} \rightarrow \underset{+2 \; -2}{2MgO(s)}$$

The oxidation states show how magnesium has been oxidised (its oxidation state has increased) and oxygen has been reduced (its oxidation number has decreased).

Aluminium

When aluminium powder is heated and then lowered into a gas jar of oxygen, it burns brightly to give aluminium oxide – a white powder. Aluminium powder also burns brightly in air (Figure 2).

$$aluminium + oxygen \rightarrow aluminium\ oxide$$
$$\underset{0}{4Al(s)} + \underset{0}{3O_2(g)} \rightarrow \underset{+3 \; -2}{2Al_2O_3(s)}$$

Aluminium is a reactive metal, but it is always coated with a strongly bonded surface layer of oxide – this protects it from further reaction. So, aluminium appears to be an unreactive metal and is used for many everyday purposes – saucepans, garage doors, window frames, and so on.

▲ **Figure 1** *Magnesium burning in oxygen from the air*

Hint

The sodium oxide formed may have a yellowish appearance due to the production of some sodium peroxide, Na_2O_2.

▲ **Figure 2** *Aluminium burning in oxygen from the air. Powdered aluminium is being sprinkled into the flame*

Hint

The sum of the oxidation states in Al_2O_3 is zero, as it is in all compounds without a charge:

$$(2 \times 3) + (3 \times -2) = 0$$

Even if the surface is scratched, the exposed aluminium reacts rapidly with the air and seals off the surface.

Silicon

Silicon will also form the oxide if it is heated strongly in oxygen:

$$\overset{0}{Si}(s) + \overset{0}{O_2}(g) \rightarrow \overset{+4\ -2}{SiO_2}(s)$$

Phosphorus

Red phosphorus must be heated before it will react with oxygen. White phosphorus spontaneously ignites in air and the white smoke of phosphorus pentoxide is given off. Red and white phosphorus are **allotropes** of phosphorus – the same element with the atoms arranged differently.

$$\overset{0}{4P}(s) + \overset{0}{5O_2}(g) \rightarrow \overset{+5\ -2}{P_4O_{10}}(s)$$

If the supply of oxygen is limited, phosphorus trioxide, P_2O_3, is also formed.

Sulfur

When sulfur powder is heated and lowered into a gas jar of oxygen, it burns with a blue flame to form the colourless gas sulfur dioxide (and a little sulfur trioxide also forms) (Figure 3).

$$sulfur + oxygen \rightarrow sulfur\ dioxide$$

$$\overset{0}{S}(s) + \overset{0}{O_2}(g) \rightarrow \overset{+4\ -2}{SO_2}(g)$$

▲ **Figure 3** *Sulfur burning in oxygen*

In all these redox reactions, the oxidation state of the Period 3 element increases and that of the oxygen decreases (from 0 to –2 in each case). The oxidation state changes are shown as small numbers above the symbol equations above. The oxidation number of the Period 3 element in the oxide increases as you move from left to right across the period.

Hint

The empirical formula of phosphorus pentoxide is P_2O_5. In the gas phase it forms molecules of P_4O_{10} and is sometimes referred to as phosphorus(V) oxide.

$$4P + 5O_2 \rightarrow P_4O_{10}$$

Phosphorus and oxygen also form phosphorus trioxide, P_2O_3. Phosphorus trioxide is also called phosphorus(III) oxide.

Summary questions

1 Metals are shiny, conduct electricity, and react with acids to give hydrogen if they are reactive. Give three more properties not mentioned in the text.

2 Non-metals do not conduct electricity. Give two more properties typical of non-metals.

3 **a** What is the oxidation state of sodium in all its compounds?

   b What is the oxidation state of oxygen in sodium peroxide, Na_2O_2? What is unusual about this?

   c Show that the sum of the oxidation states in magnesium hydroxide is zero.

4 What is the oxidation state of sulfur in sulfur trioxide?

As you move across Period 3 from left to right there are some important trends in the physical properties of the Period 3 compounds. These trends are a result of the change from metal elements on the left of the Periodic Table to non-metal elements on the right. The oxides are representative of these trends.

The metal oxides

Sodium, magnesium, and aluminium oxides are examples of compounds formed by a metal combined to a non-metal. They form giant ionic lattices where the bonding extends throughout the compound. This results in high melting points.

The bonding in aluminium oxide is ionic but has some covalent character. This is because aluminium forms a very small ion with a large positive charge and so can approach closely to the O^{2-} and distort its electron cloud. So, the bond also has some added covalent character (Figure 1).

It is possible to predict the ionic character of a bond by considering the difference in electronegativities between the two atoms. The bigger the difference, the greater the ionic character of the bond. Caesium oxide, Cs_2O, is about 80% ionic. The electronegativities are Cs = 0.7 and O = 3.5. A bond between two identical atoms, such as that in oxygen, *must* be 100% covalent.

The non-metal oxides

Silicon oxide has a giant covalent (macromolecular) structure. Again the bonding extends throughout the giant structure, but this time it is covalent (Figure 2). Again you have a compound with a high melting point because many strong covalent bonds must be broken to melt it.

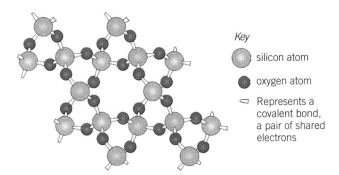

Key
- ⚪ silicon atom
- ⚫ oxygen atom
- ▱ Represents a covalent bond, a pair of shared electrons

▲ **Figure 2** *Silicon dioxide is a macromolecule*

Phosphorus and sulfur oxides exist as separate covalently bonded molecules. The phosphorus oxides are solids. The intermolecular forces are weak van der Waals and dipole–dipole forces. Their melting points are relatively low. Sulfur dioxide and sulfur trioxide are gases at 298 K.

Learning objectives:

→ Describe how the physical properties of the oxides are explained in terms of their structure and bonding.

→ State how the oxides react with water.

→ Describe how the structures of the oxides explain the trend in their reactions in water.

Specification reference: 3.2.4

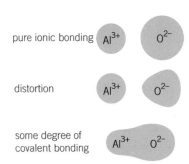

pure ionic bonding Al^{3+} O^{2-}

distortion Al^{3+} O^{2-}

some degree of covalent bonding Al^{3+} O^{2-}

▲ **Figure 1** *The covalent character of the bonding in aluminium oxide*

Synoptic link

In Topic 17.3, More enthalpy changes, you saw that the bonding in zinc selenide was not purely ionic, but had some degree of covalent character.

Synoptic link

Look back at Topic 3.5, Forces acting between molecules, to revise intermolecular forces.

The trends in the physical properties of the oxides are summarised in Table 1.

▼ **Table 1** *The trends in the physical properties of some of the oxides in Period 3*

	Na_2O	MgO	Al_2O_3	SiO_2	P_4O_{10}	SO_3	SO_2
T_m / K	1548	3125	2345	1883	573	290	200
Bonding	ionic	ionic	ionic/covalent	covalent	covalent	covalent	covalent
Structure	giant ionic	giant ionic	giant ionic	giant covalent (macromolecular)	molecular	molecular	molecular

Study tip

It is important that you understand the link between physical properties and bonding.

Note the trend in melting points: $P_4O_{10} > SO_3 > SO_2$

This is related to the increase in intermolecular forces between the larger molecules.

The structures of oxo-acids and their anions

Phosphoric(V) acid, H_3PO_4, sulfuric(VI) acid, H_2SO_4 and sulfuric (IV) acid (sulfurous acid, H_2SO_3) are called oxo acids. You can understand the structures of these acids and their anions by drawing dot-and-cross diagrams.

A phosphorus atom has five electrons in its outer shell and the bonding in phosphoric(V) acid is as shown. Note that as phosphorus is in Period 3, it can have more than eight electrons in its outer main level.

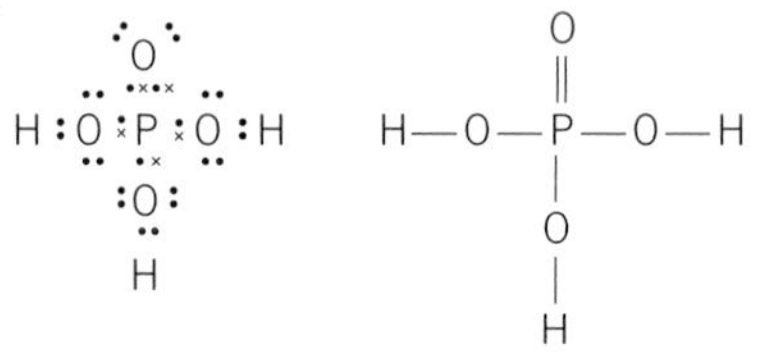

▲ **Figure 3** *Phosphoric(V) acid*

If all three hydrogens are lost as protons, the phosphate(V) ion is formed.

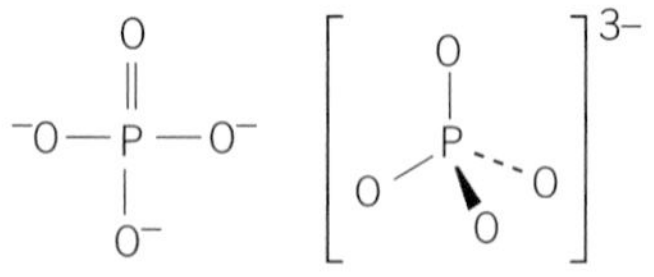

▲ **Figure 4** *Phosphate(V) ion*

At first sight this appears to have one phosphorus–oxygen double bond and three P—O single bonds.

However, it turns out that all three bonds are the same, the bonding electrons are spread out equally over all four bonds, and the shape of the ion is a perfect tetrahedron. This is an example of **delocalisation** which you have seen in graphite and in benzene.

Sulfur has six electrons in its outer shell and it, too, may have more than eight electrons in its outer main level. The bonding in sulfuric(VI) acid is as shown.

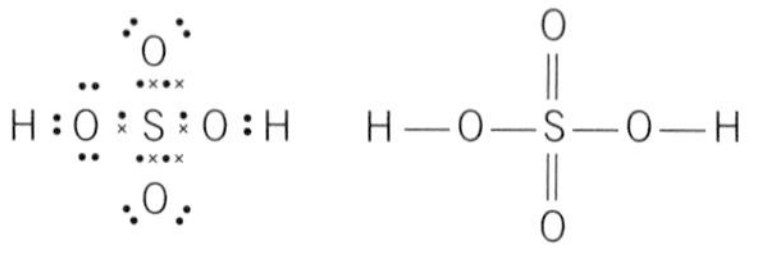

▲ **Figure 5** *Sulfuric(VI) acid*

Like the phosphate(V) ion, the sulfate(VI) anion is a perfect tetrahedron due to delocalisation.

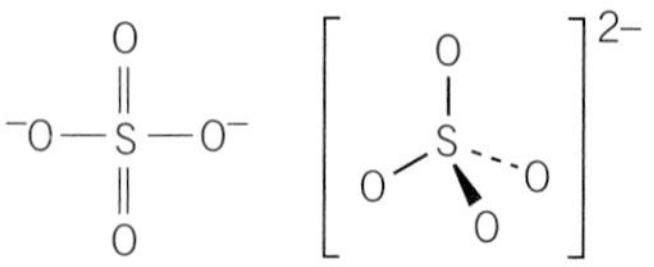

▲ **Figure 6** *Sulfate(VI) anion*

Sulfuric (IV) acid immediately dissociates in aqueous solution. The sulfate(IV) anion's bonding is as shown. Delocalisation makes the three S—O bonds the same length but the bond angle is 106°, smaller than that of a perfect tetrahedron due to the lone pair of electrons.

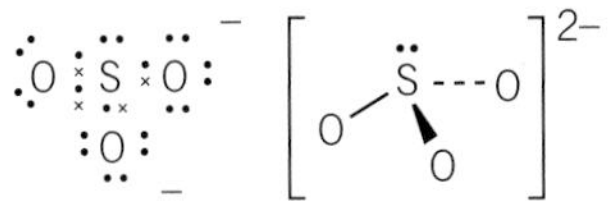

▲ **Figure 7** *Sulfate(IV) anion*

1 What is the angle of a perfect tetrahedron?
2 What can be said about the lengths of the bonds in PO_4^{3-}?
3 How many electrons does phosphorus have in its outer main level in phosphoric(V) acid?
4 The S—O bond length is 0.157 nm and the S=O bond length is 0.143 nm. Predict the sulfur–oxygen bond length in the sulfate(VI) ion.
5 How many electrons does sulfur have in its outer main level in the sulfate(IV) anion?

Reaction of oxides with water

Basic oxides

Sodium and magnesium oxides are both bases.

Sodium oxide reacts with water to give sodium hydroxide solution – a strongly alkaline solution:

$$Na_2O(s) + H_2O(l) \rightarrow 2Na^+(aq) + 2OH^-(aq) \text{ pH of solution} \sim 14$$

Magnesium oxide reacts with water to give magnesium hydroxide, which is sparingly soluble in water and produces a somewhat alkaline solution:

$$MgO(s) + H_2O(l) \rightarrow Mg(OH)_2(s) \rightleftharpoons Mg^{2+}(aq) + 2OH^-(aq)$$
$$\text{pH of solution} \sim 9$$

Insoluble oxides

Aluminium oxide is insoluble in water.
Silicon dioxide is insoluble in water.

Acidic oxides

Non-metals on the right of the Periodic Table typically form acidic oxides. For example, phosphorus pentoxide reacts quite violently with water to produce an acidic solution of phosphoric(V) acid. This ionises, so the solution is acidic.

$$P_4O_{10}(s) + 6H_2O(l) \rightarrow 4H_3PO_4(aq)$$

$H_3PO_4(aq)$ ionises in stages, the first being:

$$H_3PO_4(aq) \rightleftharpoons H^+(aq) + H_2PO_4^-(aq)$$

Sulfur dioxide is fairly soluble in water and reacts with it to give an acidic solution of sulfuric(IV) acid (sulfurous acid). This partially dissociates producing H^+ ions, which cause the acidity of the solution:

$$SO_2(g) + H_2O(l) \rightarrow H_2SO_3(aq)$$

$$H_2SO_3(aq) \rightleftharpoons H^+(aq) + HSO_3^-(aq)$$

Sulfur trioxide reacts violently with water to produce sulfuric acid (sulfuric(VI) acid):

$$SO_3(g) + H_2O(l) \rightarrow H_2SO_4(aq) \rightarrow H^+(aq) + HSO_4^-(aq)$$

> **Hint**
>
> Silicon dioxide is the main constituent of sand, which does not dissolve in water.

The overall pattern is that metal oxides (on the left of the period) form alkaline solutions in water and non-metal oxides (on the right of the period) form acidic ones, whilst those in the middle do not react.

Table 2 summarises these reactions.

▼ **Table 2** *The oxides in water*

Oxide	Bonding	Ions present after reaction with water	Acidity/ alkalinity	Approx. pH (Actual values depend on concentration)
Na_2O	ionic	$Na^+(aq), OH^-(aq)$	strongly alkaline	13–14
MgO	ionic	$Mg^{2+}(aq), OH^-(aq)$	somewhat alkaline	10
Al_2O_3	covalent/ ionic	insoluble, no reaction	—	7
SiO_2	covalent	insoluble, no reaction	—	7
P_4O_{10}	covalent	$H^+(aq) + H_2PO_4^-(aq)$	fairly strong acid	1–2
SO_2	covalent	$H^+(aq) + HSO_3^-(aq)$	weak acid	2–3
SO_3	covalent	$H^+(aq) + HSO_4^-(aq)$	strong acid	0–1

The behaviour of the oxides with water can be understood if you look at their bonding and structure, (Table 1).

- Sodium and magnesium oxides, to the left of the Periodic Table, are composed of ions.
- Sodium oxide contains the oxide ion, O^{2-}, which is a very strong base (it strongly attracts protons) and so readily reacts with water to produce hydroxide ions – a strongly alkaline solution.
- Magnesium oxide also contains oxide ions. However, its reaction with water produces a less alkaline solution than sodium oxide because it is less soluble than sodium oxide.
- Aluminium oxide is ionic but the bonding is too strong for the ions to be separated, partly because of the additional covalent bonding it has.
- Silicon dioxide is a giant macromolecule and water will not affect this type of structure.
- Phosphorus oxides and sulfur oxides are covalent molecules and react with water to form acid solutions.

General trend

Solutions of the oxides of the elements go from alkaline to acidic across the Period.

Summary questions

1 Write down an equation for the reaction of sodium oxide with water.

  a i State the oxidation number of sodium before and after the reaction.

     ii Has the sodium been oxidised, reduced, or neither?

2 a What ion is responsible for the alkalinity of the solutions formed when sodium oxide and magnesium oxide react with water?

  b What range of pH values represents an alkaline solution?

3 Phosphorus forms another oxide, P_4O_6.

  a Would you expect it to react with water to form a neutral, acidic, or alkaline solution?

  b Explain your answer.

  c Write an equation for its reaction with water.

As you saw in Topic 21.2, the general trend is alkalis to acids as you go across the period from left to right. So, you could predict that the alkaline oxides will react with acids and the acidic oxides will react with bases.

Sodium and magnesium oxides

Sodium oxide and magnesium oxide react with acids to give salt and water only.

For example, Sodium oxide reacts with sulfuric acid to give sodium sulfate:

$$Na_2O(s) + H_2SO_4(aq) \rightarrow Na_2SO_4(aq) + H_2O(l)$$

Magnesium oxide reacts with hydrochloric acid to give magnesium chloride:

$$MgO(s) + 2HCl(aq) \rightarrow MgCl_2(aq) + H_2O(l)$$

Aluminium oxide

Aluminium oxide reacts *both* with acids and alkalis. It is called an **amphoteric** oxide.

For example, with hydrochloric acid, aluminium chloride is formed:

$$Al_2O_3(s) + 6HCl(aq) \rightarrow 2AlCl_3(aq) + 3H_2O(l)$$

With hot, concentrated sodium hydroxide, sodium aluminate is formed:

$$Al_2O_3(s) + 2NaOH(aq) + 3H_2O(l) \rightarrow 2NaAl(OH)_4(aq)$$

Silicon dioxide

Silicon dioxide will react as a weak acid with strong bases, for example, with hot concentrated sodium hydroxide a colourless solution of sodium silicate is formed:

$$SiO_2(s) + 2NaOH(aq) \rightarrow Na_2SiO_3(aq) + H_2O(l)$$

Iron production

In the production of iron, at the high temperatures inside the blast furnace, calcium oxide reacts with the impurity silicon dioxide (sand) to produce a liquid slag, calcium silicate. This is also an example of the acidic silicon dioxide reacting with a base:

$$SiO_2(s) + CaO(l) \rightarrow CaSiO_3(l)$$

Learning objectives:

→ Describe how the oxides of the elements in Period 3 react with acids.

→ Describe how the oxides of the elements in Period 3 react with bases.

→ State the equations for these reactions.

Specification reference: 3.2.4

Study tip

It is important that you know the products of these reactions and that you can write equations for the reactions occurring.

Phosphorus pentoxide, P_4O_{10}

The reaction of phosphorus pentoxide with an alkali is really the reaction of phosphoric(V) acid, H_3PO_4, because this is formed when phosphorus pentoxide reacts with water. Phosphoric(V) acid has three −OH groups, and each of these has an acidic hydrogen atom. So it will react with sodium hydroxide in three stages, as each hydrogen in turn reacts with a hydroxide ion and is replaced by a sodium ion:

$$H_3PO_4(aq) + NaOH(aq) \rightarrow NaH_2PO_4(aq) + H_2O(l)$$

$$NaH_2PO_4(aq) + NaOH(aq) \rightarrow Na_2HPO_4(aq) + H_2O(l)$$

$$Na_2HPO_4(aq) + NaOH(aq) \rightarrow Na_3PO_4(aq) + H_2O(l)$$

Overall:

$$3NaOH(aq) + H_3PO_4(aq) \rightarrow Na_3PO_4(aq) + 3H_2O(l)$$

Sulfur dioxide

If you add sodium hydroxide to an aqueous solution of sulfur dioxide, first sodium hydrogensulfate(IV) is formed:

$$SO_2(aq) + NaOH(aq) \rightarrow NaHSO_3(aq)$$

Followed by sodium sulfate(IV):

$$NaHSO_3(aq) + NaOH(aq) \rightarrow Na_2SO_3(aq) + H_2O(l)$$

Calcium sulfite

Sulfur dioxide reacts with the base calcium oxide to form calcium sulfite (calcium sulfate(IV)). This is the first step of one of the methods of removing sulfur dioxide from flue gases in power stations:

$$CaO(s) + SO_2(g) \rightarrow CaSO_3(s)$$

The calcium sulfite is further converted to calcium sulfate, $CaSO_4$, and sold as gypsum for plastering.

Summary questions

1 Write balanced symbol equations for the reactions of:

   a sodium oxide with hydrochloric acid

   b magnesium oxide with sulfuric acid

   c aluminium oxide with nitric acid.

2 $SiO_2(s) + 2NaOH(aq) \rightarrow Na_2SiO_3(aq) + H_2O(l)$

   a Copy the equation above and write the oxidation numbers above each atom.

   b Is the reaction a redox reaction?

   c Explain your answer to **b**.

3 Write a balanced symbol for the reaction of phosphorus pentoxide with water.

Practice questions

1   (a)   Suggest why the melting point of magnesium oxide is much higher than the melting point of magnesium chloride.

(2 marks)

(b)   Magnesium oxide and sulfur dioxide are added separately to water. In each case describe what happens. Write equations for any reactions which occur and state the approximate pH of any solution formed.

(6 marks)

(c)   Write equations for two reactions which together show the amphoteric character of aluminium hydroxide.

(4 marks)

AQA, 2006

2   State what is observed when separate samples of sodium oxide and phosphorus(V) oxide are added to water. Write equations for the reactions which occur and, in each case, state the approximate pH of the solution formed.

(6 marks)

AQA, 2005

3   In the questions below, each of the three elements **X**, **Y**, and **Z** is one of the Period 3 elements Na, Mg, Al, Si, or P.

(a)   The oxide of element **X** has a high melting point. The oxide reacts readily with water to form a solution with a high pH.
   (i)   Deduce the type of bonding present in the oxide of element **X**.
   (ii)   Identify element **X**.
   (iii)   Write an equation for the reaction between water and the oxide of element **X**.

(3 marks)

(b)   Element **Y** has an oxide which reacts vigorously with water to form a solution containing strong acid.
   (i)   Deduce the type of bonding present in the oxide of element **Y**.
   (ii)   Identify element **Y**.
   (iii)   Identify an acid which is formed when the oxide of element **Y** reacts with water.

(3 marks)

(c)   The oxide of element **Z** is a crystalline solid with a very high melting point. This oxide is classified as an acidic oxide but it is not soluble in water.
   (i)   Deduce the type of crystal shown by the oxide of element **Z**.
   (ii)   Identify element **Z**.
   (iii)   Write an equation for a reaction which illustrates the acidic nature of the oxide of element **Z**.

(4 marks)

AQA, 2004

4   Consider the following oxides.
        Na_2O, MgO, Al_2O_3, SiO_2, P_4O_{10}, SO_3
(a)   Identify one of the oxides from the above which:
   (i)   can form a solution with a pH less than 3
   (ii)   can form a solution with a pH greater than 12.

(2 marks)

(b)   Write an equation for the reaction between:
   (i)   MgO and HNO_3
   (ii) SiO_2 and NaOH
   (iii) Na_2O and H_3PO_4.

(3 marks)

(c)   Explain, in terms of their type of structure and bonding, why P_4O_{10} can be vaporised by gentle heat but SiO_2 cannot.

(4 marks)

AQA, 2003

5   Write equations for the reactions which occur when the following compounds are added separately to water. In each case, predict the approximate pH of the solution formed when one mole of each compound is added to $1\,dm^3$ of water.
   (a) Sodium oxide
   (b) Sulfur dioxide.

(2 marks)
AQA, 2003

6   **(a)** **P** and **Q** are oxides of Period 3 elements.
      Oxide **P** is a solid with a high melting point. It does not conduct electricity when solid but does conduct when molten or when dissolved in water. Oxide **P** reacts with water forming a solution with a high pH.
      Oxide **Q** is a colourless gas at room temperature. It dissolves in water to give a solution with a low pH.
      (i) Identify **P**. State the type of bonding present in **P** and explain its electrical conductivity. Write an equation for the reaction of **P** with water.
      (ii) Identify **Q**. State the type of bonding present in **Q** and explain why it is a gas at room temperature. Write an equation for the reaction of **Q** with water.

(9 marks)

   (b) **R** is a hydroxide of a Period 3 element. It is insoluble in water but dissolves in both aqueous sodium hydroxide and aqueous sulfuric acid.
      (i) Give the name used to describe this behaviour of the hydroxide.
      (ii) Write equations for the reactions occurring.
      (iii) Suggest why **R** is insoluble in water.

(6 marks)
AQA, 2002

7   This question is about some Period 3 elements and their oxides.
   (a) Describe what you would observe when, in the absence of air, magnesium is heated strongly with water vapour at temperatures above 373 K.
      Write an equation for the reaction that occurs.

(3 marks)

   (b) Explain why magnesium has a higher melting point than sodium.

(2 marks)

   (c) State the structure of, and bonding in, silicon dioxide.
      Other than a high melting point, give **two** physical properties of silicon dioxide that are characteristic of its structure and bonding.

(4 marks)

   (d) Give the formula of the species in a sample of solid phosphorus(V) oxide. State the structure of, and describe fully the bonding in, this oxide.

(4 marks)

   (e) Sulfur(IV) oxide reacts with water to form a solution containing ions.
      Write an equation for this reaction.

(1 mark)

   (f) Write an equation for the reaction between the acidic oxide, phosphorus(V) oxide, and the basic oxide, magnesium oxide.

(1 mark)
AQA, 2014

8   Magnesium oxide, silicon dioxide and phosphorus(V) oxide are white solids but each oxide has a different type of structure and bonding.
   (a) State the type of bonding in magnesium oxide.
      Outline a simple experiment to demonstrate that magnesium oxide has this type of bonding.

(3 marks)

   (b) By reference to the structure of, and the bonding in, silicon dioxide, suggest why it is insoluble in water.

(3 marks)

(c) State how the melting point of phosphorus(V) oxide compares with that of silicon dioxide. Explain your answer in terms of the structure of, and the bonding in, phosphorus(V) oxide.

(3 marks)

(d) Magnesium oxide is classified as a basic oxide.
Write an equation for a reaction that shows magnesium oxide acting as a base with another reagent.

(2 marks)

(e) Phosphorus(V) oxide is classified as an acidic oxide.
Write an equation for its reaction with sodium hydroxide.

(1 mark)
AQA, 2013

9 White phosphorus, P_4, is a hazardous form of the element. It is stored under water.
(a) Suggest why white phosphorus is stored under water.

(1 mark)

(b) Phosphorus(V) oxide is known as phosphorus pentoxide.
Suggest why it is usually represented by P_4O_{10} rather than by P_2O_5

(1 mark)

(c) Explain why phosphorus(V) oxide has a higher melting point than sulfur(VI) oxide.

(2 marks)

(d) Write an equation for the reaction of P_4O_{10} with water to form phosphoric(V) acid. Give the approximate pH of the final solution.

(2 marks)

(e) A waste-water tank was contaminated by P_4O_{10}. The resulting phosphoric(V) acid solution was neutralised using an excess of magnesium oxide. The mixture produced was then disposed of in a lake.
   (i) Write an equation for the reaction between phosphoric(V) acid and magnesium oxide.

(1 mark)

   (ii) Explain why an excess of magnesium oxide can be used for this neutralisation.

(1 mark)

   (iii) Explain why the use of an excess of sodium hydroxide to neutralise the phosphoric(V) acid solution might lead to environmental problems in the lake.

(1 mark)
AQA, 2012

23.1 The general properties of transition metals

Learning objectives:

→ Describe the characteristic properties of the elements titanium to copper.

→ Explain these in terms of electronic structure.

Specification reference: 3.2.5

The elements from titanium to copper lie within the d-block elements.

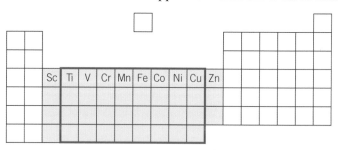

▲ **Figure 1** *The d-block elements (shaded) and the transition metals (outlined)*

Across a period, electrons are being added to a d-sub-level (3d in the case of titanium to copper). The elements from titanium to copper are metals. They are good conductors of heat and electricity. They are hard, strong, and shiny, and have high melting and boiling points.

These physical properties, together with fairly low chemical reactivity, make these metals extremely useful. Examples include iron (and its alloy steel) for vehicle bodies and to reinforce concrete, copper for water pipes, and titanium for jet engine parts that must withstand high temperatures.

Synoptic link

Look back at Topic 8.1, The Periodic Table, to revise the blocks of the Periodic Table.

Electronic configurations in the d-block elements

Figure 2 shows the electron arrangements for the elements in the first row of the d-block.

In general there are two outer 4s electrons and as you go across the period, electrons are added to the *inner* 3d sub-level. This explains the overall similarity of these elements. The arrangements of chromium, Cr, and copper, Cu, do not quite fit the pattern. The d-sub-level is full ($3d^{10}$) in Cu and half full ($3d^5$) in Cr and there is only one electron in the 4s outer level. It is believed that a half-full d-level makes the atoms more stable in the same way as a full outer main level makes the noble gas atoms stable.

Electronic configurations of the ions of d-block elements

To work out the configuration of the ion of an element, first write down the configuration of the element using the Periodic Table, from its atomic number.

		3d	4s
Sc	[Ar] $3d^1 4s^2$	↓ ☐☐☐☐	↓↑
Ti	[Ar] $3d^2 4s^2$	↓↓ ☐☐☐	↓↑
V	[Ar] $3d^3 4s^2$	↓↓↓ ☐☐	↓↑
Cr	[Ar] $3d^5 4s^{1*}$	↓↓↓↓↓	↓
Mn	[Ar] $3d^5 4s^2$	↓↓↓↓↓	↓↑
Fe	[Ar] $3d^6 4s^2$	↓↑↓↓↓↓	↓↑
Co	[Ar] $3d^7 4s^2$	↓↑↓↑↓↓↓	↓↑
Ni	[Ar] $3d^8 4s^2$	↓↑↓↑↓↑↓↓	↓↑
Cu	[Ar] $3d^{10} 4s^{1*}$	↓↑↓↑↓↑↓↑↓↑	↓
Zn	[Ar] $3d^{10} 4s^2$	↓↑↓↑↓↑↓↑↓↑	↓↑

▲ **Figure 2** *Electronic arrangements of the elements in the first d-series. [Ar] represents the electron arrangement of argon – $1s^2 2s^2 2p^6 3s^2 3p^6$*

Worked example: Electron configuration of V^{2+}

Vanadium, V, has an atomic number of 23. Its electron configuration is:

$$1s^2 2s^2 2p^6 3s^2 3p^6 3d^3 4s^2$$

The vanadium ion V^{2+} has lost the two $4s^2$ electrons and has the electron configuration:

$$1s^2 2s^2 2p^6 3s^2 3p^6 3d^3$$

Worked example: Electron configuration of the Cu^{2+} ion

The atomic number of copper is 29. The electron configuration is therefore:

$$1s^2 2s^2 2p^6 3s^2 3p^6 3d^{10} 4s^1$$

The Cu^{2+} ion has lost two electrons, so it has the electron configuration:

$$1s^2 2s^2 2p^6 3s^2 3p^6 3d^9$$

Worked example: Electron configuration of the Cr^{3+} ion

The atomic number of chromium is 24. The electron configuration is therefore:

$$1s^2 2s^2 2p^6 3s^2 3p^6 3d^5 4s^1$$

The Cr^{3+} ion has lost three electrons, so it has the electron configuration:

$$1s^2 2s^2 2p^6 3s^2 3p^6 3d^3$$

In fact, with all transition elements, the 4s electrons are lost first when ions are formed.

The definition of a transition element

The formal definition of a transition element is that it is one that forms at least one stable ion with a *part* full d-shell of electrons. Scandium only forms Sc^{3+} ($3d^0$) in all its compounds, and zinc only forms Zn^{2+} ($3d^{10}$) in all its compounds. They are therefore d-block elements but not transition elements. The transition elements are outlined in red in Figure 1.

Chemical properties of transition metals

The chemistry of transition metals has four main features which are common to all the elements:

- Variable oxidation states: Transition metals have more than one oxidation state in their compounds, for example, Cu(I) and Cu(II). They can therefore take part in many redox reactions.

- Colour: The majority of transition metal ions are coloured, for example, Cu^{2+}(aq) is blue.

- Catalysis: Catalysts affect the rate of reaction without being used up or chemically changed themselves. Many transition metals, and their compounds, show catalytic activity. For example, iron is the catalyst in the Haber process, vanadium(V) oxide in the Contact process and manganese(IV) oxide in the decomposition of hydrogen peroxide.

- Complex formation: Transition elements form complex ions. A complex ion is formed when a transition metal ion is surrounded by ions or other molecules, collectively called ligands, which are bonded to it by co-ordinate bonds. For example, $[Cu(H_2O)_6]^{2+}$ is a complex ion that is formed when copper sulfate dissolves in water.

Study tip

The electronic configuration of any element can be deduced using the Periodic Table.

▲ **Figure 3** *Some transition metals in use*

Summary questions

1 The electron arrangement of manganese is:

$1s^2\ 2s^2\ 2p^6\ 3s^2\ 3p^6\ 3d^5\ 4s^2$

Write the electron arrangement of:

a a Mn^{2+} ion

b a Mn^{3+} ion

2 The electron arrangement of iron can be written [Ar] $3d^6\ 4s^2$.

a What does the [Ar] represent?

b Which two electrons are lost to form Fe^{2+}?

c Which further electron is lost in forming Fe^{3+}?

Learning objectives:

→ State what the terms ligand, co-ordinate bond, and co-ordination number mean.

→ Explain what bidentate and multidentate ligands are.

→ Explain how the size of the ligand affects the shape of the complex ion.

→ Explain how ligand charge determines the charge on the complex ion.

Specification reference: 3.2.5

Synoptic link

You will need to understand covalent and co-ordinate bonding, shapes of simple molecules and ions studied in Topic 3.6, The shapes of molecules and ions.

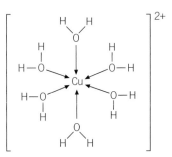

▲ **Figure 1** *Copper(II) ion surrounded by water molecules*

Hint

Some ligands are neutral and others have a negative charge.

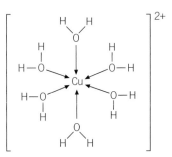

▲ **Figure 2** *Ethane-1,2-diamine*

The formation of complex ions

All transition metal ions can form co-ordinate bonds by accepting electron pairs from other ions or molecules. The bonds that are formed are **co-ordinate** (dative) bonds. An ion or molecule with a lone pair of electrons that forms a co-ordinate bond with a transition metal is called a **ligand**. Some examples of ligands are $H_2O\!:$, $:NH_3$, $:Cl^-$, $:CN^-$.

In some cases, two, four, or six ligands bond to a single transition metal ion. The resulting species is called a complex ion. The number of co-ordinate bonds to ligands that surround the d-block metal ion is called the **co-ordination number**.

- Ions with co-ordination number six are usually octahedral, for example, $[Co(NH_3)_6]^{3+}$.
- Ions with co-ordination number four are usually tetrahedral, for example, $[CoCl_4]^{2-}$.
- Some ions with co-ordination number four are square planar, for example, $[NiCN_4]^{2-}$.

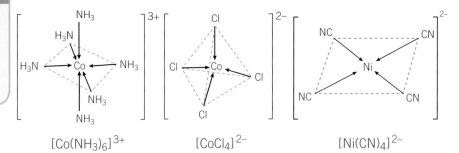

$[Co(NH3)_6]^{3+}$ $[CoCl_4]^{2-}$ $[Ni(CN)_4]^{2-}$

Aqua ions

If you dissolve the salt of a transition metal in water, for example, copper sulfate, the positively charged metal ion becomes surrounded by water molecules acting as ligands (Figure 1). Normally there are six water molecules in an octahedral arrangement. Such species are called aqua ions.

Multidentate ligands – chelation

Some molecules or ions, called **multidentate ligands**, have more than one atom with a lone pair of electrons which can bond to a transition metal ion.

Bidentate ligands include:

- Ethane-1,2-diamine, sometimes called 1,2-diaminoethane or ethylene diamine (Figure 2) Each nitrogen has a lone pair which can form a co-ordinate bond to the metal ion. The name of this ligand is often abbreviated to en, for example, $[Cr(en)_3]^{3+}$. It is a neutral ligand and the chromium ion has a 3+ charge, so the complex ion also has a 3+ charge.
- The ethanedioate (oxalate) ion, $C_2O_4^{2-}$ (Figure 3).
- Benzene-1,2-diol, sometimes called 1,2-dihydroxybenzene, is also a neutral ligand (Figure 4).

An important multidentate ligand is the ion ethylenediaminetetracetate, called EDTA^{4-} (Figure 5).

This can act as a hexadentate ligand using lone pairs on four oxygen and both nitrogen atoms. Complex ions with polydentate ligands are called **chelates**. Chelates can be used to effectively remove d-block metal ions from solution.

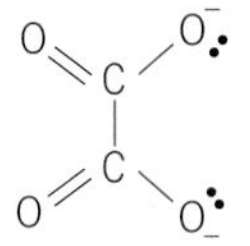

The chelate effect

If you add a hexadentate ligand such as EDTA to a solution of a transition metal salt, the EDTA will replace all six water ligands in the aqua ion $[Cu(H_2O)_6]^{2+}$ as shown:

$$[Cu(H_2O)_6]^{2+}(aq) + EDTA^{4-}(aq) \rightarrow [CuEDTA]^{2-}(aq) + 6H_2O(l)$$

In this equation, two species are replaced by seven. This increase in the number of particles causes a significant increase in entropy which drives the reaction to the right. For this reason chelate complexes with polydentate ligands are favoured over complexes with monodentate ligands and is called the **chelate effect**.

Haemoglobin

Haemoglobin is the red pigment in blood. It is responsible for carrying oxygen from the lungs to the cells of the body. The molecule consists of an Fe^{2+} ion with a co-ordination number of six. Four of the co-ordination sites are taken up by a ring system called a porphyrin which acts as a tetradentate ligand. This complex is called haem.

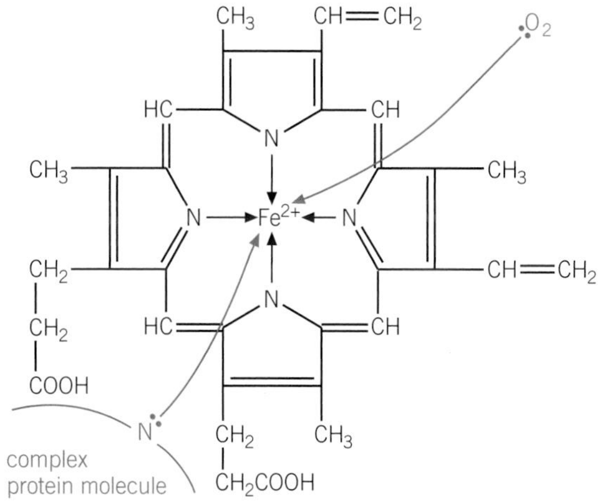

▲ **Figure 6** *Haemoglobin*

▲ **Figure 3** *Ethanedioate*

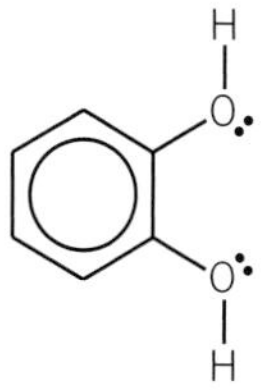

▲ **Figure 4** *Benzene-1,2-diol*

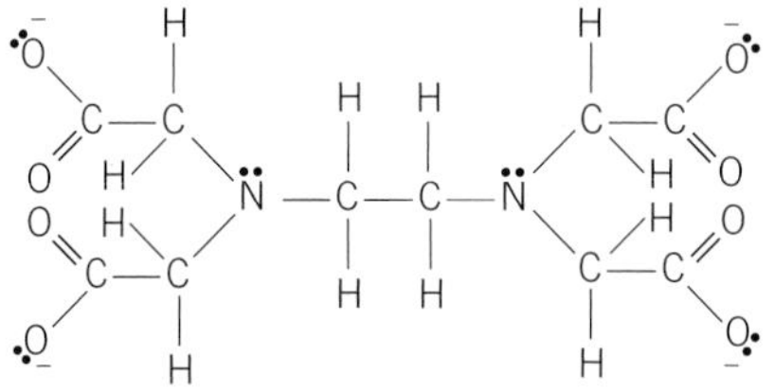

▲ **Figure 5** *EDTA*

Study tip

Both ligand sites of bidentate ligands usually bond to the same metal forming a ring. However, they can act as bridges between two metal ions.

Synoptic link

Look back at topic 17.4, Why do chemical reactions take place?, to revise entropy.

Below the plane of this ring is a fifth nitrogen atom acting as a ligand. This atom is part of a complex protein called globin. The sixth site can accept an oxygen molecule as a ligand. The Fe^{2+} to O$_2$ bond is weak, as :O$_2$ is not a very good ligand, allowing the oxygen molecule to be easily given up to cells.

Better ligands than oxygen can bond irreversibly to the iron and so destroy haemoglobin's oxygen-carrying capacity. This explains the poisonous effect of carbon monoxide, which is a better ligand than oxygen. Carbon monoxide is often formed by incomplete combustion in

faulty gas heaters. Because it binds more strongly to the iron than oxygen, it is possible to suffocate in a room with plentiful oxygen.

Anaemia is a condition which may be caused by a shortage of haemoglobin. The body suffers from a lack of oxygen and the symptoms include fatigue, breathlessness, and a pale skin colour. The causes may be loss of blood or deficiency of iron in the diet. The latter may be treated by taking 'iron' tablets which contain iron(II) sulfate.

Study tip

The chlorine atom is in Period 3 and so has an extra electron level compared with the nitrogen atom in Period 2.

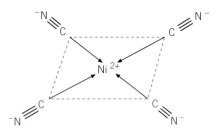

▲ **Figure 7** *A square planar complex*

Synoptic link

You will learn more about using Tollens' reagent to test for aldehydes and ketones in Topic 26.2, Reactions of the carbonyl group in aldehydes and ketones.

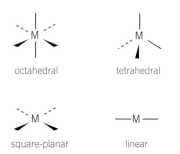

octahedral   tetrahedral

square-planar   linear

▲ **Figure 8** *The four main shapes of transition metal complexes using wedge and dotted bonds*

Hint

All six ligand positions in an octahedral complex are equivalent as are all the positions in the other three geometries.

Synoptic link

You will learn more about cisplatin in Topic 30.5, The action of anti cancer drugs.

Shapes of complex ions

As you have seen the $[Co(NH_3)_6]^{3+}$ ion, with six ligands, is an octahedral shape. An octahedron has six points but *eight* faces. The metal ion, Co^{3+}, has a charge of +3 and as the ligands are all neutral, the complex ion has an overall charge of +3.

The $[CoCl_4]^{2-}$ ion, with four ligands, is tetrahedral. The metal ion, Co^{2+}, has a charge of +2 and each of the four ligands $:Cl^-$, has a charge of −1, so the complex ion has an overall charge of −2.

The reason for this difference in shape is that the chloride ion is a larger ligand than the ammonia molecule and so fewer ligands can fit around the central metal ion.

A few complexes of co-ordination number four adopt a square planar geometry (Figure 7).

Some complexes are linear, one example being $[Ag(NH_3)_2]^+$:

$$[H_3N \rightarrow Ag \leftarrow NH_3]^+$$

A solution containing this complex ion is called Tollens' reagent and is used in organic chemistry to distinguish aldehydes from ketones. Aldehydes reduce the $[Ag(NH_3)_2]^+$ to Ag (metallic silver), while ketones do not. The silver forms a mirror on the surface of the test tube, giving the name of the test – the silver mirror test.

Complex ions may have a positive charge or a negative charge.

Representing the shapes of complex ions

Representing three-dimensional shapes on paper can be tricky. Some diagrams in this topic have thin red construction lines to help you to visualise the shapes. These are not bonds. Another way to represent shape is to use wedge bonds and dotted bonds. Wedge bonds come out of the paper and dotted bonds go in (Figure 8).

Isomerism in transition metal complexes

You have met isomerism in organic chemistry. Isomers are compounds with the same molecular formula but with different arrangements of their atoms in space. Transition metal complexes can form both geometrical isomers (*cis-trans*, or *E-Z* isomers) and optical isomers.

Geometrical isomerism

Here ligands differ in their position in space relative to one another. This type of isomerism occurs in octahedral and square planar complexes. Take the octahedral complex ion $[CrCl_2(H_2O)_4]^+$. The Cl^- ligands may be next to each other (the *cis-* or *Z-* form) or on opposite sides of the central chromium ion (the *trans-* or *E-*form) (Figure 9).

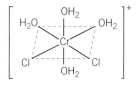

In the square planar complex platin, the Cl^- ligands may be next to each other (the *cis-* or *Z-* form) or on opposite sides of the central chromium ion (the *trans-* or *E-*form). A pair of geometrical isomers will have different chemical properties. For example, cisplatin is one of the most successful anti-cancer drugs whilst the *trans-*isomer has no therapeutic effect,

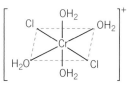

▲ **Figure 9** *The cis- or Z-isomer (top) and the trans- or E-isomer (bottom)*

Optical isomerism

Here the two isomers are non-superimposable mirror images of each other. In transition metal complexes this occurs when there are two or more bidentate ligands in a complex (Figure 10).

Look at Figure 10 and imagine rotating one of the complexes around a vertical axis until the two chlorine atoms are in the same position as in the other one. You should be able to see that the positions of the end ligands no longer match. The best way to be sure about this is to use molecular models.

Optical isomers are said to be chiral. They have identical chemical properties but can be distinguished by their effect on polarised light. One isomer will rotate the plane of polarisation of polarised light clockwise and the other anticlockwise.

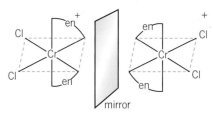

en is an abbreviation for ethane-1,2-diamine:

▲ **Figure 10** *Transition metal comlpexes that are non-identical mirror images of each other (top) and the structure of the ethane-1,2-diamine (en) ligand (bottom)*

 Ionisation isomerism

This is a third form of isomerism found in transition metal chemistry. Consider the compound of formula $CrCl_3 (H_2O)_6$. Both chloride ions, Cl^-, and water molecules can act as ligands. This compound can exist as three different isomers depending on how many of the chloride ions are bound to the chromium atom as ligands and how many are free as negative ions.

The three isomers are:

Compound 1: $[Cr(H_2O)_6]^{3+} + 3Cl^-$ violet

Compound 2: $[CrCl(H_2O)_5]^{2+} + 2Cl^- + H_2O$ light green

Compound 3: $[CrCl_2(H_2O)_6]^+ + Cl^- + 2H_2O$ dark green

1 What feature of both Cl^- and H_2O allows them to act as ligands?
2 If solutions are made of the same concentration of each compound, which would you expect to conduct electricity best? Explain your answer.
3 Only free chloride ions (and not those bonded as ligands) will react with silver nitrate. Write a balanced equation for the reaction of chloride ions with silver nitrate solution.
4 Which compound would react with most silver nitrate? Explain your answer.
5 State which other type of isomerism is shown by Compound 3.
6 The structural formula of Compound 1 is:

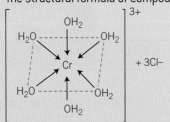

Draw the structural formula of Compound 2 in the same style.

Synoptic link

See Topic 25.2, Optical isomerism, for more detail about optical isomerism in organic chemistry.

Summary questions

1 a Predict the shapes of the following:
       i $[Cu(H_2O)_6]^{2+}$
       ii $[Cu(NH_3)_6]^{2+}$
       iii $[CuCl_4]^{2-}$
   b What is the co-ordination number of the transition metal in each complex in a?
   c Explain why the co-ordination numbers are different.

2 Benzene-1,2-dicarboxylate is shown below.

   a Suggest which atoms are likely to form co-ordinate bonds with a metal ion.
   b Mark the lone pairs.
   c Is it likely to be a mono-, bi- or hexa-dentate ligand?

23.3 Coloured ions

Learning objectives:

→ Describe the origin of the colour of a transition metal complex ion.

→ Explain what factors determine the colour of a complex ion.

Specification reference: 3.2.5

Study tip

Make sure that you know the factors which give rise to colour changes and are able to illustrate each change by an equation.

Hint $\sqrt{x}$

ΔE is also related to wavelength λ by the equation $\Delta E = \dfrac{hc}{\lambda}$, where c is the velocity of light.

▼ **Table 1** *Colours of four vanadium species*

Oxidation number	Species	Colour
5	$VO_2^+(aq)$	yellow
4	$VO^{2+}(aq)$	blue
3	$V^{3+}(aq)$	green
2	$V^{2+}(aq)$	violet

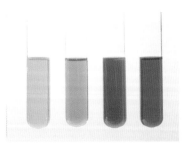

▲ **Figure 2** *Zinc ions in acid solution will reduce vanadium(V) through oxidation states V(IV) and V(III) to V(II). The final flask is normally stoppered because oxygen in the air will rapidly oxidise V(II)*

Most transition metal compounds are coloured. The colour is caused by the compounds absorbing energy that corresponds to light in the visible region of the spectrum. If a solution of a substance looks purple, it is because it absorbs all the light from a beam of white light shone at it except red and blue. The red and blue light passes through and the solution appears purple, (Figure 1).

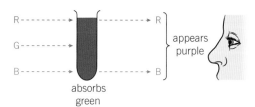

▲ **Figure 1** *Solutions look coloured because they absorb some colours and let others pass through*

Why are transition metal complexes coloured?

This is a simplified explanation, but the general principle is as follows:

- Transition metal compounds are coloured because they have part-filled d-orbitals.

- It is therefore possible for electrons to move from one d-orbital to another.

- In an isolated transition metal atom, all the d-orbitals are of exactly the same energy, but in a compound, the presence of other atoms nearby makes the d-orbitals have slightly different energies.

- When electrons move from one d-orbital to another of a higher energy level (called an excited state), they often absorb energy in the visible region of the spectrum equal to the difference in energy between levels.

- This colour is therefore missing from the spectrum and you see the combination of the colours that are not absorbed.

The frequency of the light is related to the energy difference by the expression $\Delta E = h\nu$, where E is the energy, ν the frequency, and h a constant called Planck's constant. The frequency is related to the colour of light. Violet is of high energy and therefore high frequency and red is of low energy and low frequency.

The colour of a transition metal complex depends on the energy gap ΔE, which in turn depends on the oxidation state of the metal and also on the ligands (and therefore the shape of the complex ion), so different compounds of the same metal will have different colours. For example, Table 1 shows the colours of four vanadium species each with a different oxidation state.

Some more examples of how changing the oxidation state of the metal affects the colour of the complex are given in Table 2.

▼ Table 2

Oxidation state of metal	2	3
iron complexes	$[Fe(H_2O)_6]^{2+}$ green	$[Fe(H_2O)_6]^{3+}$ pale brown
chromium complexes	$[Cr(H_2O)_6]^{2+}$ blue	$[Cr(H_2O)_6]^{3+}$ red-violet
cobalt complexes	$[Co(NH_3)_6]^{2+}$ brown	$[Co(NH_3)_6]^{3+}$ yellow

Colorimetry

A simple colorimeter uses a light source and a detector to measure the amount of light of a particular wavelength that passes through a coloured solution. The more concentrated the solution, the less light transmitted through the solution. A colorimeter is used, with a suitable calibration graph to measure the concentration of solutions of coloured transition metal compounds.

Hint

Usually the experiment is made more sensitive by using a coloured filter in the colorimeter. The filter is chosen by finding out the colour of light that the red solution absorbs most. Red absorbs light in the blue region of the visible spectrum, so a blue filter is used (Figure 3), so that only blue light passes into the sample tube.

Finding the formula of a transition metal complex using colorimetry

A colorimeter can be used to find the ratio of metal ions to ligands in a complex, which gives us the formula of the complex. Two solutions are mixed together, one containing the metal ion and one the ligand, in different proportions. When they are mixed in the same ratio as they are in the complex, there is the maximum concentration of complex in the solution. So, the solution will absorb most light.

For example, take the blood red complex formed with Fe^{3+} ions and thiocyanate ions, SCN^-.

If potassium thiocyanate is added to a solution of $Fe^{3+}(aq)$, *one* of the water molecules is replaced by a thiocyanate ion and a blood red complex forms:

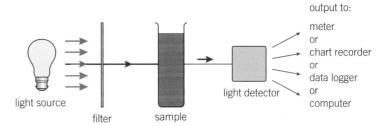

▲ **Figure 3** *Using a colorimeter to find a formula*

$$[Fe(H_2O)_6]^{3+}(aq) + SCN^- \rightarrow [Fe(SCN)(H_2O)_5]^{2+}(aq) + H_2O(l)$$

As the concentration of the red complex increases, less and less light will pass through the solution.

Start with two solutions of the same concentration, one containing $Fe^{3+}(aq)$ ions, for example, iron(III) sulfate, and one containing $SCN^-(aq)$ ions, for example, potassium thiocyanate. Mix them in the proportions shown in Table 3, adding water so that all the tubes have the same total volume of solution.

▼ **Table 3** *The absorbance of different mixtures of $Fe^{3+}(aq)$ and $SCN^-(aq)$*

Tube	1	2	3	4	5	6	7	8
Vol. of $Fe^{3+}(aq)$ solution / cm³	10.00	10.00	10.00	10.00	10.00	10.00	10.00	10.00
Vol. of $SCN^-(aq)$ solution / cm³	2.00	4.00	6.00	8.00	10.00	12.00	14.00	16.00
Vol. of water / cm³	28.00	26.00	24.00	22.00	20.00	28.00	16.00	14.00
Absorbance	0.15	0.33	0.48	0.63	0.70	0.70	0.70	0.70

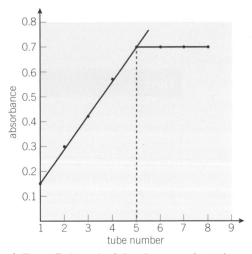

Each tube is put in the colorimeter and a reading of absorbance taken. Absorbance is a measure of the light absorbed by the solution. A graph of absorbance is plotted against tube number (Figure 5).

From the graph, the maximum absorbance occurs in tube 5 – after this, adding more thiocyanate ions makes no difference. So this shows that the ratio of SCN^- ions to Fe^{3+} ions in the complex. From Table 3, tube 5 has equal amounts of SCN^- ions and Fe^{3+} ions so their ratio in the complex must be 1 : 1. So this confirms the formula is $[Fe(SCN)(H_2O)_5]^{2+}$. (The SCN^- has substituted for one of the water molecules in the complex ion $[Fe(H_2O)_6]^{3+}$.)

▲ **Figure 5** *A graph of absorbance against tube number*

▲ **Figure 4** *The colours of many gemstones are caused by transition metal compounds*

The colour of gemstones

Transition metal ions are responsible for the colours of most gemstones. Rubies are made of aluminium oxide, Al_2O_3, which is colourless – the red colour is caused by trace amounts of Cr^{3+} ions which replace some of the Al^{3+} ions in the crystal lattice. The oxide ions, O^{2-}, are the ligands surrounding the Cr^{3+} ions.

The green colour of emeralds is also caused by Cr^{3+} ions, but in this case, the material from which the gemstone is made is beryllium aluminium silicate, $Be_3Al_2(SiO_3)_6$. The ligand surrounding the Cr^{3+} is silicate, SiO_3^{2-}, in this case. This illustrates the effect of changing the ligand on the colour of the ion.

The red of garnet and the yellow-green colour of peridot are both caused by Fe^{2+} ions – surrounded by eight silicate ligands in the case of garnet and six in peridot.

Other examples include Cu^{2+} which is responsible for the blue-green of turquoise and Mn^{2+} for the pink of tourmaline.

Summary questions

1 a Explain why copper sulfate is coloured, whereas zinc sulfate is colourless.

b A solution of copper sulfate is blue. What colour light passes through this solution?

c What happens to the other colours of the visible spectrum?

2 The graph below shows the absorbance of a series of mixtures containing different proportions of two solutions of the same concentration – one containing Ni^{2+} ions and the other containing a ligand called $EDTA^{4-}$ for short. The two solutions react together to form a coloured complex.

a Which mixture absorbs most light?

b Which mixture contains the highest concentration of the nickel EDTA complex?

c What is the simplest (empirical) formula of the complex?

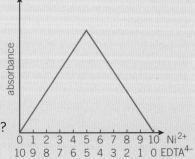

23.4 Variable oxidation states of transition elements

Group 1 metals lose their outer electron to form only +1 ions and Group 2 lose their outer two electrons to form only +2 ions in their compounds. A typical transition metal can use its 3d-electrons as well as its 4s-electrons in bonding, and this means that it can have a greater variety of oxidation states in different compounds. Table 1 shows this for the first d-series. Zinc and scandium are shown as part of the d-series although they are not transition metals.

Learning objective:

→ Describe how the concentration of iron(II) ions in aqueous solution can be found.

Specification reference: 3.2.5

▼ **Table 1** *Oxidation numbers shown by the elements of the first d-series in their compounds*

Sc	Ti	V	Cr	Mn	Fe	Co	Ni	Cu	Zn
	+I	+I	+I	+I	+I	+I	+I	+I	
	+II	+II	+II	+II	+II	+II	+II	+II	+II
+III	+III	+III	+III	+III	+III	+III	+III	+III	
	+IV	+IV	+IV	+IV	+IV	+IV	+IV		
		+V	+V	+V	+V	+V			
			+VI	+VI	+VI				
				+VII					

The most common oxidation states are shown in red, though they are not all stable.

Except for scandium and zinc all the elements show both the +1 and +2 oxidation states. These are formed by the loss of 4s electrons.

For example, nickel has the electron configuration $1s^2 2s^2 2p^6 3s^2 3p^6 3d^8 4s^2$ and Ni^{2+} is $1s^2 2s^2 2p^6 3s^2 3p^6 3d^8$.

Iron has the electron configuration $1s^2 2s^2 2p^6 3s^2 3p^6 3d^6 4s^2$ and Fe^{2+} is $1s^2 2s^2 2p^6 3s^2 3p^6 3d^6$.

Only the lower oxidation states of transition metals actually exist as simple ions, so that, for example, Mn^{2+} ions exist but Mn^{7+} ions do not. In all Mn(VII) compounds, the manganese is covalently bonded to oxygen in a compound ion as in MnO_4^- (Figure 1).

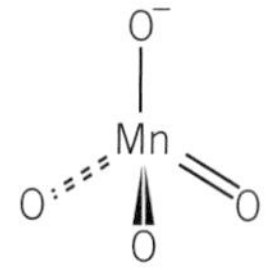

▲ **Figure 1** *Bonding in the $[MnO_4]^-$ ion*

Redox reactions in transition metal chemistry

Many of the reactions of transition metal compounds are redox reactions, in which the metals are either oxidised or reduced. For example, iron shows two stable oxidation states – Fe^{3+} and Fe^{2+}.

Fe^{2+} is the less stable state – it can be oxidised to Fe^{3+} by the oxygen in the air and also by chlorine. For example:

$$\overset{+2}{2Fe^{2+}(aq)} + \overset{0}{Cl_2(g)} \rightarrow \overset{+3}{2Fe^{3+}(aq)} + \overset{-1}{2Cl^-(aq)}$$

In this reaction, chlorine is the oxidising agent – its oxidation number drops from 0 to −1 (as it gains an electron), whilst that of the iron increases from +2 to +3 (as it loses an electron). Remember the phrase OIL RIG – oxidation is loss, reduction is gain (of electrons).

Synoptic link

You will need to understand redox equations and volumetric analysis studied in Topic 7.3, Redox equations, and Topic 2.4, Empirical and molecular formulae.

Using half equations

Potassium manganate(VII) reactions

The technique of using half equations is useful for constructing balanced equations in more complex reactions. Potassium manganate(VII) can act as an oxidising agent in acidic solution (one containing $H^+(aq)$ ions) and will, for example, oxidise Fe^{2+} to Fe^{3+}. During the reaction the oxidation number of the manganese falls from +7 to +2.

First construct the half equation for the reduction of Mn(VII) to Mn(II):

$$MnO_4^-(aq) \rightarrow Mn^{2+}(aq)$$

The oxygen atoms must be balanced using H^+ ions and H_2O molecules:

$$MnO_4^- + 8H^+(aq) \rightarrow Mn^{2+}(aq) + 4H_2O(l)$$

Then balance for charge using electrons:

$$MnO_4^-(aq) + 5e^- + 8H^+(aq) \rightarrow Mn^{2+}(aq) + 4H_2O(l)$$

The half equation for the oxidation of iron(II) to iron(III) is straightforward:

$$Fe^{2+}(aq) \rightarrow Fe^{3+}(aq) + e^-$$

To construct a balanced symbol equation for the reaction of acidified potassium manganate(VII) with $Fe^{2+}(aq)$, first multiply the Fe^{2+}/Fe^{3+} half reaction by five (so that the numbers of electrons in each half reaction are the same) and then add the two half equations:

$$5Fe^{2+}(aq) \rightarrow 5Fe^{3+}(aq) + 5e^-$$

$$MnO_4^-(aq) + 5e^- + 8H^+(aq) \rightarrow Mn^{2+}(aq) + 4H_2O(l)$$

$$5Fe^{2+}(aq) + MnO_4^-(aq) + 5e^- + 8H^+(aq) \rightarrow 5Fe^{3+}(aq) + 5e^- + Mn^{2+}(aq) + 4H_2O(l)$$

$$5Fe^{2+}(aq) + MnO_4^-(aq) + 8H^-(aq) \rightarrow 5Fe^{3+}(aq) + Mn^{2+}(aq) + 4H_2O(l)$$

This technique makes balancing complex redox reactions much easier.

Redox titrations

You may wish to measure the concentration of an oxidising or a reducing agent. One way of doing this is to do a redox titration. This is similar in principle to an acid–base titration in which you find out how much acid is required to react with a certain volume of base (or vice versa).

One example is in the analysis of iron tablets for quality control purposes. Iron tablets contain iron(II) sulfate and may be taken by patients whose diet is short of iron for some reason.

As you have seen, $Fe^{2+}(aq)$ reacts with manganate(VII) ions (in potassium manganate(VII)) in the ratio 5 : 1. The reaction does not need an indicator, because the colour of the mixture changes as the reaction proceeds (Table 2).

Using a burette, you gradually add potassium manganate(VII) solution (which contains the $MnO_4^-(aq)$ ions) to a solution containing $Fe^{2+}(aq)$ ions, acidified with excess dilute sulfuric acid. The purple colour disappears as the MnO_4^- ions are converted to pale pink $Mn^{2+}(aq)$ ions to leave a virtually colourless solution. Once just enough $MnO_4^-(aq)$ ions have been added to react with all the $Fe^{2+}(aq)$ ions, one more drop of $MnO_4^-(aq)$ ions will turn the solution purple. This is the end point of the titration.

> **Hint**
>
> H^+ ions are likely to be involved because the reaction takes place in acidic solution. Five electrons are involved because the oxidation state of each manganese atom drops by five.

> **Hint**
>
> The body needs iron compounds to make haemoglobin, the compound that carries oxygen in the blood.

▼ **Table 2** *The colours of the ions in the reaction between potassium manganate(VII) and iron(II) sulfate*

Ion	Colour
$Fe^{2+}(aq)$	pale green
$MnO_4^-(aq)$	intense purple
$Fe^{3+}(aq)$	pale violet
$Mn^{2+}(aq)$	pale pink

The apparatus used is shown in Figure 2.

You cannot use hydrochloric acid, as an alternative to sulfuric acid, to supply the $H^+(aq)$ ions in the reaction between potassium manganate(VII) and $Fe^{2+}(aq)$. You can see why this is the case by using $E^{\ominus}$ values.

Hydrochloric acid contains Cl^- ions. These are oxidised by MnO_4^- ions, as shown by the calculation of emf for the reaction below. This would affect the titration, because the manganate(VII) ions must be used only to oxidise Fe^{2+} ions. Manganate(VII) ions do not oxidise sulfate ions.

The relevant half equations with their values of $E^{\ominus}$ are:

$$MnO_4^-(aq) + 5e^- + 8H^+(aq) \rightleftharpoons Mn^{2+}(aq) + 4H_2O(l) \quad E^{\ominus} = +1.51 \text{ V}$$

$$\frac{1}{2}Cl_2 + e \rightleftharpoons Cl^- \quad\quad\quad\quad\quad\quad E^{\ominus} = +1.36 \text{ V}$$

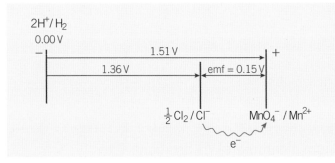

▲ **Figure 3** $E^{\ominus}$ values show that acidified MnO_4^- ions will oxidise Cl^- ions

Figure 3 shows that the electron flow is from Cl_2 to MnO_4^-, and the emf is 0.15 V, so the half reactions must be as follows:

$$MnO_4^-(aq) + 5e^- + 8H^+(aq) \rightarrow Mn^{2+}(aq) + 4H_2O(l)$$

$$Cl^- \rightarrow \frac{1}{2}Cl_2 + e^-$$

Multiplying the lower half equation by 5, to balance the electrons, and adding the equations together gives:

$$MnO_4^-(aq) + 5e^- + 8H^+(aq) + 5Cl^-(aq) \rightleftharpoons Mn^{2+}(aq) + 4H_2O(l)$$

Remember to cancel electrons $\quad\quad\quad\quad\quad\quad + 2\frac{1}{2}Cl_2(aq) + 5e^-$

$$\text{emf} = +0.15 \text{ V}$$

This reaction is feasible, so MnO_4^- ions will oxidise Cl^- ions and hydrochloric acid is not suitable for this titration.

Worked example: Iron tablets

A brand of iron tablets has this stated on the pack. 'Each tablet contains 0.200 g of iron(II) sulfate.' The following experiment was done to check this.

One tablet was dissolved in excess sulfuric acid and made up to 250 cm³ in a volumetric flask. 25.00 cm³ of this solution was pipetted into a flask and titrated with 0.001 00 mol dm⁻³ potassium manganate(VII) solution until the solution just became purple. Taking an average of several titrations, 26.30 cm³ of potassium manganate(VII) solution was needed.

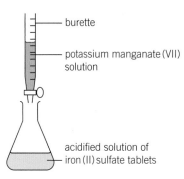

▲ **Figure 2** *Apparatus for a titration*

Synoptic link

Look back at Topic 20.2, Predicting the direction of redox reactions, to revise $E^{\ominus}$ values.

Hint

The value of $E^{\ominus}$ for the MnO_4^- / Mn^{2+} half cell will vary with pH.

Study tip

Study tip $\sqrt{x}$

Try to show all your working in calculations to demonstrate your understanding when answers are incomplete or the final answer is wrong.

Number of moles potassium manganate(VII) solution $= c \times \dfrac{V}{1000}$

where c is the concentration of the solution in $mol\,dm^{-3}$ and V is the volume of solution used in cm^3.

No. of moles potassium manganate(VII) solution $= 0.001\,00 \times \dfrac{26.30}{1000}$

$$= 2.63 \times 10^{-5}\,mol$$

$$5Fe^{2+}(aq) + MnO_4^-(aq) + 8H^+(aq) \rightarrow 5Fe^{3+}(aq) + Mn^{2+}(aq) + 4H_2O(l)$$

From the equation, 5 mol of Fe^{2+} reacts with 1 mol of MnO_4^-:

Number of moles of $Fe^{2+} = 5 \times 2.63 \times 10^{-5}\,mol = 1.315 \times 10^{-4}\,mol$

$25.00\,cm^3$ of solution contained $\dfrac{1}{10}$ tablet.

So one tablet contains $1.315 \times 10^{-4} \times 10 = 1.315 \times 10^{-3}\,mol\ Fe^{2+}$

Since 1 mol iron(II) sulfate contains 1 mol Fe^{2+}, each tablet contains $1.315 \times 10^{-3}\,mol\ FeSO_4$.

The relative formula mass of $FeSO_4$ is 151.9.

So, each tablet contains $1.315 \times 10^{-3} \times 151.9 = 0.200\,g$ of iron(II) sulfate as stated on the bottle.

➕ Potassium dichromate(VI) titrations

Acidified potassium dichromate(VI) can also be used in a titration to measure the concentration of Fe^{2+} ions. Here the half equations are:

$$Cr_2O_7^{2-}(aq) + 14H^+(aq) + 6e^- \rightarrow 2Cr^{3+}(aq) + 7H_2O(l)$$

$$Fe^{2+}(aq) \rightarrow Fe^{3+}(aq) + e^-$$

So the second half equation must be multiplied by six before adding and cancelling the electrons.

$$6Fe^{2+}(aq) \rightarrow 6Fe^{3+}(aq) + 6e^-$$

$$Cr_2O_7^{2-}(aq) + 14H^+(aq) + 6e^- \rightarrow 2Cr^{3+}(aq) + 7H_2O(l)$$

$$6Fe^{2+}(aq) + Cr_2O_7^{2-}(aq) + 14H^+(aq) + \cancel{6e^-} \rightarrow 6Fe^{3+}(aq) + 2Cr^{3+}(aq) + 7H_2O(l) + \cancel{6e^-}$$

$$6Fe^{2+}(aq) + Cr_2O_7^{2-}(aq) + 14H^+(aq) \rightarrow 6Fe^{3+}(aq) + 2Cr^{3+}(aq) + 7H_2O(l)$$

Note that although chromium is reduced from +6 to +3, the ion $Cr_2O_7^{2-}$ contains two chromium atoms so six electrons are needed.

As before, the $Fe^{2+}(aq)$ solution is placed in the flask with the dichromate in the burette with excess dilute sulfuric acid to provide the H^+ ions.

As it is not possible to see the colour change when a small volume of orange solution is added to a pale green solution, an indicator must be used – sodium diphenylaminesulfonate, which turns from colourless to purple at the end point.

Oxidation of transition metal ions in alkaline solutions

In both the above examples, a high oxidation state of a metal (Mn(VII) and Cr(VI)) are reduced in acidic solution. Oxidation of lower oxidation states of transition metal ions tends to happen in alkaline solution. This is because in alkaline solution there is a tendency to form negative ions. Since oxidation is electron loss, this is easier from negatively charged species than positively charged or neutral ones.

Typical transition metal species, where M represents a transition metal:

- Acid solution: $M(H_2O)_6^{2+}$ positively charged.
- Neutral solution: $M(H_2O)_4(OH)_2$ neutral.
- Alkaline solution: $M(H_2O)_2(OH)_4^{2-}$ negatively charged.

Low oxidation states of transition metals, such as Fe^{2+} are often stabilised against oxidation by air by keeping them in acid solution.

To oxidise a transition metal to a high oxidation state, an alkali is often added, followed by an oxidising agent.

Example – some cobalt chemistry

Many M^{2+} ions will be oxidised to M^{3+} in alkaline solution, for example cobalt(II) to cobalt(III):

$$\overset{+2}{2[Co(OH)_6]^{4-}}(aq) + H_2O_2(aq) \rightarrow \overset{+3}{2[Co(OH)_6]^{3-}}(aq) + 2OH^-(aq)$$

In ammoniacal solution, Co^{2+} ions can be oxidised by oxygen in the air.

If you add an excess of ammonia solution to an aqueous solution containing cobalt(II) ions, you get a brownish complex ion formed, $[Co(NH_3)_6]^{2+}$, containing cobalt(II) ions.

The reactions are as follows:

1   First a precipitate is formed by reaction with OH^- ions from the ammonia solution, which is alkaline:

$$[Co(H_2O)_6]^{2+} + 2OH^- \rightarrow Co(H_2O)_4(OH)_2(s) + 2H_2O(l)$$

2   Then the precipitate dissolves in excess ammonia:

$$Co(H_2O)_4(OH)_2(s) + 6NH_3(aq) \rightarrow [Co(NH_3)_6]^{2+} + 2OH^-(aq) + 4H_2O(l)$$

3   The resulting complex ion is oxidised by oxygen in air (or rapidly by hydrogen peroxide solution) to the yellow cobalt(III) ion, $[Co(NH_3)_6]^{3+}$.

You can use half equations to produce a balanced equation for the redox reaction.

The half equations are:

$$[Co(NH_3)_6]^{2+}(aq) \rightarrow [Co(NH_3)_6]^{3+}(aq) + e^-$$
$$O_2(g) + 2H_2O(l) + 4e^- \rightarrow 4OH^-(aq)$$

Multiplying the first equation by four and adding these:

$$4[Co(NH_3)_6]^{2+}(aq) \rightarrow 4[Co(NH_3)_6]^{3+}(aq) + 4e^-$$
$$O_2(g) + 2H_2O(l) + 4e^- \rightarrow 4OH^-(aq)$$

$$4[Co(NH_3)_6]^{2+}(aq) + O_2(g) + 2H_2O(l) + \cancel{4e^-} \rightarrow 4[Co(NH_3)_6]^{3+}(aq) + \cancel{4e^-} + 4OH^-(aq)$$

$$4[Co(NH_3)_6]^{2+}(aq) + O_2(g) + 2H_2O(l) \rightarrow 4[Co(NH_3)_6]^{3+}(aq) + 4OH^-(aq)$$

Summary questions

1 Zinc will reduce VO_2^+ ions to VO^{2+}; VO^{2+} to V^{3+} and V^{3+} ions to V^{2+} ions. The relevant half equations are:

$$Zn(s) \rightarrow Zn^{2+}(aq) + 2e^-$$
$$VO_2^+(aq) + 2H^+(aq) + e^- \rightarrow H_2O(l) + VO^{2+}(aq)$$
$$VO^{2+}(aq) + 2H^+(aq) + e^- \rightarrow H_2O(l) + V^{3+}(aq)$$
$$V^{3+}(aq) + e^- \rightarrow V^{2+}(aq)$$

   a Write the balanced equation for each of the reduction steps.

   b V^{2+} has to be protected from air. Suggest a reason for this.

2 A titration to determine the amount of iron(II) sulfate in an iron tablet was carried out. The tablet was dissolved in excess sulfuric acid and made up to 250 cm^3 in a volumetric flask. 25.00 cm^3 of this solution was pipetted into a flask and titrated with 0.0010 mol dm^{-3} potassium manganate(VII) solution until the solution just became purple. Taking an average of several titrations, 25.00 cm^3 of potassium manganate(VII) solution was needed. How many grams of iron are in this tablet? A_r Fe = 55.8.

3 The $E^\ominus$ value for $Cr_2O_7^{2-}(aq) + 14H^+(aq) + 6e^- \rightleftharpoons 2Cr^{3+}(aq) + 7H_2O(l)$ is +1.33 V.

   Use $E^\ominus$ values to show that acidified $Cr_2O_7^{2-}$ ions can be used in a redox titration with Fe^{2+} when Cl^- ions are present, that is, that $Cr_2O_7^{2-}$ ions will not oxidise Cl^-.

Synoptic link

Topic 20.2, Predicting the direction of redox reactions, contains the $E^\ominus$ values you will need that are not given here.

Catalysts affect the rate of a reaction without being chemically changed themselves at the end of the reaction. Catalysts play an important part in industry because they allow reactions to proceed at lower temperatures and pressures thus saving valuable resources. Modern cars have a catalytic converter in the exhaust system which is based on platinum and rhodium. This catalyses the conversion of carbon monoxide, nitrogen oxides, and unburnt petrol to carbon dioxide, nitrogen, and water.

Many catalysts used in industry are transition metals or their compounds. Catalysts can be divided into two groups:

- heterogeneous
- homogeneous.

Heterogeneous catalysts

Heterogeneous catalysts are present in a reaction in a different phase (solid, liquid, or gas) than the reactants. They are usually present as solids, whilst the reactants may be gases or liquids. Their catalytic action occurs on the solid surface. The reactants pass over the catalyst surface, which remains in place so the catalyst is not lost and does not need to be separated from the products.

Making heterogenous catalysts more efficient

Catalysts are often expensive, so the more efficiently they work, the more the costs can be minimised. Since their activity takes place on the surface you can:

- Increase their surface area – the larger the surface area, the better the efficiency.
- Spread the catalyst onto an inert support medium, or even impregnate it into one. This increases the surface-to-mass ratio so that a little goes a long way. The more expensive catalysts are often used in this way. For example, the catalytic converter in a car, has finely divided rhodium and platinum on a ceramic material.

Catalysts do not last forever.

- Over time, the surfaces may become covered with unwanted impurities. This is called poisoning. The catalytic converters in cars gradually become poisoned by substances used in fuel additives. Until a few years ago, lead-based additives were used in petrol. The lead poisoned the catalysts and so leaded fuel could not be used in cars with converters.
- The finely divided catalyst may gradually be lost from the support medium.

Learning objectives:

→ State what is meant by heterogeneous and homogeneous catalysts.

→ Describe how heterogeneous catalysts can be made more efficient.

→ Explain how a homogeneous catalyst works.

Specification reference: 3.2.5

Synoptic link

Catalysts were first introduced in Topic 5.3, Catalysts.

You will learn more about the use of catalysts in organic chemistry in Topic 26.4, Reactions of carboxylic acids and esters.

Hint

Transition metals have partly full d-orbitals which can be used to form weak chemical bonds with the reactants. This has two effects – weakening bonds within the reactant and holding the reactants close together on the metal surface in the correct orientation for reaction.

Study tip

Try to understand the factors that determine cost and catalyst efficiency.

Some important examples of heterogeneous catalysts

The Haber process

You have already met the Haber process, where ammonia is made by the reaction of nitrogen with hydrogen. The catalyst for the process is iron – present as pea-sized lumps to increase the surface area:

$$N_2(g) + 3H_2(g) \xrightleftharpoons{\text{iron catalyst}} 2NH_3(g)$$

The iron catalyst lasts about five years before it becomes poisoned by impurities in the gas stream such as sulfur compounds, and has to be replaced.

The Contact process

The Contact process produces sulfuric acid – a vital industrial chemical. Around two million tonnes are produced each year in the UK and it is involved in the manufacture of many goods.

It is made from sulfur, oxygen, and water, the key step being:

$$2SO_2 + O_2 \rightleftharpoons 2SO_3$$

This is catalysed by vanadium(V) oxide, V_2O_5, in two steps as follows:

The vanadium(V) oxide oxidises sulfur dioxide to sulfur trioxide and is itself reduced to vanadium(IV) oxide:

$$SO_2 + V_2O_5 \rightarrow SO_3 + V_2O_4$$

The vanadium(IV) oxide is then oxidised back to vanadium(V) oxide by oxygen:

$$2V_2O_4 + O_2 \rightarrow 2V_2O_5$$

The vanadium(V) oxide is regenerated unchanged. Each of the two steps has a lower activation energy than the uncatalysed single step and therefore the reaction goes faster.

This is a good example of how the variability of oxidation states of a transition metal is useful in catalysis.

The manufacture of methanol

Synthesis gas is made from methane, present in natural gas and steam:

$$CH_4(g) + H_2O(g) \rightarrow CO + 3H_2(g)$$

It is a mixture of carbon monoxide and hydrogen and is used to make methanol:

$$\underset{\text{synthesis gas}}{CO(g) + 2H_2(g)} \rightarrow \underset{\text{methanol}}{CH_3OH(g)}$$

This reaction may be catalysed by chromium oxide, Cr_2O_3. Today, the most widely used catalyst is a mixture of copper, zinc oxide, and aluminium oxide.

Methanol is an important industrial chemical (over 30 million tonnes are made each year world wide) and is used mainly as a starting material for the production of plastics such as Bakelite, Terylene, and Perspex.

Homogeneous catalysts

When the catalyst is in the same phase as the reactant, an intermediate species is formed. For example in the gas phase chlorine free radicals act as catalysts to destroy the ozone layer. The intermediate here is the ClO• free radical.

Homogeneous catalysis by transition metals

Peroxodisulfate ions, $S_2O_8^{2-}$, oxidise iodide ions to iodine. This reaction is catalysed by Fe^{2+} ions. The overall reaction is:

$$S_2O_8^{2-}(aq) + 2I^-(aq) \rightarrow 2SO_4^{2-}(aq) + I_2(aq)$$

The catalysed reaction takes place in two steps. First the peroxodisulfate ions oxidise iron(II) to iron(III):

$$S_2O_8^{2-}(aq) + 2Fe^{2+}(aq) \rightarrow 2SO_4^{2-}(aq) + 2Fe^{3+}(aq)$$

The Fe^{3+} then oxidises the I^- to I_2, regenerating the Fe^{2+} ions so that none are used up in the reaction:

$$2Fe^{3+}(aq) + 2I^-(aq) \rightarrow 2Fe^{2+}(aq) + I_2(aq)$$

So iron first gives an electron to the peroxodisulfate and later takes one back from the iodide ions.

The uncatalysed reaction takes place between two ions of the same charge (both negative), which repel, therefore giving a high activation energy. Both steps of the catalysed reaction involve reaction between pairs of oppositely charged ions. This helps to explain the increase in rate.

Figure 1 shows the reaction profile. Although there are two steps in the catalysed reaction, the overall activation energy is lower than that for the uncatalysed reaction.

Synoptic link

The destruction of the ozone layer by chlorine free radicals was introduced in Topic 12.5, The formation of halogenoalkanes.

Hint

Do not be put off by unfamiliar chemical names (such as peroxodisulfate ions). Make sure you understand the process that they are used to illustrate.

Study tip

Fe^{3+} can also act as a catalyst for this reaction. When Fe^{3+} is used, I^- is first oxidised to I_2.

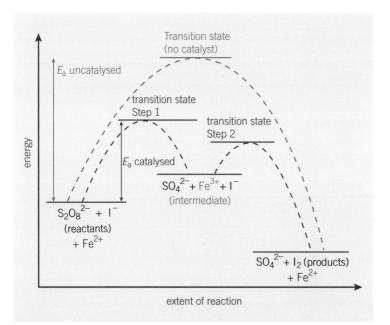

▲ **Figure 1** *Possible reaction profile for the iodine/peroxodisulfate reaction. E_a for the catalysed reaction is the energy gap between the reactants and the higher of the two transition states (transition state Step 1)*

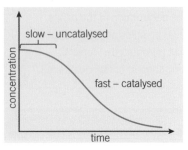

▲ **Figure 2** *A concentration/time graph for an autocatalytic reaction*

Autocatalysis

An interesting example of catalysis occurs when one of the products of the reaction is a catalyst for the reaction. Such a reaction starts slowly at the uncatalysed rate. As the concentration of the product that is also the catalyst builds up, the reaction speeds up to the catalysed rate. From then on it behaves like a normal reaction, gradually slowing down as the reactants are used up. This leads to an odd-looking rate curve (Figure 2).

The oxidation of ethanedioic acid by manganate(VII) ions

One example of an autocatalysed reaction is that between a solution of ethanedioic acid (oxalic acid) and an acidified solution of potassium manganate(VII). It is used as a titration to find the concentration of potassium manganate(VII) solution.

$$2MnO_4^-(aq) + 16H^+(aq) + 5C_2O_4^{2-}(aq) \rightarrow 2Mn^{2+}(aq) + 8H_2O(l) + 10CO_2(g)$$

| manganate(VII) ions | hydrogen ions | ethanedioate ions | manganese(II) ions | water | carbon dioxide |

The catalyst, Mn^{2+} ions, is not present at the beginning of the reaction. Once a little Mn^{2+} has formed, it can react with MnO_4^- ions to form Mn^{3+} as an intermediate species, which then reacts with $C_2O_4^{2-}$ ions to reform Mn^{2+}:

$$4Mn^{2+}(aq) + MnO_4^-(aq) + 8H^+(aq) \rightarrow 5Mn^{3+}(aq) + 4H_2O(l)$$

$$2Mn^{3+}(aq) + C_2O_4^{2-}(aq) \rightarrow 2CO_2(g) + 2Mn^{2+}(aq)$$

The reaction can easily be followed using a colorimeter to measure the concentration of MnO_4^-, which is purple. The reaction curve looks like the one in Figure 2.

Summary questions

1   a   State the difference between a homogeneous and a heterogeneous catalyst.

   b   Classify each of the examples below as homogeneous or heterogeneous.

      i   A gauze of platinum and rhodium catalyses the oxidation of ammonia gas to nitrogen monoxide during the manufacture of nitric acid.

      ii   Nickel catalyses the hydrogenation of vegetable oils.

      iii The enzymes in yeast catalyse the production of ethanol from sugar.

2   Why does a catalyst make a reaction go faster? Why is this particularly important for industry?

3   The peroxodisulfate / iodide reaction above is catalysed by Fe^{3+} ions (as well as by Fe^{2+} ions):

$$S_2O_8^{2-}(aq) + 2I^-(aq) \rightarrow 2SO_4^{2-}(aq) + I_2(aq)$$

   Write down the two equations that explain this and explain why it is slow in the absence of the catalyst.

4   Show how the overall equation for the autocatalytic reaction between MnO_4^- and $C_2O_4^{2-}$ can be obtained from the equations for the two catalytic steps.

Practice questions

1 Transition metals and their complexes have characteristic properties.
   (a) Give the electron configuration of the Zn^{2+} ion.
        Use your answer to explain why the Zn^{2+} ion is **not** classified as a transition metal
        ion.
                                                                              (2 marks)
   (b) In terms of bonding, explain the meaning of the term *complex*.
                                                                              (2 marks)
   (c) Identify **one** species from the following list that does **not** act as a ligand. Explain your
        answer.
                    H_2 O^{2-} O_2 CO   *(2 marks)*
   (d) The element palladium is in the d block of the Periodic Table. Consider the following
        palladium compound which contains the sulfate ion.

                              $[Pd(NH_3)_4]SO_4$

        (i)   Give the oxidation state of palladium in this compound.   *(1 mark)*
        (ii) Give the names of two possible shapes for the complex palladium ion in this
              compound.
                                                                              (2 marks)
                                                                              AQA, 2011

2 This question is about copper chemistry.
   (a) Aqueous copper(II) ions $[Cu(H_2O)_6]^{2+}$(aq) are blue.
        (i)   With reference to electrons, explain why aqueous copper(II) ions are blue.
                                                                              (3 marks)
        (ii) By reference to aqueous copper(II) ions, state the meaning of each of the **three**
              terms in the equation $\Delta E = hv$.
                                                                              (3 marks)
        (iii) Write an equation for the reaction, in aqueous solution, between $[Cu(H_2O)_6]^{2+}$
              and an excess of chloride ions.
              State the shape of the complex produced and explain why the shape differs from
              that of the $[Cu(H_2O)_6]^{2+}$ ion.
                                                                              (3 marks)
   (b) Draw the structure of the ethanedioate ion, $C_2O_4^{2-}$.
        Explain how this ion is able to act as a ligand.
                                                                              (2 marks)
   (c) When a dilute aqueous solution containing ethanedioate ions is added to a solution
        containing aqueous copper(II) ions, a substitution reaction occurs. In this reaction
        four water molecules are replaced and a new complex is formed.
        (i)   Write an ionic equation for the reaction. Give the co-ordination number of the
              complex formed and name its shape.
                                                                              (4 marks)
        (ii) In the complex formed, the two water molecules are opposite each other.
              Draw a diagram to show how the ethanedioate ions are bonded to a copper ion
              and give a value for one of the O—Cu—O bond angles. You are **not** required to
              show the water molecules.
                                                                              (2 marks)
                                                                              AQA, 2011

3 (a) Octahedral and tetrahedral complex ions are produced by the reaction of transition
        metal ions with ligands which form co-ordinate bonds with the transition metal ion.
        Define the term *ligand* and explain what is meant by the term *co-ordinate bond*.
                                                                              (3 marks)
   (b) (i)   Some complex ions can undergo a ligand substitution reaction in which both the
              co-ordination number of the metal and the colour change in the reaction. Write an
              equation for one such reaction and state the colours of the complex ions involved.
        (ii) Bidentate ligands replace unidentate ligands in a metal complex by a ligand
              substitution reaction.
              Write an equation for such a reaction and explain why this reaction occurs.
                                                                              (8 marks)
                                                                              AQA, 2005

Learning objectives:

→ Describe metal aqua ions.

→ State what determines the acidity of metal aqua ions in aqueous solution.

Specification reference: 3.2.6

If you dissolve a salt of a transition metal–such as iron(II) nitrate, $Fe(NO_3)_2$ – in water, water molecules cluster around the Fe^{2+} ion so it actually exists as the complex ion $[Fe(H_2O)_6]^{2+}$. Six water molecules act as ligands bonding to the metal ion in an octahedral arrangement. They each use one of their lone pairs of electrons to form a co-ordinate (dative) bond with the metal ion. A similar situation occurs with an iron(III) salt – here the complex formed is $[Fe(H_2O)_6]^{3+}$. These complexes are called **aqua ions**.

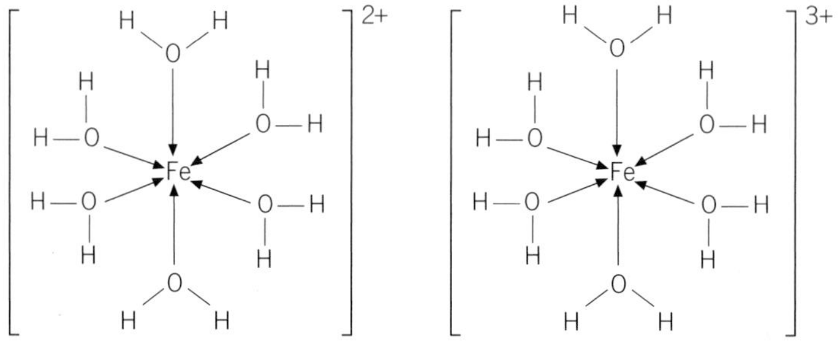

▲ **Figure 1** $[Fe(H_2O)_6]^{2+}$ (left) and $[Fe(H_2O)_6]^{3+}$ (right)

However, there is a significant difference in the acidity of these two complexes.

Solutions of $Fe^{2+}(aq)$ are not noticeably acidic, whereas a solution of $Fe^{3+}(aq)$ ($pK_a = 2.2$) is a stronger acid than ethanoic acid ($pK_a = 4.8$). Why is $Fe^{3+}(aq)$ acidic at all and why the difference with $Fe^{2+}(aq)$? This is because the Fe^{3+} ion is both smaller and more highly charged than Fe^{2+} (it has a higher charge density) making it more strongly polarising. So in the $[Fe(H_2O)_6]^{3+}(aq)$ ion the iron strongly attracts electrons from the oxygen atoms of the water ligands, so weakening the O—H bonds in the water molecules. This complex ion will then readily release an H^+ ion making the solution acidic (Figure 2). Fe^{2+} is less polarising and so fewer O—H bonds break in solution.

> **Hint**
>
> When drawing co-ordinate bonds, the arrow → represents the donated pair of electrons.

> **Synoptic link**
>
> pK_a was discussed in Topic 21.3, Weak acids and bases. It is a measure of the strength of an acid. The *smaller* the value of pK_a, the *stronger* the acid.

> **Hint**
>
> In a dilute solution of iron(II) nitrate there will be many times more water molecules than nitrate ions so these are far more likely to act as ligands.

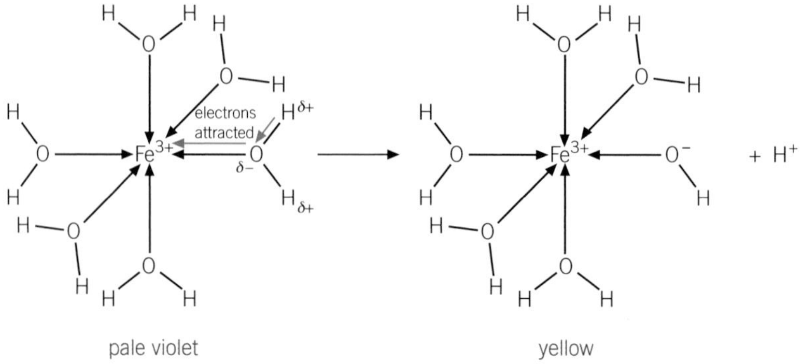

▶ **Figure 2** *The acidity of $Fe^{3+}(aq)$ ions*

pale violet   yellow

Written as an equation:

$$[Fe(H_2O)_6]^{3+}(aq) \rightleftharpoons [Fe(H_2O)_5(OH)]^{2+}(aq) + H^+(aq)$$

With transition metals, there is a general rule that aqua ions of M^{3+} are significantly more acidic than those of M^{2+}.

A similar situation occurs in solutions of $Al^{3+}(aq)$, although aluminium is not a transition metal.

Reactions such as the above are often called **hydrolysis** (reaction with water) because they may also be represented as:

$$[Fe(H_2O)_6]^{3+}(aq) + H_2O(l) \rightleftharpoons [Fe(H_2O)_5(OH)]^{2+}(aq) + H_3O^+(aq)$$

This stresses the fact that the $[Fe(H_2O)_6]^{3+}$ ion is donating a proton, H^+, to a water molecule and behaving as Brønsted–Lowry acid.

Study tip

A hydrolysis reaction is one in which O—H bonds of water are broken and new species are formed.

Synoptic link

You will need to understand Brønsted–Lowry acids and bases studied in Topic 21.1, Defining an acid.

Lewis acids and bases

The Brønsted–Lowry theory of acidity describes acids as proton (H^+ ion) donors, and bases, such as OH^- ions, as proton acceptors.

Another theory (the Lewis theory) is also used to describe acids. This theory defines acids as electron pair acceptors, and bases as electron pair donors in the formation of co-ordinate (dative) covalent bonds. For example:

F—B + :N—H ⟶ F—B←N—H (boron trifluoride / ammonia)

boron   ammonia
trifluoride

Here, boron trifluoride is acting as a Lewis acid (electron pair acceptor) and ammonia as a Lewis base (electron pair donor). The Lewis definition of acids is wider than the Brønsted–Lowry one. Boron trifluoride contains no hydrogen and so cannot be an acid under the Brønsted–Lowry definition. H^+ ions have no electrons at all and so can *only* form bonds by accepting an electron pair.

$$H^+ + \,^-:O—H \longrightarrow H \leftarrow O—H$$

A water molecule has two lone pairs of electrons and it can use one of these to accept a proton (acting as a Lewis base and as a Brønsted–Lowry base) or, for

example, to form a co-ordinate bond with a metal ion (acting as a Lewis base).

$$H^+ + :O—H \longrightarrow [H \leftarrow O—H]^+$$

Lewis base and Brønsted–Lewis base

$$Cu^{2+} + 6:O—H \longrightarrow$$

Lewis base

All Brønsted–Lowry acids are also Lewis acids. Ligands which form bonds to transition metal ions using lone pairs are acting as Lewis bases and the metal ions as Lewis acids.

Which of the following can act as Lewis acids and which as Lewis bases?
$AlCl_3$, AlF_3, V^{3+}, Zn^{2+}

Acids: $AlCl_3$, AlF_3 Bases: V^{3+}, Zn^{2+}

Theories of acidity over the years

Acids are a group of compounds with similar properties, for example, neutralising bases, producing hydrogen with the more-reactive metals, releasing carbon dioxide from carbonates. They were probably first recognised as a group by their sour taste. Today, of course, no one would dream of tasting a newly synthesised compound before it had been thoroughly tested for toxicity (which could take some time), but in old chemical papers it is not uncommon to find the taste of new compounds reported along with colour, crystal form, melting point, and so on.

Many theories of acidity have been proposed, and these have been discarded or modified as new facts have come along. This is how scientific understanding progresses. Theories of acidity include:

Lavoisier (1777) proposed that all acids contain oxygen. This is fine for many acids, for example, nitric, HNO_3, sulfuric, H_2SO_4, and ethanoic (acetic), CH_3COOH, and was a good working theory. However, once the formula of hydrochloric acid, HCl, was worked out, it became clear that this theory could not be correct.

Davy (1816) suggested that all acids contain hydrogen. This looks better – all the above acids fit and the theory has no problem including HCl. However, it does not explain why the hydrogen is important.

Liebig (1838) defined acids as substances containing hydrogen which could be replaced by a metal. This is an improvement on Davy's theory as it explains why not all hydrogen-containing compounds are acidic, for example, ammonia, NH_3, is not acidic. There must be something special about that hydrogen that makes it replaceable by a metal. This is a theory that is not far from one that could be used today.

Arrhenius (1887) thought of acids as producing hydrogen ions, H^+. This is a development of Liebig's theory. It tells us what exactly is special about the hydrogen – it must be able to become an H^+ ion.

The **Brønsted–Lowry** description of acidity (developed in 1923 by Thomas Lowry and Johannes Brønsted independently) is the most generally useful current theory. This defines an acid as a substance which can donate a proton (an H^+ ion) and a base as a substance which can accept a proton. However, this theory has difficulty with acids that do not contain hydrogen – aluminium chloride, $AlCl_3$, or boron trifluoride, BF_3, for example.

Another theory (the **Lewis theory**) is also used today to describe acids. This theory regards acids as electron pair acceptors and bases as electron pair donors in the formation of co-ordinate covalent bonds.

Acid–base reactions of M^{2+}(aq) and M^{3+}(aq) ions

If you add a base (such as OH^-) it will remove protons from the aqueous complex. This takes place in a series of steps.

> **Hint**
>
> The boron in boron trifluoride has only six electrons in its outer shell and is therefore able to accept an electron pair from ammonia, for example.

In the case of M³⁺

$$[M(H_2O)_6]^{3+}(aq) + OH^-(aq) \rightarrow [M(H_2O)_5(OH)]^{2+}(aq) + H_2O \; (l)$$

$$[M(H_2O)_5(OH)]^{2+}(aq) + OH^-(aq) \rightarrow [M(H_2O)_4(OH)_2]^+(aq) + H_2O \; (l)$$

$$[M(H_2O)_4(OH)_2]^+(aq) + OH^-(aq) \rightarrow M(H_2O)_3(OH)_3(s) + H_2O \; (l)$$

The neutral metal(III) hydroxide, $M(H_2O)_3(OH)_3$, is in effect $M(OH)_3$, which is uncharged and insoluble and forms as a precipitate.

In the case of M²⁺

$$[M(H_2O)_6]^{2+}(aq) + OH^-(aq) \rightarrow [M(H_2O)_5(OH)]^+(aq) + H_2O \; (l)$$

$$[M(H_2O)_5(OH)]^+(aq) + OH^-(aq) \rightarrow M(H_2O)_4(OH)_2(s) + H_2O \; (l)$$

The neutral metal(II) hydroxide, $M(H_2O)_4(OH)_2$, is in effect $M(OH)_2$, which is uncharged and insoluble and forms a precipitate.

Ammonia, which is basic, has the same effect as OH^- ions in removing protons.

$$[M(H_2O)_6]^{3+}(aq) + 3NH_3(aq) \rightarrow M(H_2O)_3(OH)_3 \; (s) + 3NH_4^+$$

$$[M(H_2O)_6]^{2+}(aq) + 2NH_3(aq) \rightarrow M(H_2O)_4(OH)_2(s) + 2NH_4^+$$

Reactions with the base CO_3^{2-}, the carbonate ion

The greater acidity of the aqueous Fe^{3+} ion explains why iron(III) carbonate does not exist, but iron(II) carbonate does. The carbonate ion is able to remove protons from $[Fe(H_2O)_6]^{3+}(aq)$ to form hydrated iron(III) hydroxide but cannot do so from $[Fe(H_2O)_6]^{2+}(aq)$.

$$[Fe(H_2O)_6]^{3+}(aq) + 3CO_3^{2-}(aq) \rightleftharpoons Fe(OH)_3(H_2O)_3(s) + 3HCO_3^-(aq)$$

The overall reaction is:

$$2[Fe(H_2O)_6]^{3+}(aq) + 3CO_3^{2-}(aq) \rightarrow 2[Fe(H_2O)_3(OH)_3](aq) + 3CO_2(g) + 3H_2O(l)$$

The reaction can be derived as a combination of the following:

$$2H_3O^+(aq) + CO_3^{2-}(aq) \rightarrow 3H_2O(l) + CO_2(g)$$

With the removal of H_3O^+ displacing the hydrolysis equilibrium below to the right.

$$[Fe(H_2O)_6]^{3+}(aq) + 3H_2O(l) \rightleftharpoons Fe(H_2O)_3(OH)_3 \; (s) + 3H_3O^+(aq)$$

In the case of the aqueous Fe^{2+} ion, which is less acidic than $Fe^{3+}(aq)$, insoluble iron(II) carbonate is formed:

$$[Fe(H_2O)_6]^{2+}(aq) + CO_3^{2-}(aq) \rightarrow FeCO_3 \; (s) + 6H_2O(l)$$

In general, carbonates of transition metal ions in oxidation state +2 exist, whilst those of ions in the +3 state do not.

Distinguishing iron ions

As you have seen, both Fe^{2+} and Fe^{3+} exist in aqueous solution as octahedral hexa-aqua ions. $[Fe(H_2O)_6]^{2+}$ is pale green and $[Fe(H_2O)_6]^{3+}$ is pale brown, and dilute solutions are hard to tell

> **Synoptic link**
>
> You will need to understand bond polarity studied in Topic 3.4, Electronegativity – bond polarity in covalent bonds, and equilibria in Chapter 6, Equilibria.

> **Hint**
>
> If a solution of sodium carbonate is added to a solution containing Fe^{3+} ions (e.g., $Fe(NO_3)_3$) it will fizz due to carbon dioxide being released. a solution containing Fe^{2+} ions will form a precipitate of the carbonate.

apart. A simple test to distinguish the two is to add dilute alkali, which precipitates the hydroxides whose colours are more obviously different.

$$[Fe(H_2O)_6]^{3+}(aq) + 3OH^-(aq) \rightarrow Fe(H_2O)_3(OH)_3(s) + 3H_2O(l)$$
iron(III) hydroxide (brown)

$$[Fe(H_2O)_6]^{2+}(aq) + 2OH^-(aq) \rightarrow Fe(H_2O)_4(OH)_2(s) + 2H_2O(l)$$
iron(II) hydroxide (green)

▲ **Figure 3** *Iron(III) hydroxide precipitate*

▲ **Figure 4** *Iron(II) hydroxide precipitate*

Amphoteric hydroxides

Amphoteric means showing both acidic and basic properties. Aluminium hydroxide is an example of this – it will react with both acids and bases. For example:

$$Al(H_2O)_3(OH)_3 + 3HCl \rightarrow Al(H_2O)_6^{3+} + 3Cl^-$$

This is what you would expect from a normal metal hydroxide – it reacts with acid and is therefore basic.

But aluminium hydroxide also shows acidic properties – it will react with the base sodium hydroxide to give a colourless solution of sodium tetrahydroxoaluminate:

$$Al(H_2O)_3(OH)_3 + OH^- \rightarrow [Al(OH)_4]^- + 3H_2O$$

> ### Study tip
>
> An excess of a strong base (e.g. NaOH) is needed to redissolve the $Al(H_2O)_3(OH)_3$ precipitate.

Summary questions

1 What are the oxidation states of the metal atoms in these ions?

   a MnO_4^- b CrO_4^{2-} c $Cr_2O_7^{2-}$

2 Classify the reaction as redox or acid–base. Explain your answer.

   $$CrO_4^{2-}(aq) + 2H^+(aq) \rightleftharpoons Cr_2O_7^{2-}(aq) + H_2O(l)$$

The water molecules that act as ligands in metal aqua ions can be replaced by other ligands – either because the other ligands form stronger co-ordinate bonds or because they are present in higher concentration and an equilibrium is displaced.

Replacing water as a ligand

There are a number of possibilities:

- The water molecules may be replaced by other neutral ligands, such as ammonia.
- The water molecules may be replaced by negatively charged ligands, such as chloride ions.
- The water molecules may be replaced by bi- or multidentate ligands – this is called chelation.
- Replacement of the water ligands may be complete or partial.

Replacement by neutral ligands – no change in co-ordination number

In general for an M^{2+} ion, water molecules may be replaced one at a time by ammonia. Both ligands are uncharged and are of similar size, so there is no change in co-ordination number or charge on the ion:

$$[M(H_2O)_6]^{2+} + NH_3 \rightleftharpoons [M(NH_3)(H_2O)_5]^{2+} + H_2O$$

$$[M(NH_3)(H_2O)_5]^{2+} + NH_3 \rightleftharpoons [M(NH_3)_2(H_2O)_4]^{2+} + H_2O$$

$$[M(NH_3)_2(H_2O)_4]^{2+} + NH_3 \rightleftharpoons [M(NH_3)_3(H_2O)_3]^{2+} + H_2O$$

$$[M(NH_3)_3(H_2O)_3]^{2+} + NH_3 \rightleftharpoons [M(NH_3)_4(H_2O)_2]^{2+} + H_2O$$

$$[M(NH_3)_4(H_2O)_2]^{2+} + NH_3 \rightleftharpoons [M(NH_3)_5(H_2O)]^{2+} + H_2O$$

$$[M(NH_3)_5(H_2O)]^{2+} + NH_3 \rightleftharpoons [M(NH_3)_6]^{2+} + H_2O$$

Overall:

$$[M(H_2O)_6]^{2+} + 6NH_3 \rightleftharpoons [M(NH_3)_6]^{2+} + 6H_2O$$

There is a complication, because ammonia is a base as well as a ligand, and therefore contains OH^- ions, a precipitate may form and then redissolve

$$[M(H_2O)_6]^{2+} + 2OH^-(aq) \rightarrow M(H_2O)_4(OH)_2(s) + 2H_2O(l)$$

$$M(H_2O)_4(OH)_2(s) + 6NH_3(aq) \rightleftharpoons [M(NH_3)_6]^{2+}(aq) + 4H_2O(l) + 2OH^-(aq)$$

Cobalt(II)

When M is cobalt, Co, the first step is the formation of a blue precipitate of hydrated cobalt(II) hydroxide when ammonia, NH_3, is added. This is produced by the loss of a proton from each of two of the six water molecules co-ordinated to the Co^{2+} ion:

$$[Co(H_2O)_6]^{2+}(aq) + 2NH_3(aq) \rightarrow [Co(H_2O)_4(OH)_2](s) + 2NH_4^+(aq)$$
$$\text{hydrated cobalt(II)}$$
$$\text{hydroxide}$$

Here ammonia is acting as a base.

Learning objectives:

→ Explain the changes in the co-ordination numbers and charges of complexes when different ligands are substituted.

→ Explain why complexes formed with multidentate ligands are more stable than those with monodentate ligands.

Specification reference: 3.2.6

Synoptic link

Chelation was covered in Topic 23.2, Complex formation and the shape of complex ions.

Synoptic link

You will need to understand free energy change and entropy studied in 17.4, Why do chemical reactions take place?

Study tip

Because NH_3 and H_2O ligands are similar in size and both are uncharged, ligand exchange occurs without a change in charge or co-ordination number.

Hint

Ammonia is a better ligand than water because the lone pair on the nitrogen atom of ammonia is less strongly held than that on the more electronegative oxygen atom of water. It is therefore more readily donated to the Co^{2+}.

▲ **Figure 1** *Pale blue solution of $[Cu(H_2O)_6]^{2+}$, the pale blue precipitate of $[Cu(OH)_2(H_2O)_4]$, and the deep blue solution of $[Cu(NH_3)_4(H_2O)_2]^{2+}$*

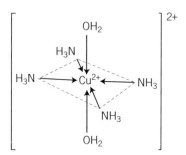

▲ **Figure 2** *The shape of the $[Cu(NH_3)_4(H_2O)_2]^{2+}$ ion. The dotted lines are not bonds, they are construction lines to show the square-planar arrangement of the NH_3 ligands*

Synoptic link

There are some further examples of ligand substitution reactions of other metal ions, both M^{2+} and M^{3+}, summarised in Topic 24.3, A summary of acid–base and substitution reactions of some metal ions.

If you add more of the concentrated ammonia, then both OH^- and all four water ligands are replaced by ammonia. This is for two reasons:

1 Ammonia is a better ligand than water.
2 The high concentration of ammonia displaces equilibria like those above to the right, thus displacing water and OH^-.

Overall:

$$[Co(H_2O)_4(OH)_2](s) + 6NH_3(aq) \rightleftharpoons [Co(NH_3)_6]^{2+}(aq) + 4H_2O(l) + 2OH^-(aq)$$

The blue precipitate dissolves to form a pale yellow solution (which is oxidised by oxygen in air to a brown mixture containing $Co(III)$).

Copper(II)

When aqueous copper ions react with ammonia in aqueous solution, ligand replacement is only partial – only four of the water ligands are replaced. The overall reaction is:

$$[Cu(H_2O)_6]^{2+} + 4NH_3 \rightleftharpoons [Cu(NH_3)_4(H_2O)_2]^{2+} + 4H_2O$$

$[Cu(H_2O)_6]^{2+}$ is pale blue whilst $[Cu(NH_3)_4(H_2O)_2]^{2+}$ is a very deep blue.

The steps are similar to those above for Co^{2+}. The ammonia first acts as a base removing protons from two of the water molecules in $[Cu(H_2O)_6]^{2+}$ to form $[Cu(OH)_2(H_2O)_4](s)$. The first thing we see is a pale blue precipitate of copper hydroxide. When more of the concentrated ammonia is added, the precipitate dissolves to form a deep blue solution containing $[Cu(NH_3)_4(H_2O)_2]^{2+}$ (Figure 1). The ammonia has replaced both OH^- ligands, and two of the H_2O ligands:

$$[Cu(OH)_2(H_2O)_4](s) + 4NH_3(aq) \rightleftharpoons [Cu(NH_3)_4(H_2O)_2]^{2+}(aq) + 2H_2O(l) + 2OH^-(aq)$$

The shape of the $[Cu(NH_3)_4(H_2O)_2]^{2+}$ ion

The $[Cu(NH_3)_4(H_2O)_2]^{2+}$ is octahedral, as expected for a six co-ordinate ion. The four ammonia molecules exist in a square-planar arrangement around the metal ion with the two water molecules above and below the plane (Figure 2).

The Cu—O bonds are longer (and therefore weaker) than the Cu—N bonds, as would be expected because water is a poorer ligand than ammonia. The octahedron is slightly distorted.

Replacement by chloride ions – change in co-ordination number

When aqueous copper ions react with concentrated hydrochloric acid there is a change in both charge and co-ordination number. Concentrated hydrochloric acid provides a high concentration of Cl^- ligands:

$$[Cu(H_2O)_6]^{2+} + 4Cl^- \rightleftharpoons [CuCl_4]^{2-} + 6H_2O$$

The pale blue colour of the $[Cu(H_2O)_6]^{2+}$ ion is replaced by the yellow $[CuCl_4]^{2-}$ ion. (Although the solution may look green as some $[Cu(H_2O)_6]^{2+}$ will remain.) Again, the actual replacement takes place in steps. The co-ordination number of the ion is four and the ion is tetrahedral.

$[Cu(H_2O)_6]^{2+}$ is six co-ordinate and $[CuCl_4]^{2-}$ is four co-ordinate (Figure 3), because Cl^- is larger than H_2O and fewer ligands can physically fit around the central copper ion.

Chelation

Chelation is the formation of complexes with multidentate ligands. These are ligands with more than one lone pair so they can form more than one co-ordinate bond. Examples include ethylene diamine, benzene-1,2-diol, and $EDTA^{4-}$. These complexes are usually more stable than those with monodentate ligands. This increased stability is mainly due to the entropy change of the reaction.

Ethylene diamine (often represented as en for short) is a bidentate ligand and can be thought of as two ammonia ligands linked by a short hydrocarbon chain. Each en can replace two water molecules:

$$[Cu(H_2O)_6]^{2+}(aq) + 3en \rightarrow [Cu(en)_3]^{2+}(aq) + 6H_2O(l)$$
$$\text{four entities} \qquad\qquad \text{seven entities}$$

In this reaction, *three* molecules of ethylene diamine release *six* of water. The larger number of entities on the right in this reaction means that there is a significant entropy increase as the reaction goes from left to right. This favours the reaction.

A single hexadentate ligand $EDTA^{4-}$ can displace all six water ligands from $[M(H_2O)_6]^{2+}$. For example:

$$[Cu(H_2O)_6]^{2+}(aq) + EDTA^{4-}(aq) \rightarrow [CuEDTA]^{2-}(aq) + 6H_2O(l)$$
$$\text{two entities} \qquad\qquad \text{seven entities}$$

In this reaction, *one* ion of $EDTA^{4-}$ releases six water ligands. The larger number of entities on the right in this reaction means that there is a significant entropy increase as the reaction goes from left to right. This entropy increase favours the formation of chelates (complexes with polydentate ligands) over complexes with monodentate ligands.

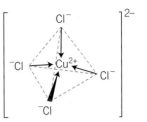

▲ **Figure 3** *The shape of the $[CuCl_4]^{2-}$ ion*

Study tip

Practise writing equations for the step-by-step replacement of one ligand with another. They are straightforward but can be tedious, and it is easy to make mistakes with charges. Make sure each step balances for both ligands and charge.

Hint

Ethylene diamine (en) is also called ethane-1,2-diamine and 1,2-diaminoethane. Its formula is $H_2NCH_2CH_2NH_2$. Each nitrogen atom has a lone pair which makes it a bidentate ligand.

Synoptic link

Entropy is a measure of the degree of disorder (Topic 17.4, Why do chemical reactions take place?) so the more molecules that are released and are free to move around, the greater the entropy.

Summary questions

1   In the stepwise conversion of $[Cu(H_2O)_6]^{2+}$ to $[CuCl_4]^{2-}$, one of the species formed is neutral. Suggest two possible formulae that it could have.

2   **a** Draw the shape of $[Cu(H_2O)_6]^{2+}$ and **b** predict the shape of $[CuBr_4]^{2-}$. Explain your answer.

3   Write equations for the step-by-step replacement of all the water ligands in $[Cu(H_2O)_6]^{2+}$ by en. How many entities are there on each side of each equation? Predict the likely sign of the entropy change for each reaction.

4   When concentrated hydrochloric acid is added to an aqueous solution containing Co(II) ions, the following change takes place:

$[Co(H_2O)_6]^{2+} \rightarrow [CoCl_4]^{2-}$ and the colour changes from pink to blue.

a   Is there any change in the oxidation state of the cobalt?

b   Give the shapes of the two ions concerned.

c   Suggest two possible reasons for the colour change.

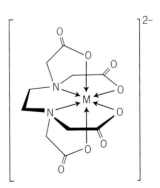

▲ **Figure 4** *A metal(II)–EDTA complex*

24.3 A summary of acid–base and substitution reactions of some metal ions

Learning objective:

→ Describe products of the reactions between bases and metal aqua ions.

Specification reference: 3.2.6

Synoptic link

You will need to understand co-ordinate bonding studied in, Topic 3.2, Covalent bonding.

The products of many of the reactions of transition metal compounds can be identified by their colours.

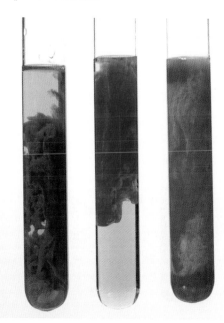

▲ **Figure 1** *Precipitates of (left to right) iron(III) hydroxide, copper(II) hydroxide, and a suspension of chromium(III) hydroxide*

Table 1 and Table 2 show a number of examples of reactions involving different bases. You should be able to rationalise these observations using the principles explained in Topic 23.2. In each case the effect of adding the base shown on the left to solutions of aqua ions is shown in the following tables.

▼ **Table 1** *Adding a base to $M^{2+}(aq)$ complexes*

	$[Fe(H_2O)_6]^{2+}(aq)$ pale green	$[Cu(H_2O)_6]^{2+}(aq)$ pale blue
OH^- little	green gelatinous ppt of $[Fe(H_2O)_4(OH)_2]$*	pale blue ppt of $[Cu(H_2O)_4(OH)_2]$
OH^- excess	green gelatinous ppt of $[Fe(H_2O)_4(OH)_2]$*	pale blue ppt of $[Cu(H_2O)_4(OH)_2]$
NH_3 little	green gelatinous ppt of $[Fe(H_2O)_4(OH)_2]$*	pale blue ppt of $[Cu(H_2O)_4(OH)_2]$
NH_3 excess	green gelatinous ppt of $[Fe(H_2O)_4(OH)_2]$*	deep blue solution of $[Cu(NH_3)_4(H_2O)_2]^{2+}$
CO_3^{2-}	green ppt of $FeCO_3$	blue-green ppt of $CuCO_3$

*Pale green $[Fe(H_2O)_4(OH)_2]$ is soon oxidised by air to brown $[Fe(H_2O)_3(OH)_3]$

▼ **Table 2** *Adding a base to M³⁺(aq) complexes*

	$[Fe(H_2O)_6]^{3+}(aq)$ purple/yellow/brown	$[Al(H_2O)_6]^{3+}(aq)$ colourless
OH⁻ little	brown gelatinous ppt of $[Fe(H_2O)_3(OH)_3]$	white ppt of $[Al(H_2O)_3(OH)_3]$
OH⁻ excess	brown gelatinous ppt of $[Fe(H_2O)_3(OH)_3]$	colourless solution of $[Al(OH)_4]^-$
NH₃ little	brown gelatinous ppt of $[Fe(H_2O)_3(OH)_3]$	white ppt of $[Al(H_2O)_3(OH)_3]$
NH₃ excess	brown gelatinous ppt of $[Fe(H_2O)_3(OH)_3]$	white ppt of $[Al(H_2O)_3(OH)_3]$
CO_3^{2-}	brown gelatinous ppt of $[Fe(H_2O)_3(OH)_3]$ and bubbles of CO_2	white ppt of $[Al(H_2O)_3(OH)_3]$ and bubbles of CO_2

In the case of the M^{2+} ions, precipitates of the metal carbonates form when carbonate ions are added. In the case of the M^{3+} ions, bubbles of carbon dioxide are produced instead. This is a reflection of the greater acidity of $[M(H_2O)_6]^{3+}$ compared with $[M(H_2O)_6]^{2+}$, see Topic 23.1.

Summary questions

1   Why are M^{3+} aqua ions more acidic than M^{2+} aqua ions?

2   Explain why all the compounds of aluminium are colourless.

3   Explain why $[Co(H_2O)_6]^{2+}$ and $[Co(NH_3)_6]^{2+}$ both have a co-ordination number of six and have the same charge.

4   a   Write the equations for the reactions of:

   i $[Fe(H_2O)_6]^{3+}(aq)$ with sodium hydroxide solution

   ii $[Cu(H_2O)_6]^{2+}(aq)$ with excess ammonia.

   b   What colour changes would you expect to see in **a i** and **ii**?

Practice questions

1 Consider the reaction scheme below and answer the questions which follow.

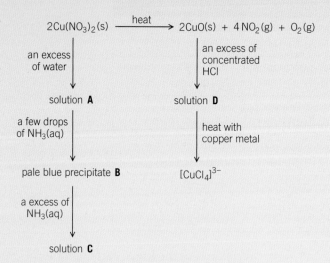

$$2Cu(NO_3)_2(s) \xrightarrow{\text{heat}} 2CuO(s) + 4NO_2(g) + O_2(g)$$

an excess of water

an excess of concentrated HCl

solution **A**

solution **D**

a few drops of $NH_3(aq)$

heat with copper metal

pale blue precipitate **B**

$[CuCl_4]^{3-}$

a excess of $NH_3(aq)$

solution **C**

(a) A redox reaction occurs when $Cu(NO_3)_2$ is decomposed by heat. Deduce the oxidation state of nitrogen in $Cu(NO_3)_2$ and in NO_2 and identify the product formed by oxidation in this decomposition.

(3 marks)

(b) Identify and state the shape of the copper-containing species present in solution **A**.

(2 marks)

(c) Identify the pale blue precipitate **B** and write an equation, or equations, to show how **B** is formed from the copper-containing species in solution **A**.

(2 marks)

(d) Identify the copper-containing species present in solution **C**. State the colour of this copper-containing species and write an equation for its formation from precipitate **B**.

(3 marks)

(e) Identify the copper-containing species present in solution **D**. State the colour and shape of this copper-containing species.

(3 marks)

(f) The oxidation state of copper in $[CuCl_4]^{3-}$ is +1.
   (i) Give the electron arrangement of a Cu^+ ion.
   (ii) Deduce the role of copper metal in the formation of $[CuCl_4]^{3-}$ from the copper-containing species in solution **D**.

(2 marks)
AQA, 2005

2 (a) State what is observed when aqueous ammonia is added dropwise, until present in excess, to a solution of cobalt(II) chloride, and the mixture obtained is then left to stand in air.
   Give the formula of each cobalt-containing species formed. Explain the change which occurs when the mixture is left to stand in air.

(8 marks)

   (b) Explain why separate solutions of iron(II) sulfate and iron(III) sulfate of equal concentration have different pH values.
   State what is observed when sodium carbonate is added separately to solutions of these two compounds. Give the formula of each iron-containing species formed.

(9 marks)
AQA, 2003

3 Consider the following reaction scheme that starts from aqueous $[Cu(H_2O)_6]^{2+}$ ions.

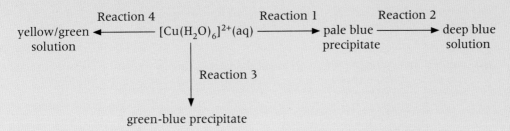

For each of the reactions 1 to 4, identify a suitable reagent, give the formula of the copper-containing species formed and write an equation for the reaction.

(a) Reaction 1   *(3 marks)*

(b) Reaction 2   *(3 marks)*

(c) Reaction 3   *(3 marks)*

(d) Reaction 4   *(3 marks)*

AQA, 2014

4 The scheme below shows some reactions of copper(II) ions in aqueous solution. **W**, **X**, **Y**, and **Z** are all copper-containing species.

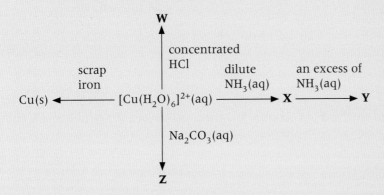

(a) Identify ion **W**. Describe its appearance and write an equation for its formation from $[Cu(H_2O)_6]^{2+}(aq)$ ions.

(3 marks)

(b) Identify compound **X**. Describe its appearance and write an equation for its formation from $[Cu(H_2O)_6]^{2+}(aq)$ ions.

(3 marks)

(c) Identify ion **Y**. Describe its appearance and write an equation for its formation from **X**.

(3 marks)

(d) Identify compound **Z**. Describe its appearance and write an equation for its formation from $[Cu(H_2O)_6]^{2+}(aq)$ ions.

(3 marks)

(e) Copper metal can be extracted from a dilute aqueous solution containing copper(II) ions using scrap iron.
   (i) Write an equation for this reaction and give the colours of the initial and final aqueous solutions.

(3 marks)

   (ii) This method of copper extraction uses scrap iron. Give **two** other reasons why this method of copper extraction is more environmentally friendly than reduction of copper oxide by carbon.

(2 marks)

AQA, 2010

Section 2 practice questions

1   Due to their electron arrangements, transition metals have characteristic properties including catalytic action and the formation of complexes with different shapes.

   (a)   Give **two other** characteristic properties of transition metals. For each property, illustrate your answer with a transition metal of your choice.

(4 marks)

   (b)   Other than octahedral, there are several different shapes shown by transition metal complexes. Name **three** of these shapes and for each one give the formula of a complex with that shape.

(6 marks)

   (c)   It is possible for Group 2 metal ions to form complexes. For example, the $[Ca(H_2O)_6]^{2+}$ ion in hard water reacts with $EDTA^{4-}$ ions to form a complex ion in a similar manner to hydrated transition metal ions. This reaction can be used in a titration to measure the concentration of calcium ions in hard water.

      (i)   Write an equation for the equilibrium that is established when hydrated calcium ions react with $EDTA^{4-}$ ions.

(1 mark)

      (ii)   Explain why the equilibrium in part **(c)(i)** is displaced almost completely to the right to form the EDTA complex.

(3 marks)

      (iii)   In a titration, $6.25\ cm^3$ of a $0.0532\ mol\ dm^{-3}$ solution of EDTA reacted completely with the calcium ions in a $150\ cm^3$ sample of a saturated solution of calcium hydroxide. Calculate the mass of calcium hydroxide that was dissolved in $1.00\ dm^3$ of the calcium hydroxide solution.

(3 marks)

AQA, 2012

2   Iron is an important element in living systems. It is involved in redox and in acid–base reactions.

   (a)   Explain how and why iron ions catalyse the reaction between iodide ions and $S_2O_8^{2-}$ ions. Write equations for the reactions that occur.

(5 marks)

   (b)   Iron(II) compounds are used as moss killers because iron(II) ions are oxidised in air to form iron(III) ions that lower the pH of soil.

      (i)   Explain, with the aid of an equation, why iron(III) ions are more acidic than iron(II) ions in aqueous solution.

(3 marks)

      (ii)   In a titration, $0.321\ g$ of a moss killer reacted with $23.60\ cm^3$ of acidified $0.0218\ mol\ dm^{-3}\ K_2Cr_2O_7$ solution.
Calculate the percentage by mass of iron in the moss killer. Assume that all of the iron in the moss killer is in the form of iron(II).

(5 marks)

   (c)   Some sodium carbonate solution was added to a solution containing iron(III) ions. Describe what you would observe and write an equation for the reaction that occurs.

(3 marks)

AQA, 2011

3 **(a)** State what is meant by the term *homogeneous* as applied to a catalyst.

(b) **(i)** State what is meant by the term *autocatalysis*. *(1 mark)*

(ii) Identify the species which acts as an autocatalyst in the reaction between ethanedioate ions and manganate(VII) ions in acidic solution.

(2 marks)

(c) When petrol is burned in a car engine, carbon monoxide, carbon dioxide, oxides of nitrogen and water are produced. Catalytic converters are used as part of car exhaust systems so that the emission of toxic gases is greatly reduced.

(i) Write an equation for a reaction which occurs in a catalytic converter between two of the toxic gases. Identify the reducing agent in this reaction.

(ii) Identify a transition metal used in catalytic converters and state how the converter is constructed to maximise the effect of the catalyst.

(5 marks)

AQA, 2004

4 **(a)** Using complex ions formed by Co^{2+} with ligands selected from H_2O, NH_3, Cl^-, $C_2O_4^{2-}$ and $EDTA^{4-}$, give an equation for each of the following.

(i) A ligand substitution reaction which occurs with no change in either the co-ordination number or in the charge on the complex ion.

(ii) A ligand substitution reaction which occurs with both a change in the co-ordination number and in the charge on the complex ion.

(iii) A ligand substitution reaction which occurs with no change in the co-ordination number but a change in the charge on the complex ion.

(iv) A ligand substitution reaction in which there is a large change in entropy.

(8 marks)

(b) An aqueous solution of iron(II) sulfate is a pale-green colour. When aqueous sodium hydroxide is added to this solution a green precipitate is formed. On standing in air, the green precipitate slowly turns brown.

(i) Give the formula of the complex ion responsible for the pale-green colour.

(ii) Give the formula of the green precipitate.

(iii) Suggest an explanation for the change in the colour of the precipitate.

(4 marks)

AQA, 2004

5 This question is about a product, Kwik Kleen, sold for unblocking clogged waste pipes of sinks in the home. The product consists of pure, powdered, solid sodium hydroxide, NaOH, a base.

The instructions for use state 'Wearing gloves and eye protection, add 100 g of Kwik Kleen to 1 litre (1 dm^3) of cold water and pour into the blocked drain. The solution will get warm. Always add Kwik Kleen to water rather than water to Kwik Kleen.'

(a) Write an ionic equation for the reaction that occurs when sodium hydroxide is added to water. Give the appropriate state symbols.

(2 marks)

(b) The pack says that Kwik Kleen is 'super strength'. Comment on the use of the term 'strength' in relation to sodium hydroxide dissolving in water.

(1 mark)

(c) Is it realistic to claim that solid sodium hydroxide is 'super strength' compared with other products consisting of solid sodium hydroxide?

(2 marks)

(d) Give the sign of ΔH for this reaction. Is it exothermic or endothermic?

(2 marks)

(e) Explain why it is safer to add sodium hydroxide to water than *vice versa*.

(2 marks)

(f) Calculate the concentration in $mol\,dm^{-3}$ of a solution containing 100 g of sodium hydroxide in 1 dm^3 of solution.

(1 mark)

(g) Part of the action unblocking action depends on the reaction of the sodium hydroxide solution with solid fat (resulting from cooking) to form soluble products. Fat contains esters such as:

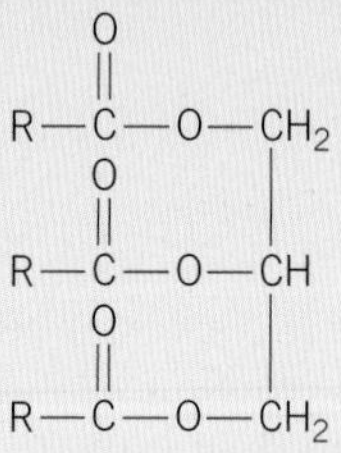

$$R-\overset{\overset{\displaystyle O}{\|}}{C}-O-CH_2$$
$$R-\overset{\overset{\displaystyle O}{\|}}{C}-O-CH$$
$$R-\overset{\overset{\displaystyle O}{\|}}{C}-O-CH_2$$

Where R is a long chain alkyl group, such as $C_{17}H_{35}$.
Calculate the relative molecular mass of the fat.

(1 mark)

(h) The equation for the reaction of the fat with sodium hydroxide is

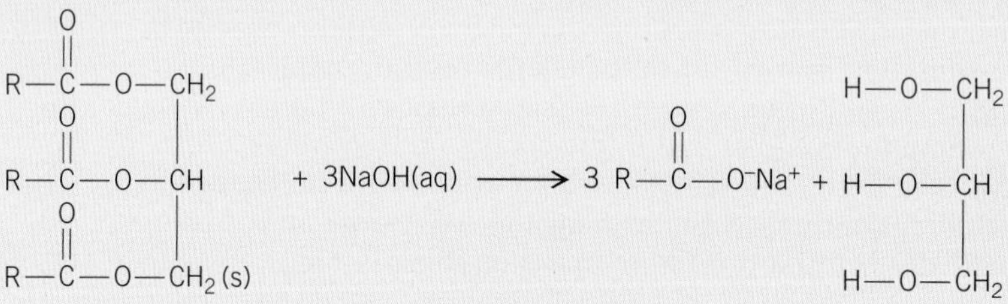

$$R-\overset{\overset{\displaystyle O}{\|}}{C}-O-CH_2$$
$$R-\overset{\overset{\displaystyle O}{\|}}{C}-O-CH \quad + \; 3NaOH(aq) \longrightarrow 3 \; R-\overset{\overset{\displaystyle O}{\|}}{C}-O^-Na^+ + \quad \begin{matrix} H-O-CH_2 \\ H-O-CH \\ H-O-CH_2 \end{matrix}$$
$$R-\overset{\overset{\displaystyle O}{\|}}{C}-O-CH_2 \, (s)$$

(i) Calculate the maximum number of grams of fat that could be dissolved by $1\,dm^3$ of the sodium hydroxide solution.

(2 marks)

(ii) Why is this unlikely to be the case in practice?

(2 marks)

(j) Explain why $R-\overset{\overset{\displaystyle O}{\|}}{C}-O^-Na^+$ is more soluble in water than was the original fat.

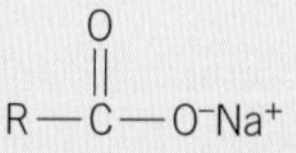

(1 mark)

6 A co-ordinate bond is formed when a transition metal ion reacts with a ligand.
(a) Explain how this co-ordinate bond is formed.

(2 marks)

(b) Describe what you would observe when dilute aqueous ammonia is added dropwise, to excess, to an aqueous solution containing copper(II) ions.
Write equations for the reactions that occur.

(4 marks)

(c) When the complex ion $[Cu(NH_3)_4(H_2O)_2]^{2+}$ reacts with 1,2-diaminoethane, the ammonia molecules but not the water molecules are replaced.
Write an equation for this reaction.

(1 mark)

(d) Suggest why the enthalpy change for the reaction in part (c) is approximately zero.

(2 marks)

(e) Explain why the reaction in part (c) occurs despite having an enthalpy change that is approximately zero.

(2 marks)

AQA, Specimen paper 1

7   **Table 1** shows observations of changes from some test-tube reactions of aqueous solutions of compounds **Q, R** and **S** with five different aqueous reagents. The initial colours of the solutions are not given.

▼ **Table 1**

	$BaCl_2$ + HCl	$AgNO_3$ + HNO_3	NaOH	Na_2CO_3	HCl (conc)
Q	no change observed	pale cream precipitate	white precipitate	white precipitate	no change observed
R	no change observed	white precipitate	white precipitate, dissolves in excess of NaOH	white precipitate, bubbles of a gas	no change observed
S	white precipitate	no change observed	brown precipitate	brown precipitate, bubbles of a gas	yellow solution

(a)   Identify each of compounds **Q, R** and **S**.
You are **not** required to explain your answers.

(6 marks)

(b)   Write ionic equations for each of the positive observations with **S**.

(4 marks)

AQA, Specimen paper 1

8   (a)   Define the term transition metal.

(1 mark)

(b)   Explain why scandium, Sc, is classified as a d-block element but not as a transition metal element.   *(2 marks)*

9   (a)   Explain what is meant by the terms complex ion and ligand.

(2 marks)

(b)   Complete the electron configuration of:
(i)   Cu atom: $1s^2$, $2s^2$

(1 mark)

(ii) Cu^{2+} ion: $1s^2$, $2s^2$

(1 mark)

(c)   Consider the hexaaquachromium(III), $[Cr(H_2O)_6]^{3+}$, complex.
(i)   Draw the shape of this complex.

(1 mark)

(ii)   Name the shape of this complex.

(1 mark)

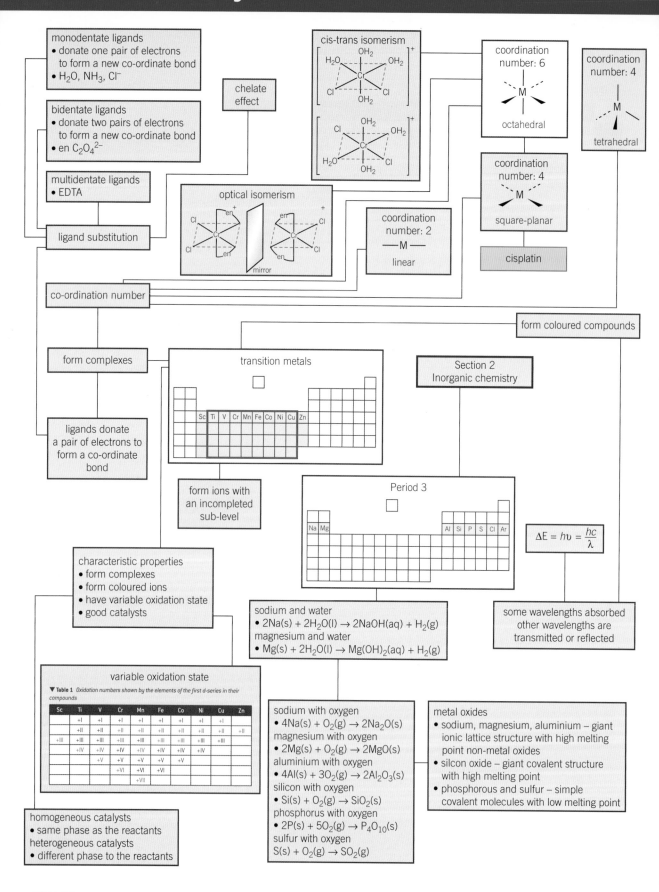

monodentate ligands
- donate one pair of electrons to form a new co-ordinate bond
- H_2O, NH_3, Cl^-

bidentate ligands
- donate two pairs of electrons to form a new co-ordinate bond
- en $C_2O_4^{2-}$

multidentate ligands
- EDTA

ligand substitution

chelate effect

cis-trans isomerism

optical isomerism

mirror

coordination number: 6

octahedral

coordination number: 4

square-planar

cisplatin

coordination number: 4

tetrahedral

coordination number: 2
— M —
linear

co-ordination number

form coloured compounds

form complexes

transition metals

| Sc | Ti | V | Cr | Mn | Fe | Co | Ni | Cu | Zn |

Section 2 Inorganic chemistry

ligands donate a pair of electrons to form a co-ordinate bond

form ions with an incompleted sub-level

Period 3

| Na | Mg | | | | | | | | | | | | Al | Si | P | S | Cl | Ar |

$$\Delta E = h\upsilon = \frac{hc}{\lambda}$$

characteristic properties
- form complexes
- form coloured ions
- have variable oxidation state
- good catalysts

sodium and water
- $2Na(s) + 2H_2O(l) \rightarrow 2NaOH(aq) + H_2(g)$

magnesium and water
- $Mg(s) + 2H_2O(l) \rightarrow Mg(OH)_2(aq) + H_2(g)$

some wavelengths absorbed other wavelengths are transmitted or reflected

variable oxidation state

▼ **Table 1** *Oxidation numbers shown by the elements of the first d-series in their compounds*

Sc	Ti	V	Cr	Mn	Fe	Co	Ni	Cu	Zn
+I	+I	+I	+I	+I	+I	+I	+I	+I	
	+II	+II	+II	+II	+II	+II	+II	+II	+II
+III	+III	+III	+III	+III	+III	+III	+III	+III	
	+IV	+IV	+IV	+IV	+IV	+IV			
		+V	+V	+V	+V				
			+VI	+VI	+VI				
				+VII					

sodium with oxygen
- $4Na(s) + O_2(g) \rightarrow 2Na_2O(s)$

magnesium with oxygen
- $2Mg(s) + O_2(g) \rightarrow 2MgO(s)$

aluminium with oxygen
- $4Al(s) + 3O_2(g) \rightarrow 2Al_2O_3(s)$

silicon with oxygen
- $Si(s) + O_2(g) \rightarrow SiO_2(s)$

phosphorus with oxygen
- $2P(s) + 5O_2(g) \rightarrow P_4O_{10}(s)$

sulfur with oxygen
$S(s) + O_2(g) \rightarrow SO_2(g)$

metal oxides
- sodium, magnesium, aluminium – giant ionic lattice structure with high melting point non-metal oxides
- silcon oxide – giant covalent structure with high melting point
- phosphorous and sulfur – simple covalent molecules with low melting point

homogeneous catalysts
- same phase as the reactants

heterogeneous catalysts
- different phase to the reactants

Practical skills

In this section you have met the following ideas:

- Investigating the reactions of period 3 oxides.
- Investigating ligand substitution reactions.
- Finding the concentration of a solution using colorimetry.
- Investigating the reduction of vanadate(V) ions.
- Finding out how Tollens' reagent can be used to distinguish between aldehydes and ketones.
- Investigating redox titrations.
- Finding out about autocatalysis.
- Investigating metal-aqua reactions.
- Identifying positive and negative ions and finding the identity of unknown substances.

Maths skills

In this section you have met the following maths skills:

- Using information about ligands to draw the shapes of complex ions.
- Working out how to draw cis and trans and optical isomers of complexes.
- Calculating the concentration of a solution from a graph of absorption versus concentration.

Extension

Produce a report exploring how the electronic configuration of transition metals affects their reactivity and properties.

Suggested resources:

Winter, M[2015], *d-Block Chemistry: Oxford Chemistry Primers*. Oxford University Press, UK. ISBN 978-0-19-870096-8

McCleverty, J[1999], *Chemistry of the First Row Transition Metals. Oxford Chemistry Primers*. Oxford University Press, UK. ISBN 978-0-19-850151-0

Section 3
Organic Chemistry

Chapters in this section

25 Nomenclature and isomerism in organic chemistry

26 Compounds containing the carbonyl group

27 Aromatic chemistry

28 Amines

29 Polymerisation

30 Amino acids, proteins, and DNA

31 Organic synthesis and analysis

32 Structure determination

33 Chromatography

Nomenclature and isomerism in organic chemistry revisits the IUPAC naming system introduced earlier and applies it to further families of organic compounds. A further type of isomerism, optical isomerism based on mirror image molecules, is introduced.

Compounds containing the carbonyl group introduces the chemistry of aldehydes, ketones, carboxylic acids, and esters, all of which contain the carbonyl group, C=O.

Aromatic chemistry looks at the chemistry of compounds based on the benzene ring, which have unexpected properties due to their system of electrons delocalised over a hexagonal ring of carbon atoms.

Amines are organic compounds based on the ammonia molecule. Their nitrogen atom has a lone pair of electrons which explains their reactivity as bases and nucleophiles

Polymerisation looks at two types of long chain molecules based on smaller repeating units – condensation and addition polymers. It describes their synthesis and uses and also their biodegradability (or lack of it).

Amino acids, proteins, and DNA are two groups of biologically important groups of polymers which are vital for life. The chapter looks at how proteins are built up from amino acids and explains how DNA molecules hold the 'blueprint' for living things.

Organic synthesis and analysis shows how a series of organic reactions can be linked together to make a target molecule from a given starting material.

Structure determination explains the techniques of proton nuclear magnetic resonance (NMR) and carbon-13 NMR and shows how these techniques can be used to help deduce the structures of organic compounds.

Chromatography describes a group of techniques used for separating mixtures of organic compounds and shows how they can be linked with mass spectrometry to help identify the components.

What you already know:

The material in this unit builds on knowledge and understanding that you have built up at AS level. In particular the following:

- ○ Organic compounds are based on chains and rings of carbon atoms along with hydrogen.
- ○ Organic compound exist in families called homologous series.
- ○ Different families of organic compounds have different functional groups.
- ○ There is a systematic naming system for organic compounds.
- ○ Organic compounds can exist as isomers with the same molecular formula but different arrangement of atoms.
- ○ Some important groups of organic compound include alkanes, alkenes, halogenoalkanes, and alcohols.
- ○ Characteristic reactions and infra-red spectroscopy can be used to help identify organic compounds.

25 Nomenclature and isomerism
25.1 Naming organic compounds

Learning objective:

→ Describe how IUPAC rules are used for naming organic compounds.

Specification reference: 3.3.1

Synoptic link

You will need to know the nomenclature and isomerism studied in Topic 11.1, Carbon compounds, and Topic 11.2, Nomenclature - naming organic compounds, and the shapes and bond angles in simple molecules in Topic 3.6, The shapes of molecules and ions.

The IUPAC systematic naming system for organic chemistry is based on a root, which describes the length of the carbon chain. Functional groups have names with numbers to show where they occur on the chain. You can use Table 1 to revise this system.

▼ **Table 1** *Examples of systematic naming of organic compounds*

Structural formula	Name	Notes
	2,2-dibromopropane	The di tells you there are two bromine atoms. The 2,2 says where the two bromine atoms are on the chain.
	3-bromobutan-1-ol	The suffix -ol defines the compound as an alcohol. The −OH group defines the end that the chain is counted from, so the bromine is attached to carbon 3
	butan-2-ol	Not butan-3-ol as the smallest locant possible is used.
	but-1-ene	Not but-2-ene, but-3-ene, or but-4-ene as the smallest locant possible is used
	cyclohexane	Cyclo- is used to indicate a ring
	methylbutane	There is no need to use a number to locate the side chain because it must be on carbon number 2

Structural formula	Name	Notes
	3-methylpentane	This is not 2-ethylbutane. The rule is to base the name on the longest unbranched chain, in this case pentane (picked out in red). Remember the bond angles are 109.5° not 90°
	2,3-dimethylpentane	Again remember the root is based on the longest unbranched chain

More functional groups

Table 2 shows how to name organic compound and includes the functional groups that you will meet in the following chapters, as well as those you have already met.

▼ **Table 2** *The suffixes and prefixes of some functional groups*

Family	Formula	Suffix	Prefix	Example
alkenes	$RCH=CH_2$	-ene		propene, CH_3CHCH_2
alkynes	$RC\equiv CH$	-yne		propyne, CH_3CCH
halogenoalkanes	R—X (X is F, Cl, Br, or I)		halo- (fluoro-, chloro-, bromo-, iodo-)	chloromethane, CH_3Cl
carboxylic acids	RCOOH	-oic acid		ethanoic acid, CH_3COOH
anhydrides	RCOOCOR′	-anhydride		ethanoic anhydride, $CH_3COOCOCH_3$
esters	RCOOR′	-oate (Esters are named from their parent alcohol and acid, so propyl ethanoate is derived from propanol and ethanoic acid.)		propyl ethanoate, $CH_3COOC_3H_7$
acyl chlorides	RCOCl	-oyl chloride		ethanoyl chloride, CH_3COCl
amides	$RCONH_2$	-amide		ethanamide, CH_3CONH_2
nitriles	$RC\equiv N$	-nitrile		ethanenitrile, $CH_3C\equiv N$
aldehydes	RCHO	-al		ethanal, CH_3CHO
ketones	RCOR′	-one		propanone, CH_3COCH_3
alcohols	ROH	-ol	hydroxy-	ethanol, C_2H_5OH 2-hydroxyethanal, $HOCH_2CHO$
amines	RNH_2	-amine	amino-	ethylamine, $CH_3CH_2NH_2$
arenes	C_6H_5R			methylbenzene, $C_6H_5CH_3$

Some functional groups may be identified by either a prefix or a suffix. For example, alcohols have the suffix -ol and the prefix hydroxy-. The suffix-ol is used if it is the only functional group. When there are two (or more) functional groups, the IUPAC rules have a comprehensive system of priority. In Table 2 the higher in the list is a suffix and the lower a prefix. So the amino acid alanine (see below), which is both a carboxylic acid and an amine, has the systematic name 2-aminopropanoic acid.

$$H_3C - \underset{\underset{H}{|}}{\overset{\overset{NH_2}{|}}{C}} - COOH$$

The International Union of Pure and Applied Chemistry

According to its website, the International Union of Pure and Applied Chemistry (IUPAC) 'serves to advance the worldwide aspects of the chemical sciences and to contribute to the application of chemistry in the service of mankind. As a scientific, international, non-governmental and objective body, IUPAC can address many global issues involving the chemical sciences.'

One of its services to the world of chemistry is to have developed a systematic naming system for organic chemicals, the full rules for which are held in a publication known as the Red Book. There is a companion Blue Book for inorganic chemistry.

This means that any organic chemical can be given a name which can be recognised and used by chemists throughout the world. This can obviously reduce confusion. For example, one well-respected database lists a total of 28 names that are in use for the compound with the IUPAC name butanone, a relatively simple compound:

$$H_3C \underset{\underset{H_2}{C}}{\diagdown} \overset{\overset{O}{||}}{C} \diagup CH_3$$

butanone

IUPAC also rules on the naming of newly discovered elements. An element 'can be named after a mythological concept, a mineral, a place or country, a property, or a scientist' and the discoverer has the right to propose a name and symbol.

Study tip

In chemical names, strings of numbers are separated by commas. A hyphen is placed between words and numbers.

Hint

Sometimes hydrocarbon molecules form rings. These are indicated by the prefix cyclo-, for example, cyclohexane:

The molecular formula of cyclohexane is C_6H_{12}. It is different from the molecular formula for hexane, C_6H_{14}, as there are no $-CH_3$ groups at the end of the carbon chain as the carbons are joined in a ring.

Study tip

Compounds are named according to the longest hydrocarbon chain. When the molecule is branched, look at it carefully and count the longest chain before you start numbering the carbon atoms.

Summary questions

1 Draw the displayed formula of:

   a 3-ethyl-3-methylhexane

   b 2,4-dimethylpentane.

2 What is the name of :

   a $CH_3CH_2CH_2OH$

   b $CH_3CH(Cl)CH_3$

   c $CH_3CH_2CH{=}CHCH_2CH_3$

   d CH_3CH_2OH

   e C_4H_9COOH?

Isomers are compounds with the same molecular formula but have different molecular structures or a different arrangement of atoms in space. Organic chemistry provides many examples of isomerism. Structural isomers:

- have different functional groups (Figure 1a) or
- have functional groups attached to the main chain at different points (Figure 1b) or
- have a different arrangement of carbon atoms in the skeleton of the molecule (Figure 1c).

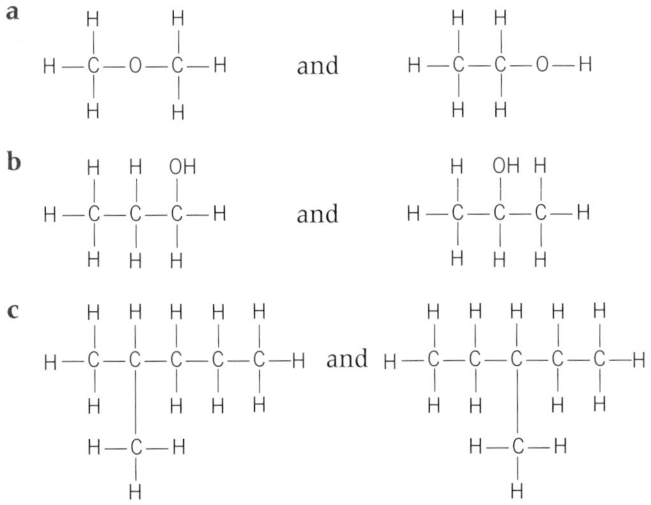

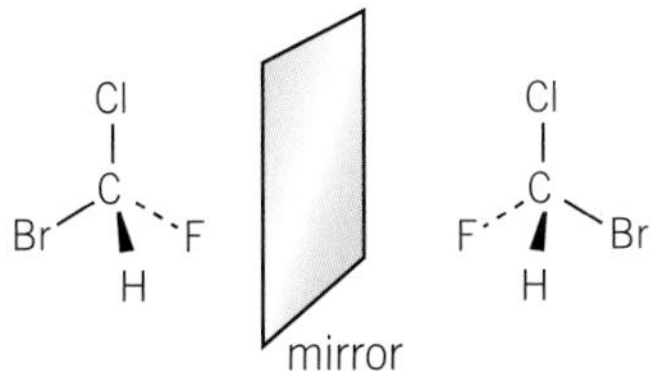

Learning objective:

→ Describe what type of molecules show optical isomerism.

Specification reference: 3.3.7

a

b

c

▲ **Figure 1** *Pairs of the different types of structural isomers*

Stereoisomerism

Stereoisomerism is where two (or more) compounds have the same structural formula. They differ in the *arrangement* of the bonds in space. There are two types:

- *E-Z* isomerism
- optical isomerism.

Optical isomerism

Optical isomers occur when there are four different substituents attached to one carbon atom. This results in two isomers that are non-superimposable mirror images of one another, but are not identical. For example, bromochlorofluoromethane exists as two mirror image forms:

Link

You met with structural isomerism and *E-Z* isomerism in Topic 11.3, Isomerism. *E-Z* involves the arrangement of atoms around double bonds.

mirror

Hint

It really helps to use models to understand optical isomerism. Even matchsticks and Plasticine will do.

Study tip

Drawn structures of optical isomers should show a tetrahedral arrangement of groups around a central carbon atom.

The ball and stick models of bromochlorofluoromethane in Figure 2 may help you to see that these are not identical.

Imagine rotating one of the molecules about the C—Cl bond (pointing upwards) until the two bromine atoms (in red) are in the same position.

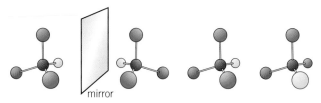

▲ **Figure 2** *Bromochlorofluoromethane has a pair of mirror isomers which are not identical*

The positions of the hydrogen (blue) and fluorine atoms (yellow) will not match – you cannot superimpose one molecule onto the other.

This is just like a pair of shoes. A left shoe and a right shoe are mirror images, but they are not identical, that is, they cannot be superimposed.

Pairs of molecules like this are called **optical isomers** because they differ in the way they rotate the plane of polarisation of polarised light – either clockwise ((+)-isomer) or anticlockwise ((−)-isomer).

Chirality

Optical isomers are said to be **chiral** and the two isomers are called a pair of **enantiomers**. The carbon bonded to the four different groups is called the **chiral centre** or the **asymmetric carbon atom**, and is often indicated on formulae by *. You can easily pick out a chiral molecule because it contains at least one carbon atom that has four different groups attached to it.

- All α-amino acids, except aminoethanoic acid (glycine) the simplest one, (Figure 3) have a chiral centre. For example, the chiral centre of α-aminopropanoic acid (2-aminopropanoic acid) is:

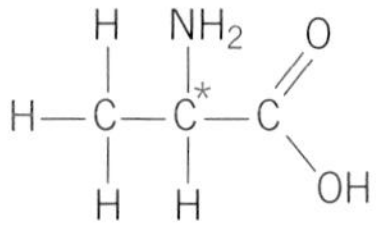

α-aminopropanoic acid

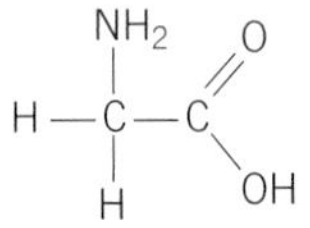

glycine

▲ **Figure 3** *Aminoethanoic acid (glycine)*

- 2-hydroxypropanoic acid (non-systematic name lactic acid) is also chiral. Although the chiral carbon is bonded to two other carbon atoms, these carbons are part of different groups and you must count the whole group.

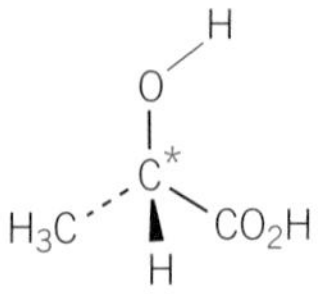

2-hydroxypropanoic acid

Synoptic link

There is more about α-amino acids in Topic 30.1, Introduction to amino acids.

Optical isomerism happens because the isomers have three-dimensional structures so it can only be shown by three-dimensional representations or by models.

Optical activity

Light consists of vibrating electric and magnetic fields. You can think of it as waves with vibrations occurring in all directions at right angles to the direction of motion of the light wave. If the light passes through a special filter, called a **polaroid** (as in polaroid sunglasses) all the vibrations are cut out except those in one plane, for example, the vertical plane (Figure 4).

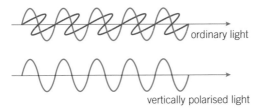

ordinary light

vertically polarised light

▲ **Figure 4** *Polarised light*

The light is now vertically polarised and it will be affected differently by different optical isomers of the same substance.

Optical rotation can be measured using a **polarimeter** (Figure 5).

1   Polarised light is passed through two solutions of the same concentration, each containing a different optical isomer of the same substance.

2   One solution will rotate the plane of polarisation through a particular angle, clockwise. This is the (+)-isomer.

3   The other will rotate the plane of polarisation by the same angle, anticlockwise. We call this the (−)-isomer.

(There are several other systems in use for distinguishing pairs of isomers, as well as (+) and (−) you may see R and S, D and L, or d and l.)

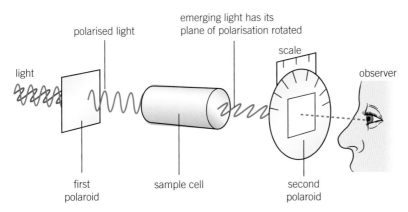

polarised light

emerging light has its plane of polarisation rotated

scale

light

observer

first polaroid

sample cell

second polaroid

▲ **Figure 5** *A polarimeter measures rotation of the plane of polarisation of polarised light*

Summary questions

1   Which of the following compounds show optical isomerism?

a
$$Cl—\overset{\displaystyle H}{\underset{\displaystyle H}{C}}—Br$$

b
$$Cl—\overset{\displaystyle H}{\underset{\displaystyle CH_3}{C}}—Br$$

c
$$Cl—\overset{\displaystyle H}{\underset{\displaystyle H}{C}}—H$$

d
$$H—\overset{\displaystyle H}{\underset{\displaystyle H}{C}}—H$$

2   Mark the chiral centre on this molecule with a *:

$$CH_3—\overset{\displaystyle Cl}{\underset{\displaystyle H}{C}}—\overset{\displaystyle O}{\underset{}{C}}\diagdown_{OH}$$

3   The molecule below has two chiral centres. Mark them both.

$$HO_2C—\overset{\displaystyle H}{\underset{\displaystyle OH}{C}}—\overset{\displaystyle H}{\underset{\displaystyle OH}{C}}—CO_2H$$

Learning objectives:

→ State what a racemate is

→ Describe how a racemate is formed by synthesis.

Specification reference: 3.3.7

A great many of the reactions that are used in organic synthesis to produce optically active compounds actually produce a 50:50 mixture of two optical isomers. This is called a racemic mixture or **racemate** and is not optically active because the effects of the two isomers cancel out.

The synthesis of 2-hydroxypropanoic acid (lactic acid)

2-hydroxypropanoic acid (lactic acid) has a chiral centre, marked by * in the structure.

Study tip

If a mixture of equal amounts of the two enantiomers is formed. This is optically inactive as their effects cancel out.

2-hydroxypropanoic acid

The synthesis below produces a mixture of optical isomers. It can be made in two stages from ethanal.

Stage 1

Hydrogen cyanide is added across the C=O bond to form 2-hydroxypropanenitrile.

Synoptic link

You will cover nucleophilic addition reactions in Topic 26.2, Reactions of the carbonyl group in aldehydes and ketones.

This is a nucleophilic addition reaction in which the nucleophile is the CN⁻ ion. It takes place as follows:

ethanal   2-hydroxypropanenitrile

2-hydroxypropanenitrile has a chiral centre, the starred carbon, which has four different groups ($-CH_3$, $-H$, $-OH$, and $-CN$). The reaction used does not favour one of these isomers over the other (i.e., the $-CN$ group could add on with equal probability from above or below the CH_3CHO which is planar) so you get a racemic mixture.

Hint

The carbon chain length of the product is one greater than that of the starting material. You started with *ethanal* (two carbon atoms) and ended with 2-hydroxy*propane*nitrile (three carbon atoms). This type of reaction is important in synthesis, whenever a carbon chain needs to be lengthened.

Stage 2

The nitrile group is converted into a carboxylic acid group.

The 2-hydroxypropanenitrile is reacted with water acidified with a dilute solution of hydrochloric acid. This is a hydrolysis reaction:

2-hydroxypropanoic acid

The balanced equations for the two steps are shown below:

Step 1

$$H_3C\!\!\diagdown \!\!\!\!\!\!\! \underset{H}{\overset{}{C}}\!\!=\!\!O \;+\; HCN \;\longrightarrow\; H_3C\!\!\diagdown\!\!\!\!\!\underset{H}{\overset{CN}{C}}\!\!\diagup OH$$

Step 2

$$H_3C\!\!\diagdown\!\!\!\underset{H}{\overset{CN}{C}}\!\!\diagup OH \;+\; HCl \;+\; 2H_2O \;\longrightarrow\; H_3C\!\!\diagdown\!\!\!\underset{H}{\overset{COOH}{C}}\!\!\diagup OH \;+\; NH_4Cl$$

The 2-hydroxypropanoic acid that is produced still has a chiral centre – this has not been affected by the hydrolysis reaction, which only involves the –CN group. So you still have a racemic mixture of two optical isomers (Figure 1).

$$\underset{H_3C}{\overset{OH}{\underset{H}{|}}}\overset{}{\underset{}{\overset{*}{C}}}COOH \qquad HOOC\overset{OH}{\underset{H}{\overset{|}{\underset{}{\overset{*}{C}}}}}CH_3$$

▲ **Figure 1** *The optical isomers of 2-hydroxypropanoic acid*

It is often the case that a molecule with a chiral centre that is made synthetically ends up as a mixture of optical isomers. However, the same molecule produced naturally in living systems will often be present as only one optical isomer. Amino acids are a good example of this. All naturally occurring amino acids (except aminoethanoic acid, glycine) are chiral, but in every case only one of the isomers is formed in nature. This is because most naturally occurring molecules are made using enzyme catalysts, which only produce one of the possible optical isomers.

Optical isomers in the drug industry

Some drugs are optically active molecules. For some purposes, a racemic mixture of the two optical isomers will do. For other uses, only one isomer is required. Many drugs work by a molecule of the active ingredient fitting an area of a cell (called a receptor) or an enzyme's active site like a piece in a jigsaw puzzle. Because receptors have a three-dimensional structure, only one of a pair of optical isomers will fit. In some cases, one optical isomer is an effective drug and the other is inactive. This is a problem and there are three options:

1 Separate the two isomers – this may be difficult and expensive as optical isomers have very similar properties.

2 Sell the mixture as a drug – this is wasteful because half of it is inactive.

3 Design an alternative synthesis of the drug that makes only the required isomer.

The over-the-counter painkiller and anti-inflammatory drug ibuprofen (sold as Nurofen and Calprofen) is an example.

▲ **Figure 2** *Lactic acid is produced naturally in sour milk. In muscle tissue a build-up of lactic acid causes cramp. The two situations produce different optical isomers*

Synoptic link

You will look at enzymes in more detail in Topic 30.3, Enzymes.

▲ **Figure 3** *Nurofen contains ibuprofen, which is a mixture of two optical isomers*

The structure of ibuprofen is:

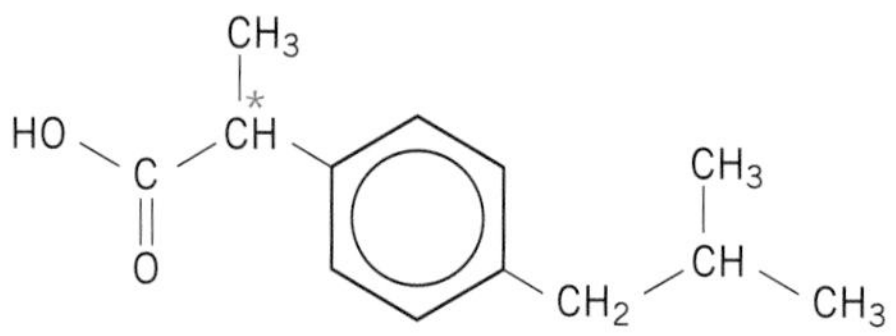

$$HO - \underset{\underset{O}{\|}}{C} - \overset{\overset{CH_3}{|}}{\underset{*}{CH}} - \text{(benzene ring)} - CH_2 - \overset{CH_3}{\underset{|}{CH}} - CH_3$$

The starred carbon is the chiral centre. At present, most ibruprofen is made and sold as a racemate.

In some cases one of the optical isomers is an effective drug and the other is toxic or has unacceptable side effects. For example, naproxen has one isomer that is used to treat arthritic pain, whilst the other causes liver poisoning. In this case it is vital that only the effective optical isomer is sold.

The structure of ibuprofen

Ibuprofen is a popular remedy for mild pain and inflammation that is available over-the-counter under a variety of trade names such as Nurofen and Cuprofen. The skeletal formula of ibuprofen in shown in Figure 4.

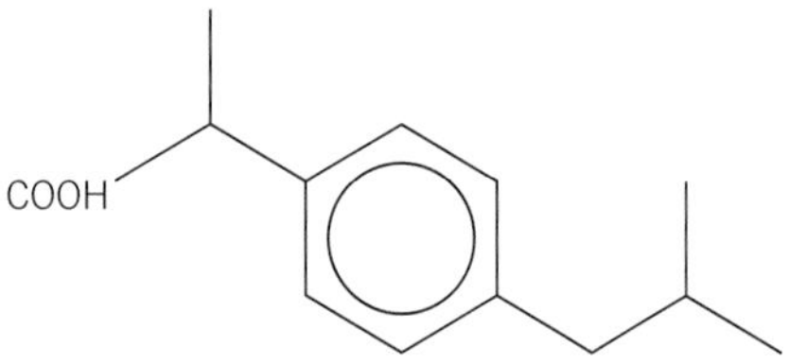

▲ **Figure 4** *Skeletal formula of ibuprofen*

Optical activity of ibuprofen

Ibuprofen can exist as a pair of optical isomers that are mirror images of each other. These mirror images are non-superimposable. This mirror image property occurs in molecules that have a carbon atom to which four different groups are bonded. The two optical isomers of ibuprofen are identified by the prefixes *R*– and *S*+,

Mirror image isomers are identical in many properties such as solubility, melting point, and boiling point. They can be distinguished by the fact that they rotate the plane of polarisation of polarised light in different directions– the (+)-isomer clockwise as the observer looks at the light, and the (–)-isomer anticlockwise. The symbols *R* and *S* refer to the 3D arrangement of the atoms in space. The two isomers do, however, behave differently when they interact with other 'handed' molecules such as the prostaglandins, which are involved in the process of inflammation. Of the two optical isomers of ibuprofen it is the *S*+ form which has the anti-inflammatory and pain-killing effect.

However, it has been found that there is an enzyme in the body that converts the *R*– form into the *S*+. In fact 60% of the *R*– form is converted into *S*+. This means that in a typical dose of ibuprofen of 400 mg, 200 mg is *S*+ and 200 mg *R*–. Of the 200 mg of *R*–, 60% (i.e., 120 mg) is converted into the active *S*+ form, giving a total of active form of 320 mg. Therefore there is little point in going to the trouble of synthesising the *S*+ form only, and ibuprofen is sold as a racemic mixture (one initially containing equal amounts of both optical isomers). However, a synthesis is possible that produces a pure sample of just one of the isomers.

▲ **Figure 5** *S+ibuprofen (top) and R–ibuprofen (bottom) showing their mirror image relationship*

Identify the functional group in ibuprofen.

carboxylic acid

The thalidomide tragedy

Around the late 1950s there was a spate of incidents of children born with serious and unusual birth defects – missing, shortened, or deformed limbs. There were over 10 000 of these world wide, almost 500 of them in the UK. Eventually it was realised that the common factor was that their mothers had all taken a drug called thalidomide in early pregnancy. The drug had been prescribed to relieve the symptoms of morning sickness. It had been tested on animals and considered safe (although the testing regime was much more relaxed than it would be today) but, crucially, there had been no tests on pregnant animals. In 1961 the drug was withdrawn.

Thalidomide exists as a pair of optical isomers. They differ in how they interact with other chiral molecules, which are common in living things. The isomers are extremely hard to separate and thalidomide was supplied as a racemic mixture produced when the the drug was synthesised. Apparently no one thought to test the two isomers separately.

Figure 6 shows one of the two enantiomers of thalidomide with the chiral carbon marked. The other isomer would have the positions of the C—H and the C—N bonds reversed, that is, the C—H going into the paper and the C—N coming out.

One of the enantiomers, called the *S*-form, is the one that caused the birth defects whilst the other, the *R*-form, is a safe sedative. It has been suggested that if just the *R*-form had been used, the tragedy would have been averted. However, there is evidence that in the human body, *R*-thalidomide is converted into *S*-thalidomide and so even if the pure *R*-form had been produced and taken,

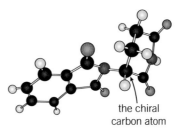

the chiral
carbon atom

the chiral carbon –
remember, there is a hydrogen atom bonded
to it which is not drawn in skeletal notation

▲ **Figure 6** *One of the enantiomers of thalidomide with the chiral carbon atom marked.*
Key: black = carbon, red = oxygen, pale blue = hydrogen, dark blue = nitrogen

patients would have ended up with some of the *S*-form in their bodies.

Even after the ban in the early 1960s, pharmacologists continued to work with thalidomide. It appears that so long as it is not given in pregnancy, it is a safe and potentially useful drug and it is now being investigated as a treatment for a number of conditions including leprosy. Chemists have also produced a number of related compounds that are up to 4000 times more effective and have fewer side effects.

Summary questions

1 a What would be the product of the reaction of propanal with hydrogen cyanide followed by reaction with dilute hydrochloric acid?

   b Does this molecule have a chiral centre? Explain your answer.

2 a What would be the product of the reaction of propanone with hydrogen cyanide followed by reaction with dilute hydrochloric acid?

   b Does this molecule have a chiral centre? Explain your answer.

3 Explain why the carbon marked ** in the formula of ibuprofen is *not* a chiral centre.

Study tip

Appreciate that different optical isomers can have very different drug action.

Synoptic link

The symbol ⬡ represents an aryl group – see Topic 27.1, Introduction to arenes.

Practice questions

1 The amino acid alanine is shown below.

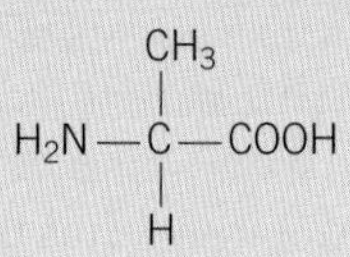

$$H_2N-\underset{\underset{H}{|}}{\overset{\overset{CH_3}{|}}{C}}-COOH$$

Give the systematic name for alanine.

(1 mark)
AQA, 2007

2 Phenylethanone, $C_6H_5COCH_3$, reacts with HCN according to the equation below.

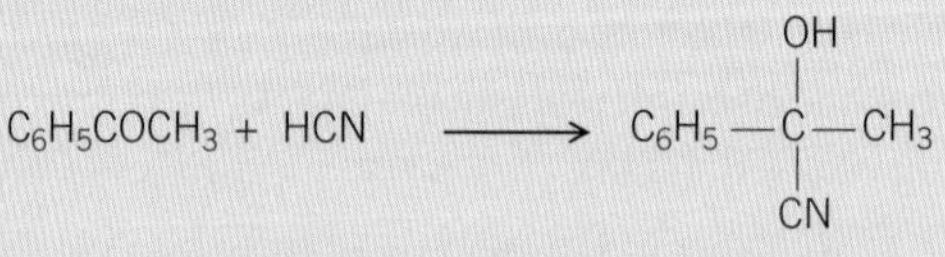

$$C_6H_5COCH_3 + HCN \longrightarrow C_6H_5-\underset{\underset{CN}{|}}{\overset{\overset{OH}{|}}{C}}-CH_3$$

The product formed exists as a racemic mixture. State the meaning of the term racemic mixture and explain why such a mixture is formed in this reaction.

(3 marks)
AQA, 2007

3 The reaction of but-2-ene with hydrogen chloride forms a racemic mixture of the stereoisomers of 2-chlorobutane.
   (a) Name the type of stereoisomerism shown by 2-chlorobutane and give the meaning of the term *racemic mixture*. State how separate samples of the stereoisomers could be distinguished.

(4 marks)

   (b) By considering the shape of the reactive intermediate involved in the mechanism of this reaction, explain how a racemic mixture of the two stereoisomers of 2-chlorobutane is formed.

(3 marks)
AQA, 2006

4 Consider the reactions shown below.

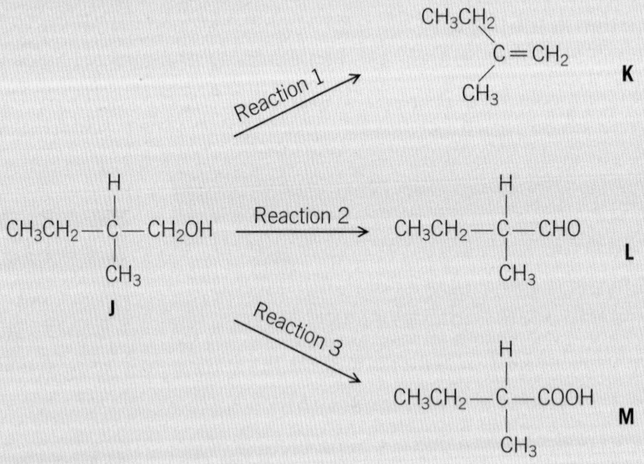

$$CH_3CH_2-\underset{\underset{CH_3}{|}}{\overset{\overset{H}{|}}{C}}-CH_2OH$$

J

Reaction 1 → $\underset{\underset{CH_3}{}}{\overset{\overset{CH_3CH_2}{}}{C}}=CH_2$ **K**

Reaction 2 → $CH_3CH_2-\underset{\underset{CH_3}{|}}{\overset{\overset{H}{|}}{C}}-CHO$ **L**

Reaction 3 → $CH_3CH_2-\underset{\underset{CH_3}{|}}{\overset{\overset{H}{|}}{C}}-COOH$ **M**

   (a) Name compound **J**.
   (b) Compound **J** exists as a pair of stereoisomers. Name this type of stereoisomerism.
   (c) Draw the structure of an isomer of **K** which shows stereoisomerism.

(3 marks)
AQA, 2007

5 **(a)** State the meaning of the term stereoisomerism.

(2 marks)

(b) Draw the structure of an isomer of C_5H_{10} which shows *E–Z* isomerism and explain how this type of isomerism arises. Name the structure you have drawn.

(3 marks)

(c) Name the structure below and state the type of isomerism it shows.

(2 marks)

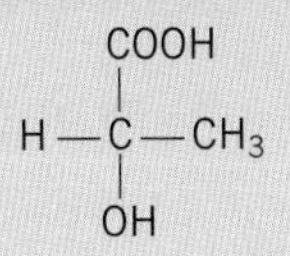

$$\begin{array}{c} COOH \\ | \\ H-C-CH_3 \\ | \\ OH \end{array}$$

(d) State how the different isomers of this structure can be distinguished from each other.

(2 marks)

6 **(a)** Define the term stereoisomer.

(1 mark)

(b) **(i)** Draw the display formula of but-2-ene.

(1 mark)

(ii) What type of isomerism is shown by but-2-ene?

(1 mark)

(iii) What are the conditions necessary for this type of isomerism?

(2 mark)

7 Consider molecule A which is optically active.

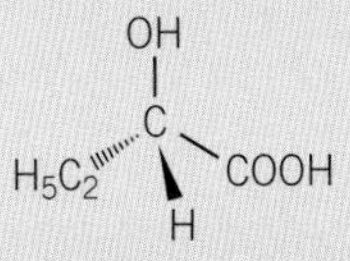

$$\begin{array}{c} OH \\ | \\ C \\ H_5C_2\overset{|}{\underset{H}{\diagdown}}COOH \end{array}$$

(a) Define the term optically active.

(1 mark)

(b) Draw the optical isomer of molecule A.

(1 mark)

(c) **(i)** What is a racemic mixture?

(1 mark)

(ii) Why is a racemic mixture not optically active?

(1 mark)

8 The display formula of α-aminopropanoic acid is shown below.

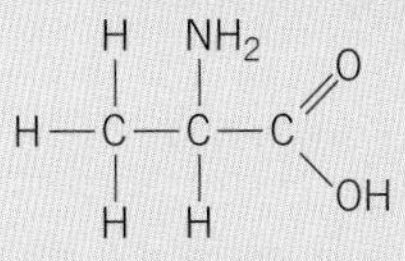

$$\begin{array}{c} H \quad NH_2 \quad\quad O \\ | \quad\quad | \quad\quad \diagup\!\!\diagup \\ H-C-C-C \\ | \quad\quad | \quad\quad \diagdown \\ H \quad H \quad\quad OH \end{array}$$

(a) Circle the chiral centre.

(1 mark)

(b) Use the display formula above to explain why this molecule is optically active.

(1 mark)

The carbonyl group consists of a carbon–oxygen double bond: $C=O$

The group is present in **aldehydes** and **ketones**.

In aldehydes, the carbon bonded to the oxygen (the carbonyl carbon) has at least one hydrogen atom bonded to it, so the general formula of an aldehyde is:

$$\begin{array}{c} R \\ \diagdown \\ C=O \\ \diagup \\ H \end{array}$$

This is sometimes written as RCHO.

In ketones, the carbonyl carbon has two organic groups, which can be represented by R and R', so the formula of a ketone is:

$$\begin{array}{c} R' \\ \diagdown \\ C=O \\ \diagup \\ R \end{array}$$

The R groups in both aldehydes and ketones may be alkyl or aryl.

How to name aldehydes and ketones

Aldehydes are named using the suffix -al. The carbon of the aldehyde functional group is counted as part of the carbon chain of the root. So:

$$H-C\diagup_{H}^{O} \text{ or HCHO is methanal and } H-\underset{H}{\overset{H}{C}}-C\diagup_{H}^{O} \text{ or CH}_3\text{CHO is ethanal.}$$

The aldehyde group can *only* occur at the end of a chain, so a numbering system is not needed to show its location.

or C_6H_5CHO, is counted as a derivative of benzene (not of methylbenzene) and is called benzenecarbaldehyde. It is often still called by the old name benzaldehyde.

Ketones are named using the suffix -one. In the same way as aldehydes, the carbon atom of the ketone functional group is counted as part of the root. So the simplest ketone:

H—C—C—C—H or CH_3COCH_3, is called propanone.

No ketone with fewer than three carbon atoms is possible.

You do not need to number the carbon in propanone or in butanone:

H—C—C—C—C—H,

$C_2H_5COCH_3$, because the carbonyl group can only be in one position. With larger numbers of carbon atoms, numbers are needed to locate the carbonyl group on the chain, for example, pentanone could be pentan-3-one, $CH_3CH_3COCH_2CH_3$, or pentan-2-one, $CH_3COCH_2CH_2CH_3$.

Physical properties of carbonyl compounds

The carbonyl group is strongly polar, $C^{\delta+}=O^{\delta-}$, so there are permanent dipole–dipole forces between the molecules. These forces mean that boiling points are higher than those of alkanes of comparable relative molecular mass but not as high as those of alcohols, where hydrogen bonding can occur between the molecules (Table 1).

▼ **Table 1** *Boiling point data*

Name	Formula	M_r	T_b / K
butane	$CH_3CH_2CH_2CH_3$	60	273
propanone	CH_3COCH_3	58	359
propan-1-ol	$CH_3CH_2CH_2OH$	60	370

Solubility in water

Shorter chain aldehydes and ketones mix completely with water because hydrogen bonds form between the oxygen of the carbonyl compound and water (Figure 1). As the length of the carbon chain increases, carbonyl compounds become less soluble in water.

Methanal, HCHO, is a gas at room temperature. Other short chain aldehydes and ketones are liquids, with characteristic smells (propanone, sometimes known as acetone, is found in many brands of nail varnish remover). Benzenecarbaldehyde smells of almonds and is used to scent soaps and flavour food.

The reactivity of carbonyl compounds

The C=O bond in carbonyl compounds is strong (Table 2) and you might think that the C=O bond would be the least reactive bond. However, almost all reactions of carbonyl compounds involve the C=O bond.

This is because the big difference in electronegativity between carbon and oxygen makes the $C^{\delta+}=O^{\delta-}$ strongly polar. So, nucleophilic reagents can attack the $C^{\delta+}$. Also, since they contain a double bond, carbonyl compounds are unsaturated and addition reactions are possible.

In fact the most typical reactions of the carbonyl group are nucleophilic additions.

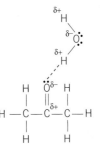

▲ **Figure 1** *Hydrogen bonding between propanone and water*

▼ **Table 2** *Comparison of bond strengths*

Bond	Mean bond enthalpy / kJ mol^{-1}
C=O	740
C=C	612
C—O	358
C—C	347

Summary questions

1 Name the following compounds:

   a $CH_3CH_2CCH_2CH_3$ (O double bond)

   b CH_3CH_2C (O double bond, H)

2 Explain why:

   a No ketone with fewer than three carbons is possible.

   b No numbering system is needed in the ketone butanone.

   c No numbering is ever needed to locate the position of the C=O group when naming aldehydes.

3 Explain why there are no hydrogen bonds between propanone molecules.

4 Explain why hydrogen bonds can form between propanone and water molecules.

Learning objectives:

→ Describe the mechanism of nucleophilic addition reactions of carbonyl compounds.

→ Describe how these compounds react when oxidised or reduced.

Specification reference: 3.3.8

Many of the reactions of carbonyl compounds are nucleophilic addition reactions.

They also undergo redox reactions.

Nucleophilic addition reactions

By representing the nucleophile as $:Nu^-$, the general reaction is:

The addition of hydrogen cyanide is a good example of a nucleophilic addition.

Addition of hydrogen cyanide

Sodium cyanide or potassium cyanide is used as a source of cyanide ions followed by the addition of dilute hydrochloric acid. You will not carry out this reaction in the laboratory because of the toxic nature of the CN^- ion. The products are called hydroxynitriles. Hydroxynitriles are useful in synthesis because both the –OH and –CN groups are reactive and can be converted into other functional groups. Here the nucleophile is $:CN^-$

With a ketone:

Or with an aldehyde:

The overall balanced equation for the reaction with an aldehyde is:

$$RCHO + HCN \longrightarrow R-\overset{\overset{\textstyle CN}{|}}{\underset{\underset{\textstyle OH}{|}}{C}}-H$$

This reaction is important in organic synthesis because it increases the length of the carbon chain by one carbon. The products are called hydroxynitriles. (This is an example where the –OH group is named using the prefix hydroxy- rather than the suffix -ol.)

This reaction will produce a racemic mixture of two optical isomers (enantiomers) when carried out with an aldehyde or an unsymmetrical ketone, because the $:CN^-$ ion may attack from above or below the flat C=O group (Figure 1).

> **Study tip**
>
> Nucleophiles have a lone pair to attack the $C^{\delta+}$. Some nucleophiles are negatively charged, others use the negative end of a dipole to attack $C^{\delta+}$.

> **Study tip** ⚗
>
> In theory hydrogen cyanide could be used as the nucleophile. However, this is toxic and, being a gas, it is hard to stop it escaping into the laboratory.

> **Synoptic link**
>
> Optical isomerism and racemic mixtures were covered in Topic 25.3, Synthesis of optically active compounds.

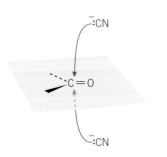

▲ **Figure 1** *The :CN⁻ ion may attack from above or below the C=O group*

Redox reactions

Oxidation

Aldehydes can be oxidised to carboxylic acids. Remember that [O] is used to represent the oxidising agent.

One oxidising agent commonly used is acidified (with dilute sulfuric acid) potassium dichromate(VI), $K_2Cr_2O_7/H^+$.

Ketones *cannot* be oxidised easily to carboxylic acids because, unlike aldehydes, a C—C bond must be broken. Stronger oxidising agents break the hydrocarbon chain of the ketone molecule resulting in a shorter chain molecule, carbon dioxide, and water.

Distinguishing aldehydes from ketones

Weak oxidising agents can oxidise aldehydes but not ketones. This is the basis of two tests to distinguish between them.

Fehling's test

Fehling's solution is made from a mixture of two solutions – Fehling's A which contains the Cu^{2+} ion and is therefore coloured blue, and Fehling's B which contains an alkali and a complexing agent.

- When an aldehyde is warmed with Fehling's solution, a brick red precipitate of copper(I) oxide is produced as the copper(II) oxidises the aldehyde to a carboxylic acid, and is itself reduced to copper(I).

- Ketones give no reaction to this test.

▲ **Figure 2** *When an aldehyde is warmed with Fehling's solution, the blue colour will turn green then a brick-red precipitate forms*

The silver mirror test

Tollens' reagent contains the complex ion $[Ag(NH_3)_2]^+$ which is formed when aqueous ammonia is added to an aqueous solution of silver nitrate.

- When an aldehyde is warmed with Tollens' reagent, metallic silver is formed. Aldehydes are oxidised to carboxylic acids by Tollens' reagent. The Ag^+ is reduced to metallic silver. A silver mirror will be formed on the inside of the test tube (which has to be spotlessly clean).

$$RCHO + [O] \rightarrow RCOOH \qquad \text{The aldehyde is oxidised.}$$

$$[Ag(NH_3)_2]^+ + e^- \rightarrow Ag + 2NH_3 \qquad \text{The silver is reduced.}$$

- Ketones give no reaction to this test.

Synoptic link

You met with aldehydes and ketones in Topic 15.3, The reactions of alcohols. They are formed from the oxidation of primary and secondary alcohols, respectively.

Synoptic link

Remember what was covered on the Fehling's test in Topic 15.3, The reactions of alcohols.

Synoptic link

A complexing agent can form co-ordinate (dative bonds) with metal atoms or ions, see 23.2, Complex formation and the shape of complex ions.

Hint

Benedict's solution is similar to Fehling's solution but is more convenient as it does not have to be prepared by mixing. It also contains Cu^{2+} ions but has a different complexing agent.

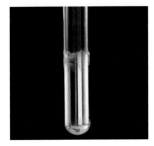

▲ **Figure 3** *The reaction of aldehydes with Ag^+ ions was once used as a method of silvering mirrors.*

Study tip

Compounds containing the carbonyl group have a strong absorption in the infra-red spectrum at around 1700 cm^{-1}. This can be used to show the presence of this bond.

Hint

Sodium tetrahydridoborate(III) is sometimes called sodium borohydride.

Summary questions

1   Which of the following is a nucleophile?

    H^+, Cl^-, $Cl \cdot$

2   Sodium tetrahydridoborate(III) generates the nucleophile $:H^-$ and converts aldehydes and ketones to alcohols.

    a   Would you expect this reagent to reduce

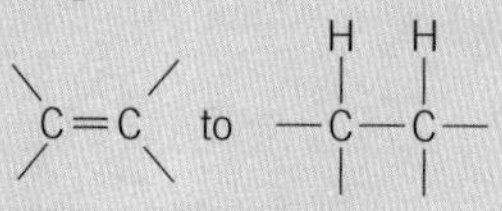

    b   Explain your answer.

    c   Predict the product when sodium tetrahydridoborate(III) reacts with:

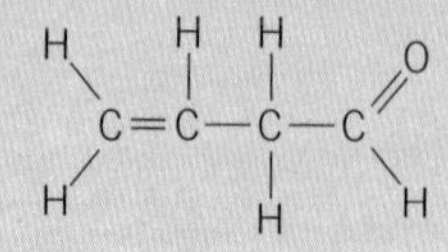

3   Hydrogen with a suitable catalyst will add on to C=C bonds as well as reducing the carbonyl group to an alcohol. Predict the product when hydrogen reacts with the compound in question **2c** in the presence of a suitable catalyst.

4   Explain why the reaction of CH_3CHO with HCN forms a racemic mixture, whilst that with CH_3COCH_3 forms a single compound.

Reduction

Many reducing agents will reduce both aldehydes and ketones to alcohols. One such reducing agent is sodium tetrahydridoborate(III), $NaBH_4$, in aqueous solution. This generates the nucleophile $:H^-$, the hydride ion.

This reduces $C^{\delta+}{=}O^{\delta-}$ but not C=C as it is repelled by the high electron density in the C=C bond, but is attracted to the $C^{\delta+}$ of the C=O bond.

Reducing an aldehyde

Aldehydes are reduced to primary alcohols by the following mechanism in which H^- acts as a nucleophile:

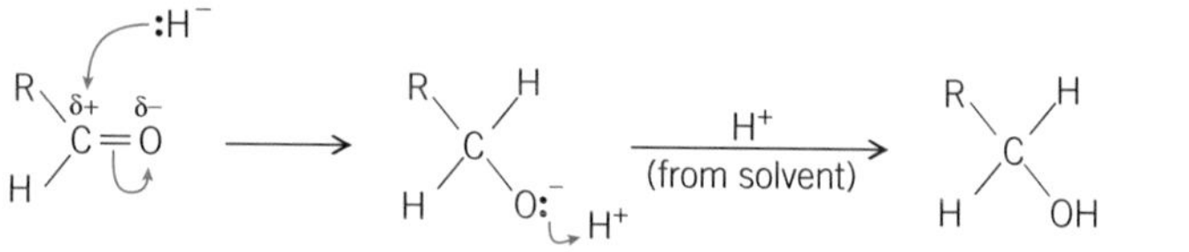

[H] is used to represent reduction in equations.

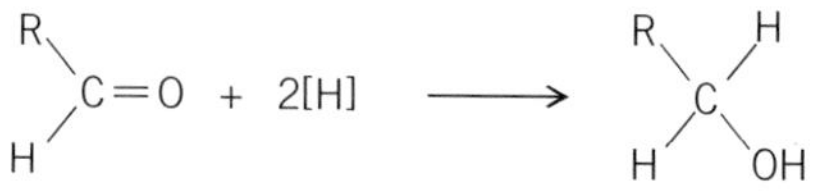

Reducing a ketone

Ketones are reduced to secondary alcohols in a similar way.

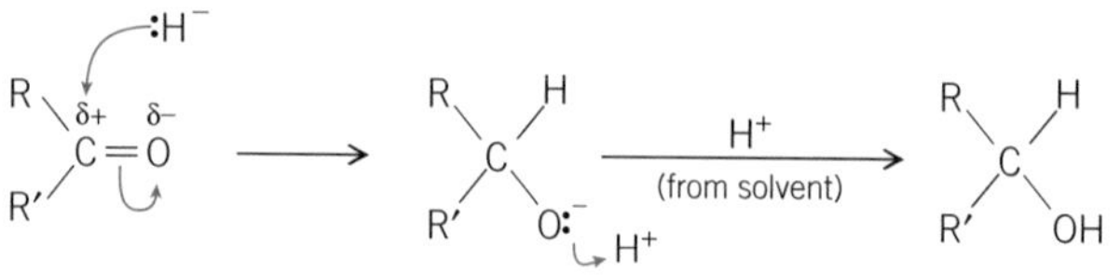

Using [H]:

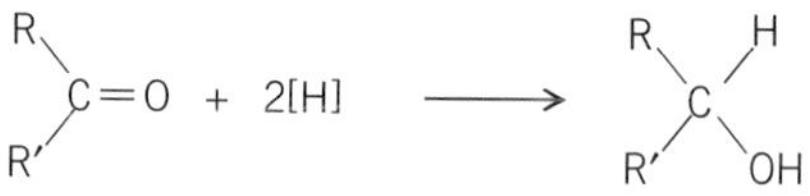

These reactions are **nucleophilic addition reactions**, (because the H^- ion is a nucleophile).

Carboxylic acids and esters

Carboxylic acids

The carboxylic acid functional group is

This is sometimes written as –COOH or as –CO_2H. This group can only be at the end of a carbon chain.

Carboxylic acids have two functional groups that you have seen before:

- the carbonyl group, C=O, found in aldehydes and ketones
- the hydroxy group, –OH, found in alcohols.

Having two groups on the same carbon atom changes the properties of each group. The most obvious difference is that the –OH group in carboxylic acids is much more acidic than the –OH group in alcohols.

The most familiar carboxylic acid is ethanoic acid (acetic acid), which is the acid in vinegar.

How to name carboxylic acids

Carboxylic acids are named using the suffix -oic acid. The carbon atom of the functional group is counted as part of the carbon chain of the root. So, HCOOH is methanoic acid, CH_3COOH is ethanoic acid, and so on.

methanoic acid   ethanoic acid

Where there are substituents or side chains on the carbon chain, they are numbered, counting from the carbon of the carboxylic acid as carbon number one. So, $CH_3CHBrCOOH$ is 2-bromopropanoic acid and $CH_3CH(CH_3)CH_2COOH$, is 3-methylbutanoic acid.

2-bromopropanoic acid   3-methylbutanoic acid

When the functional group is attached to a benzene ring, the suffix -carboxylic acid is used and the carbon of the functional group is *not* counted as part of the root. So, C_6H_5COOH is benzenecarboxylic acid.

benzenecarboxylic acid (This is still often called benzoic acid).

Learning objectives:

→ Describe carboxylic acids and esters.

→ State how they are named.

Specification reference: 3.3.9

Synoptic link

You will need to know: the principles of the IUPAC naming system covered in Topic 11.2, Nomenclature – naming organic compounds, and Topic 25.1, Naming organic compounds.

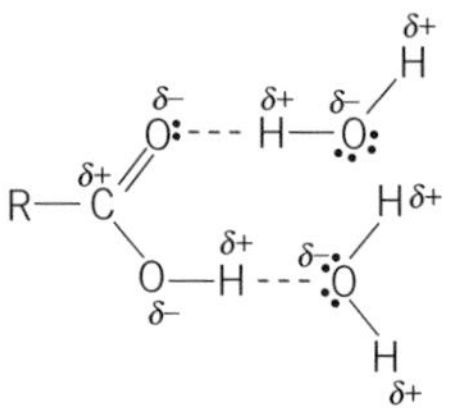

▲ **Figure 1** *A molecule of a carboxylic acid forming hydrogen bonds with water*

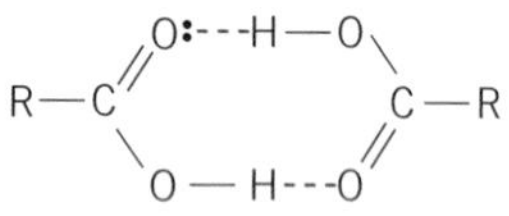

▲ **Figure 2** *Two carboxylic acid molecules can hydrogen bond together to form a pair called a dimer.*

▲ **Figure 4** *Modern electrical melting point apparatus*

Physical properties of carboxylic acids

The carboxylic acid group can form hydrogen bonds with water molecules (Figure 1). For this reason carboxylic acids up to, and including, four carbons (butanoic acid) are completely soluble in water.

The acids also form hydrogen bonds with one another in the solid state (Figure 2). They therefore have much higher melting points than the alkanes of similar relative molecular mass. Ethanoic acid ($M_r = 60$) melts at 290 K whilst butane ($M_r = 58$) melts at 135 K.

One way of identifying a carboxylic acid is to measure its melting point and compare it with tables of known melting points. A Thiele tube may be used (Figure 3), or the melting point can be found electrically (Figure 4).

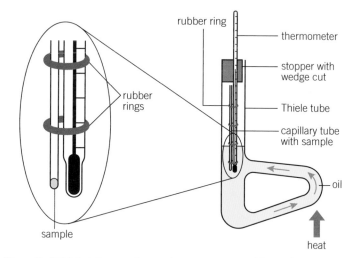

▲ **Figure 3** *A Thiele tube may be used to measure a melting point*

Pure ethanoic acid is sometimes called glacial ethanoic acid because it may freeze on a cold day – its freezing point is 13 °C (260 K).

The acids have characteristic smells. You will recognise the smell of ethanoic acid as vinegar, whilst butanoic acid causes the smell of rancid butter.

The non-systematic names of hexanoic and octanoic acids are caproic and capryllic acid respectively, from the same derivation as Capricorn the goat. They are present in goat fat and cause its unpleasant smell.

Esters

Esters are derived from carboxylic acids. The hydrogen (from the –OH group of the acid) is replaced by a hydrocarbon group – an alkyl or aryl group (OH is replaced by OR). So the general formula is:

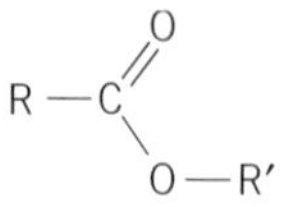

$$R-C\begin{smallmatrix}O\\\\O-R'\end{smallmatrix} \quad \text{or RCOOR}'.$$

How to name esters

The names of esters are based on that of the parent acid, for example, all esters from ethanoic acid are called ethanoates. But, the name always begins with the alkyl (or aryl) group that has replaced the hydrogen of the acid, rather than the name of the acid.

For example:

ethanoate

$$CH_3-C \overset{O}{\underset{O-CH_3 \text{ methyl}}{\big\backslash}}$$

is called methyl ethanoate.

methanoate

$$H-C \overset{O}{\underset{O-C_2H_5 \text{ ethyl}}{\big\backslash}}$$

is called ethyl methanoate.

Short chain esters are fairly volatile and have pleasant fruity smells, so that they are often used in flavourings and perfumes. For example, 3-methylbutyl ethanoate has the smell of pear drops, octyl ethanoate is orange flavoured, whilst pentyl pentanoate smells and tastes of apples. They are also used as solvents and plasticisers. Fats and oils are esters with longer carbon chains.

Study tip

Take care with the names of esters. It is easy to get them the wrong way round. The part of the name relating to the acid comes last. Also, remember that the acid is named from the number of carbon atoms – including the carbon of the functional group.

Study tip

Esters such as ethyl ethanoate are ingredients of many brands of nail varnish remover. They dissolve the nitrocellulose-based polymer in the varnish.

Hint

Plasticisers are added to plastics such as PVC to make them softer and more flexible. The small molecules of the plasticiser get in between the long chain molecules and allow them to slide across one another more easily.

Summary questions

1   Give the name of:

$$H-\overset{\overset{\displaystyle H}{|}}{\underset{\underset{\displaystyle H}{|}}{C}}-\overset{\overset{\displaystyle Br}{|}}{\underset{\underset{\displaystyle H}{|}}{C}}-\overset{\overset{\displaystyle H}{|}}{\underset{\underset{\displaystyle H}{|}}{C}}-C\overset{O}{\underset{OH}{\big\backslash}}$$

2   Write the displayed formula for 3-chloropropanoic acid.

3   Why is it not necessary to call propanoic acid 1-propanoic acid?

4   Give the names of the following esters:

$$CH_3-C \overset{O}{\underset{O-C_2H_5}{\big\backslash}}$$

$$CH_3CH_2-C \overset{O}{\underset{O-CH_3}{\big\backslash}}$$

Learning objectives:
→ Describe how carboxylic acids react.
→ State how esters are formed from carboxylic acids.
→ Describe how esters are hydrolysed.
→ Describe how esters are used.
Specification reference: 3.3.9

Reactivity of carboxylic acids

The carboxylic acid group is polarised as shown:

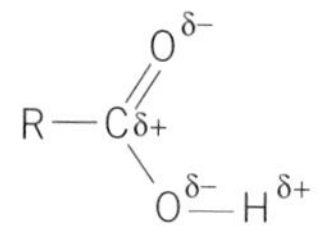

- The $C^{\delta+}$ is open to attack from nucleophiles.
- The $O^{\delta-}$ of the $C{=}O$ may be attacked by positively charged species (like H^+, in which case you say it has been protonated).
- The $H^{\delta+}$ may be lost as H^+, in which case the compound is behaving as an acid.

Loss of a proton

If the hydrogen of the –OH group is lost, a negative ion – a carboxylate ion – is left.

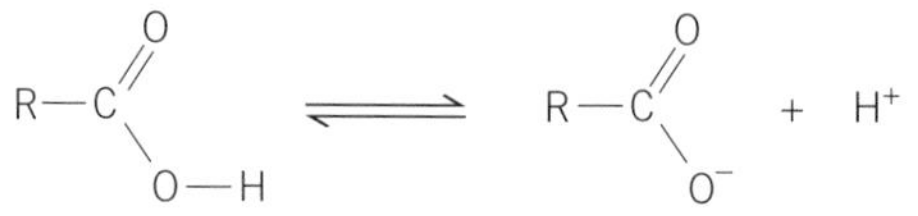

a carboxylate ion

The negative charge is shared over the whole of the carboxylate group.

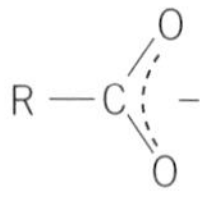

This **delocalisation** makes the resulting ion more stable. Carboxylic acids are weak acids, so the equilibrium is well over to the left:

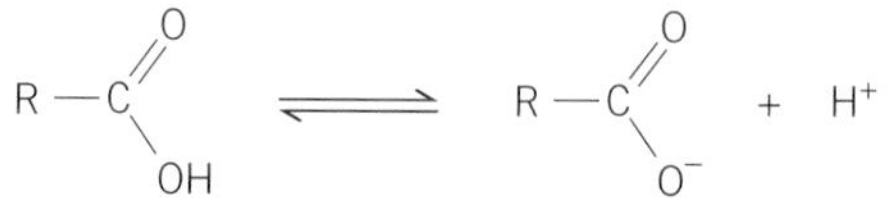

Even so, they are strong enough to react with sodium hydrogencarbonate, $NaHCO_3$, to release carbon dioxide. This distinguishes them from other organic compounds that contain the –OH group, such as alcohols.

$$CH_3COOH(aq) + NaHCO_3(aq) \rightarrow CH_3COONa\ (aq) + H_2O(l) + CO_2(g)$$

ethanoic acid · sodium hydrogencarbonate · sodium ethanoate · water · carbon dioxide

Reactions of acids

Carboxylic acids are proton donors and show the typical reactions of acids.

They form ionic salts with the more reactive metals, alkalis, metal oxides, or metal carbonates in the usual way. The salts that are formed have the general name carboxylates, and are named from the particular acid. Methanoic acid gives methanoates, ethanoic acid gives ethanoates, propanoic acid gives propanoates, and so on.

Study tip

The carboxylic acid group contain both the carbonyl group and the alcohol group. However, the two groups react differently when they are next to each other in a molecule.

Study tip

The stability of the carboxylate ion is what allows the H^+ ion to be released and makes the molecules acidic.

Study tip

Carboxylic acids give CO_2 with $NaHCO_3(aq)$, solid Na_2CO_3, and solid $NaHCO_3$.

For example, ethanoic acid reacts with aqueous sodium hydroxide:

$$CH_3COOH(aq) + NaOH(aq) \rightarrow CH_3COONa(aq) + H_2O(l)$$

ethanoic acid   sodium hydroxide   sodium ethanoate   water

Ethanoic acid reacts with aqueous sodium carbonate:

$$2CH_3COOH(aq) + Na_2CO_3(aq) \rightarrow 2CH_3COONa(aq) + H_2O(l) + CO_2(g)$$

ethanoic acid   sodium carbonate   sodium ethanoate   water   carbon dioxide

Esters

Formation of esters

Esters, general formula RCOOR′, are **acid derivatives**.

Carboxylic acids react with alcohols to form esters. This reaction is speeded up by a strong acid catalyst. This is a reversible reaction and forms an equilibrium mixture of reactants and products. For example:

▲ **Figure 1** *Carboxylic acids fizz with sodium carbonate*

ethanoic acid ethanol   ethyl ethanoate   water

Hydrolysis of esters

The carbonyl carbon atom of an ester has a δ+ charge and is therefore attacked by water acting as a weak nucleophile. This is the reverse of the reaction above. The equation is:

ester   carboxylic acid   alcohol

The hydrolysis (reaction with water) of esters does not go to completion. It produces an equilibrium mixture containing the ester, water, acid, and alcohol. The acid is a catalyst so it affects only the rate at which equilibrium is reached, not the composition of the equilibrium mixture.

An ester can be hydolysed at room temperature when a strong acid catalyst is used. The balanced equation for the acid catalysed hydrolysis of ethyl ethanoate is:

Bases also catalyse hydrolysis of esters. In this case, the salt of the acid is produced rather than the acid itself. This removes the acid from the reaction mixture so an equilibrium is not established and the reaction goes to completion, so there is more product in the mixture.

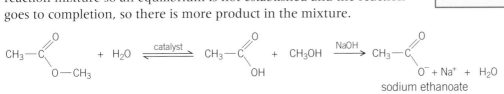

sodium ethanoate

  Esters

In the laboratory, esters are made by warming the appropriate acid and alcohol with concentrated sulfuric acid. The ester will be more volatile than the original alcohol and carboxylic acid and may be distilled off the reaction mixture.

Synoptic link

Equilibrium is covered in Topic 6.4, The equilibrium constant K_c.

Study tip

When an acid catalyst is used in the hydrolysis of an ester an equilibrium mixture of reactants and products is obtained.

The mechanism of base hydrolysis

The mechanism of base hydrolysis of esters can be explained using 'curly arrows' to show the movement of electron pairs:

$$R - C \overset{O^{\delta-}}{\underset{O-R}{\overset{\delta+}{\|}}} + :\bar{O}H \xrightarrow{\text{Step 1}} R - \overset{:\bar{O}}{\underset{O-R}{\overset{|}{C}}} - OH \xrightarrow{\text{Step 2}} R - C \overset{O}{\underset{O-H}{\|}} \xrightarrow{\text{Step 3}} R - C \overset{O}{\underset{O:^-}{\|}}$$

$$\bar{O} - R \qquad H - O - R$$

1   Describe Step 1.
2   What is the leaving group in Step 2.
3   How is RO⁻ acting in Step 3.

1 nucleophilic attack   2 RO⁻   3 As a base

Uses of esters

Animal and vegetable oils and fats are the esters of the alcohol propane-1,2,3-triol, (non-systematic name is glycerol). The only difference between a fat and an oil is that oils are liquid at room temperature, whilst fats are solid. Oils and fats contain three molecules of long chain (around 12–18 carbons) carboxylic acids called **fatty acids**. Since they are based on glycerol, fats and oils are referred to as **triglycerides** (Figure 2).

Fats and oils can be hydrolysed in acid conditions to give a mixture of glycerol and the component fatty acids.

They can also be hydrolysed by boiling with sodium hydroxide. Both the products are useful – glycerol and a mixture of sodium salts of the three acids which formed part of the ester. These salts are soaps. Soap can be a mixture containing many different salts. The type of soap depends on the fatty acids initially present in the ester.

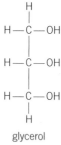

glycerol

glycerol   3 fatty acids

a triglyceride

▲ **Figure 2** *Glycerol and a triglyceride*

$$\begin{array}{c} H \\ | \\ H - C - O - \overset{O}{\overset{\|}{C}} - R \\ | \\ H - C - O - \overset{O}{\overset{\|}{C}} - R \\ | \\ H - C - O - \overset{O}{\overset{\|}{C}} - R \\ | \\ H \end{array} \; + \; 3NaOH \; \longrightarrow \; \begin{array}{c} H \\ | \\ H - C - OH \\ | \\ H - C - OH \\ | \\ H - C - OH \\ | \\ H \end{array}$$

$$+$$

$$3\,Na - O - \overset{O}{\overset{\|}{C}} - R$$

These sodium salts are ionic and dissociate to form Na⁺ and RCOO⁻. RCOO⁻ has two distinct ends – a long hydrocarbon chain which is non-polar and the COO⁻ group which is polar and ionic. The hydrocarbon will mix with grease, while the COO⁻ mixes with water (Figure 3). So, these tadpole-shaped molecules allow grease and water to mix and therefore are used as cleaning agents.

▼ **Table 1** *Some common fatty acids*

Name	Formula	Details
stearic acid	$CH_3(CH_2)_{16}CO_2H$	present in most animal fats
palmitic acid	$CH_3(CH_2)_{14}CO_2H$	used in making soaps
oleic acid	$CH_3(CH_2)_7CH=CH(CH_2)_7$ CO_2H	monounsaturated – it has one double bond, present in most fats and in olive oil
linoleic acid	$CH_3(CH_2)_4(CH=CHCH_2)_2$ $(CH_2)_6CO_2H$	polyunsaturated, present in many vegetable oils

Propane-1,2,3-triol (glycerol)

Glycerol has three O—H bonds, so it readily forms hydrogen bonds and is very soluble in water. It is a very important chemical in many industries and has a really wide range of uses.

- It is used extensively in many pharmaceutical and cosmetic preparations. Because it attracts water, it is used to prevent ointments and creams from drying out.

- It is used as a solvent in many medicines, and is present in toothpastes.

- It is used as a solvent in the food industry, for example, for food colourings.

- It is used to plasticise various materials like sheets and gaskets, cellophane, and special quality papers. Plasticisers are introduced between the molecules of the polymer which makes up the material and by allowing the molecules to slip over each other, the material becomes flexible and smooth. PVC may contain up to 50% plasticiser, such as esters of hexanedioic acid. Over time, the plasticiser leaks away, leaving the plastic brittle and inflexible.

Biodiesel

One possible solution to the reliance on crude oil as a source of fuel for motor vehicles is **biodiesel**. This is a renewable fuel, as it is made from oils derived from crops such as rape seed. Rape seed oil is a triglyceride ester. To make biodiesel the oil is reacted with methanol (with a strong alkali as a catalyst) to form a mixture of methyl esters, which can be used as a fuel in diesel vehicles with little or no modification. This process is being introduced commercially, but as the chemistry is relatively simple, some people are making their own biodiesel at home starting with used chip-shop oil, for example. Germany has thousands of filling stations supplying biodiesel, and it is cheaper there than ordinary diesel fuel. All fossil diesel fuel sold in France contains between 2% and 5% biodiesel.

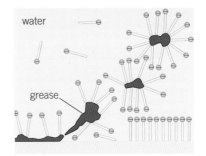

water

grease

▲ **Figure 3** *The action of soap. The hydrocarbon ends of the ions, in yellow, mix with grease and the COO– ends, in blue, lift it into aqueous solution*

Plasticiser

Without plasticiser, PVC is rigid and is used for drain pipes, for example. With plasticiser, it is flexible and can be used as a waterproof fabric in tablecloths and aprons, for example.

▲ **Figure 4** *Many vehicles now run of fuels which contain biodiesel*

Summary questions

1. Carboxylic acids, being acidic, will react with the reactive metals. Give three other reactions that are typical of acids.

2. Name the acid and the alcohol that would react together to give the ester methyl ethanoate.

3. Name the acid and the alcohol that would react together to give the ester ethyl methanoate.

4. Methyl ethanoate and ethyl methanoate are a pair of isomers. Explain what this means.

26.5 Acylation

Learning objectives:

→ Describe acylation reactions.

→ Explain the nucleophilic addition–elimination mechanism for acylation reactions.

Specification reference: 3.3.9

Acylation is the process by which the acyl group is introduced into another molecule. The acyl group is:

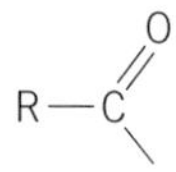

There is a group of compounds called **acid derivatives**, which all have the acyl group as part of their structure. Two important acid derivatives are acid chlorides and acid anhydrides. Acid derivatives are derived from carboxylic acids and have the general formula:

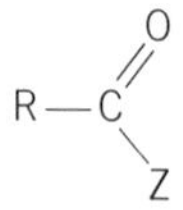

Z may be a variety of groups (Table 1).

If R is CH_3, the group is called **ethanoyl**.

▼ **Table 1** *Some acid derivatives*

−Z	Name of acid derivative	General formula	Example
−OR′	ester	RCOOR′	ethyl ethanoate, $CH_3COOC_2H_5$
−Cl	acid chloride	RCOCl	ethanoyl chloride, CH_3COCl
−OCOR′	acid anhydride	RCOOCOR′	ethanoic anhydride, $CH_3COOCOCH_3$

The carbonyl group of an acid derivative is polarised as shown:

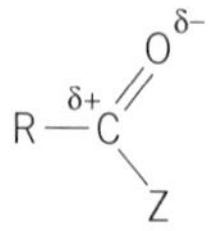

It is attacked by nucleophiles at the $C^{\delta+}$ and, in the process, the nucleophile replaces Z and the nucleophile therefore acquires an acyl group. So the nucleophile has been acylated.

The general reaction is:

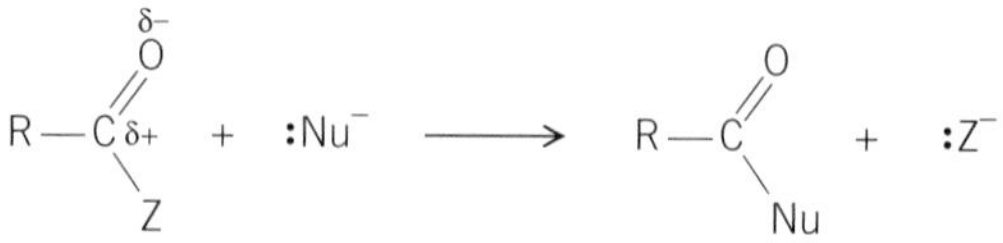

How readily the reaction occurs depends on three factors:

1 The magnitude of the δ+ charge on the carbonyl carbon, which in turn depends on the electron-releasing or attracting power of Z.
2 How easily Z is lost. (Z is called the leaving group.)
3 How good the nucleophile is.

Synoptic link

You will need to know the bond polarity and shapes of molecules studied in Topic 3.4, Electronegatvity – bond polarity in covalent bonds, and Topic 3.6, The shapes of molecules and ions, and nucleophilic substitution reactions of halogenoalkanes studied in Topic 13.2, Nucleophilic substitution in halogenoalkanes

Study tip

A nucleophile has a lone pair of electrons and attacks positively charged carbon atoms.

Factors 1 and 2 tend to be linked – groups which strongly attract electrons tend to form stable negative ions, Z⁻, and are good leaving groups.

acid chlorides   anhydrides

The Z groups of acyl chlorides and acid anhydrides *withdraw* electrons from the carbonyl carbon. This makes the carbon more positive and makes these compounds reactive towards nucleophiles. So, acyl chlorides and acid anhydrides are both good acylating agents. Acyl chlorides are somewhat more reactive than acid anhydrides.

Nucleophiles

Nucleophiles must have a lone pair of electrons which they use to attack an electron-deficient carbon, $C^{\delta+}$. The best nucleophiles are the ones that are best at donating their lone pair.

Acyl chlorides and acid anhydrides will both react with all the following nucleophiles, listed in order of reactivity:

primary amine   ammonia   alcohol   water

The products of the reactions of these nucleophiles with acyl chlorides and acid anhydrides are shown in Table 2. These reactions are called **addition–elimination reactions**. These nucleophiles are all neutral so they must lose a hydrogen ion during the reaction.

One way of looking at these reactions is that a hydrogen atom of the nucleophile ('the active hydrogen') has been replaced by an acyl group.

- If the nucleophile is ammonia the product is an amide.

- If the nucleophile is a primary amine, the product is an N-substituted amide.

- If the nucleophile is the –OH group of an alcohol the product is an ester.
- If the nucleophile is OH from water, the product is a carboxylic acid.

▼ **Table 2** *The products of the reactions of acid derivatives with nucleophiles. All reactions take place at room temperature*

The mechanism of the reactions

The mechanism of these reactions follows the same pattern, shown below.

1   Ethanoyl chloride and water (called hydrolysis).

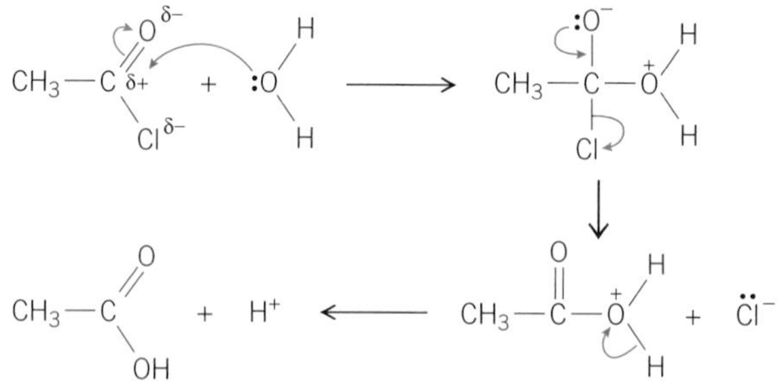

The overall equation may be written:

$$CH_3COCl + H_2O \rightarrow CH_3COOH + HCl$$

2 Ethanoyl chloride and ethanol.

The overall equation may be written:

$$CH_3COCl + C_2H_5OH \rightarrow CH_3COOC_2H_5 + HCl$$

3 Ethanoyl chloride and ammonia. (The H^+ ion that is lost then reacts with a second molecule of ammonia to form NH_4^+.)

The overall equation may be written:

$$CH_3COCl + 2NH_3 \rightarrow CH_3CONH_2 + NH_4Cl$$

4 Ethanoyl chloride and methylamine.

The overall equation may be written:

$$CH_3COCl + CH_3NH_2 \rightarrow CH_3CONHCH_3 + HCl$$

Uses of acylation reactions

Ethanoic anhydride is manufactured on a large scale. Its advantages over ethanoyl chloride as an acylating agent are:

- it is cheaper
- it is less corrosive
- it does not react with water as readily
- it is safer, as the by-product of its reaction is ethanoic acid rather than hydrogen chloride.

One important use is in the production of aspirin.

> **Study tip**
>
> Many students find that it is best to work out the products of these reactions from the mechanism rather than to remember them.

> **Study tip**
>
> A second NH_3 removes a proton from the intermediate ion to give the final product.

Aspirin

Aspirin (systematic name 2-ethanoyloxybenzenecarboxylic acid) is probably the most used medicine of all time — it must have been used to treat millions of headaches. It is often thought of as an over-the-counter remedy for moderate pain, which also reduces fever. However, it has more recently been shown to have many other effects such as reducing the risk of heart attacks and some cancers. It is not without risks itself, for example, it can cause intestinal bleeding, and it has been suggested that if it were to be introduced as a new drug today it would be prescription-only.

Aspirin has a long history. Compounds related to it were originally extracted from willow bark. One old theory held that cures to diseases could be found near the cause, and willow, which grows in damp places, was suggested as a cure for rheumatism, which is made worse by dampness.

In 1890, the German chemist Felix Hofmann produced the ethanoyl (or acyl) derivative of salicylic acid (2-hydroxybenzenecarboxylic acid), from willow bark extract, and used it to treat his father's rheumatism. This derivative is what is now used for aspirin.

The reagent used is ethanoyl anhydride:

The by-product is ethanoic acid.

1 Work out the atom economy for this reaction. You will probably need to draw out the displayed formulae first.
2 An alternative acylating agent for this reaction would be ethanoyl chloride.
   a Write the equation for the reaction of salicylic acid (2-hydroxybenzenecarboxylic acid) with ethanoyl chloride.
   b What is the by-product in this case?
   c Work out the atom economy for the reaction.

1 $(180 \div 240) \times 100 = 25.0\%$ 2 a $HOOCC_6H_4OH + CH_3COCl \rightarrow HOOCC_6H_4OOCCH_3 + HCl$ b HCl c $(180 \div 216.5) \times 100 = 83.1\%$

Synoptic link

See Topic 2.6, Balanced equations, atom economies, and percentage yield.

Summary questions

1 Why is ethanoyl chloride a good acylating agent?
2 Which of the following could be acylated? **A** NH_4^+ **B** OH^- **C** CH_4? Explain your answers.
3 Why is acylation called an addition–elimination reaction?
4 Write down the equation for the formation of propanamide from the reaction between ammonia and propanoyl chloride. Give the mechanism for this reaction.

Practice questions

1 **(a)** Write an equation for the formation of methyl propanoate, $CH_3CH_2COOCH_3$, from methanol and propanoic acid.

(1 mark)

(b) Name and outline a mechanism for the reaction between methanol and propanoyl chloride to form methyl propanoate.

(5 marks)

(c) Propanoic anhydride could be used instead of propanoyl chloride in the preparation of methyl propanoate from methanol. Draw the structure of propanoic anhydride.

(1 mark)

(d) (i) Give **one** advantage of the use of propanoyl chloride instead of propanoic acid in the laboratory preparation of methyl propanoate from methanol.

(ii) Give **one** advantage of the use of propanoic anhydride instead of propanoyl chloride in the industrial manufacture of methyl propanoate from methanol.

(2 marks)

AQA, 2006

2 Consider the sequence of reactions below

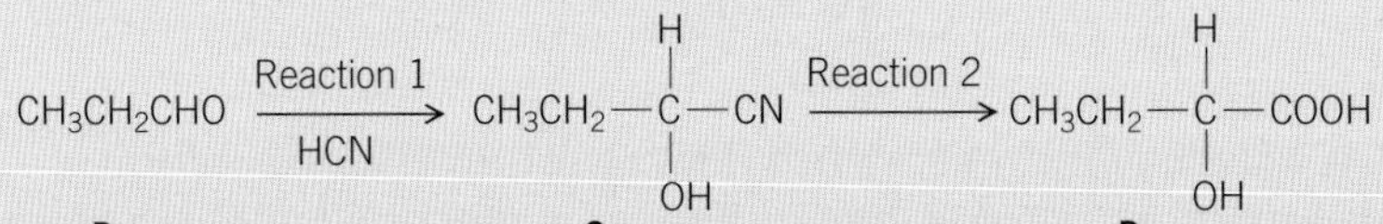

$$CH_3CH_2CHO \xrightarrow[HCN]{Reaction\ 1} CH_3CH_2-\underset{\underset{OH}{|}}{\overset{\overset{H}{|}}{C}}-CN \xrightarrow{Reaction\ 2} CH_3CH_2-\underset{\underset{OH}{|}}{\overset{\overset{H}{|}}{C}}-COOH$$

P **Q** **R**

(a) Name and outline a mechanism for Reaction 1.

(5 marks)

(b) (i) Name compound **Q**.

(ii) The molecular formula of **Q** is C_4H_7NO. Draw the structure of the isomer of **Q** which shows geometrical isomerism and is formed by the reaction of ammonia with an acyl chloride.

(3 marks)

(c) Draw the structure of the main organic product formed in each case when **R** reacts separately with the following substances:

(i) methanol in the presence of a few drops of concentrated sulfuric acid;

(ii) acidified potassium dichromate(VI);

(iii) concentrated sulfuric acid in an elimination reaction.

(3 marks)

AQA, 2006

3 **(a)** Name and outline a mechanism for the reaction between propanoyl chloride, CH_3CH_2COCl, and methylamine, CH_3NH_2.

(5 marks)

(b) Draw the structure of the organic product.

(1 mark)

AQA, 2005

4 A naturally-occurring triester, shown below, was heated under reflux with an excess of aqueous sodium hydroxide and the mixture produced was then distilled. One of the products distilled off and the other was left in the distillation flask.

$$CH_3(CH_2)_{16}COOCH_2$$
$$CH_3(CH_2)_{16}COOCH$$
$$CH_3(CH_2)_{16}COOCH_2$$

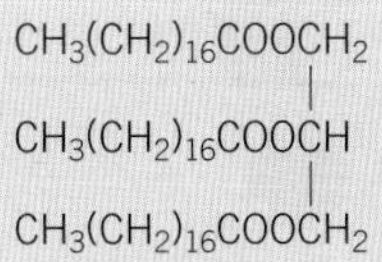

(a) Draw the structure of the product distilled off and give its name.

(2 marks)

(b) Give the formula of the product left in the distillation flask and give a use for it.

(2 marks)

AQA, 2005

27 Aromatic chemistry

27.1 Introduction to arenes

Learning objectives:

- ➔ Explain why the benzene ring is a planar hexagon.
- ➔ Explain why the carbon–carbon bonds are of equal length.

Specification reference: 3.3.10

Hint

A linear six-carbon alkane would have the formula C_6H_{14} – more than twice the number of hydrogens compared with benzene.

Synoptic link

You will need to know the covalent bonding studied in Topic 3.2, Covalent bonding, and bonding in alkenes studied in 14.1, Alkenes.

Arenes are hydrocarbons based on benzene, C_6H_6, which is the simplest one. Although benzene is an unsaturated molecule, it is very stable. It has a hexagonal (six-sided) ring structure with a special type of bonding. Arenes were first isolated from sweet-smelling oils, such as balsam, and this gave them the name aromatic compounds. Arenes are still called aromatic compounds, but this now refers to their structures rather than their aromas. Benzene and other arenes have characteristic properties.

Benzene is given the special symbol:

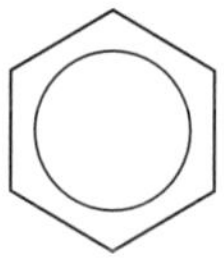

This is a skeletal formula, which does not show the carbon or hydrogen atoms. There is one carbon atom and one hydrogen atom at each point of the hexagon.

An arene can have other functional groups (substituents) replacing one or more of the hydrogen atoms in its structure.

Bonding and structure of benzene

The bonding and structure of benzene, C_6H_6, was a puzzle for a long time to organic chemists, because:

- in spite of being unsaturated, it does not readily undergo addition reactions
- all the carbon atoms were equivalent, which implied that all the carbon–carbon bonds are the same.

Benzene consists of a flat, regular hexagon of carbon atoms, each of which is bonded to a single hydrogen atom. The geometry of benzene is shown in Figure 1. Notice the difference between the flat benzene ring (Figure 2) and the puckered cyclohexane ring (Figure 3).

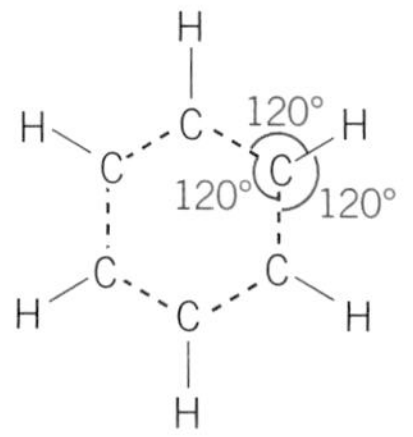

▲ **Figure 1** *The geometry of benzene (the dashed lines show the shape and do not represent single bonds)*

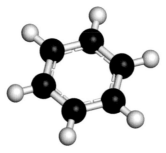

▲ **Figure 2** *The flat benzene ring*

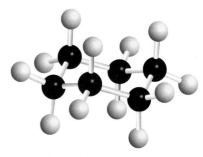

▲ **Figure 3** *The puckered cyclohexane ring*

▼ **Table 1** *Carbon–carbon bond lengths*

Bond	Length / nm
C—C	0.154
C⋯C (in benzene)	0.140
C═C	0.134
C≡C	0.120

The C—C bond lengths in benzene are intermediate between those expected for a carbon–carbon single bond and a carbon–carbon double bond (Table 1). So, each bond is intermediate between a single and a double bond.

The symbol 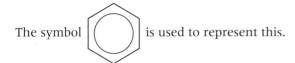 is used to represent this.

This can be explained by using the idea that some of the electrons are delocalised. Delocalisation means that electrons are spread over more than two atoms – in this case the six carbon atoms that form the ring.

Each carbon has three covalent bonds – one to a hydrogen atom and the other two to carbon atoms. The fourth electron of each carbon atom is in a p-orbital, and there are six of these – one on each carbon atom. The p-orbitals overlap and the electrons in them are **delocalised**. They form a region of electron density above and below the ring (Figure 4).

Overall, each carbon–carbon bond is intermediate between a single and a double bond. The delocalised system is very important in the chemistry of benzene and its derivatives. It makes benzene unusually stable. This is sometimes called **aromatic stability**.

The thermochemical evidence for stability

The enthalpy change for the hydrogenation of cyclohexene is:

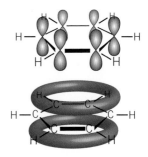

▲ **Figure 4** *Delocalisation of p-electrons to form areas of electron density above and below the ring*

> **Study tip**
>
> Benzene is more stable than the hypothetical molecule cyclohexa-1,3,5-triene because of delocalisation.
>
> HC⎓CH / HC⎓CH ring structure with C and H labels

$$\text{cyclohexene} + H_2 \longrightarrow \text{cyclohexane} \qquad \Delta H^{\ominus} = -120\,\text{kJ mol}^{-1}$$

So the hydrogenation of a ring with alternate double bonds would be expected to be three times this:

$$\text{hypothetical non-delocalised benzene} + 3H_2 \longrightarrow \text{cyclohexane} \qquad \Delta H^{\ominus} = -360\,\text{kJ mol}^{-1}$$

The enthalpy change for benzene is in fact:

$$\text{benzene} + 3H_2 \longrightarrow \text{cyclohexane} \qquad \Delta H^{\ominus} = -208\,\text{kJ mol}^{-1}$$

If these values are put on an enthalpy diagram (Figure 3), you can see that benzene is $152\,\text{kJ mol}^{-1}$ more stable than the unsaturated ring structure.

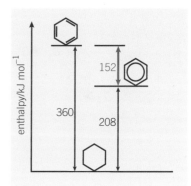

▲ **Figure 5** *Enthalpy diagram for the hydrogenation of cyclohexa-1,3,5-triene, cyclohexane, and benzene*

The most important dream in history?

This is how Friedrich August von Kekulé's insight into a chemical mystery – the structure of benzene – has been described. Benzene, C_6H_6, had been discovered by Michael Faraday but its structure was a puzzle, as the proportion of carbon to hydrogen seemed to be too great for conventional theories. In 1865, the Belgian chemist Friedrich August von Kekulé published a paper in which he suggested that benzene's structure was based on a ring of carbon atoms with alternating double and single bonds.

His idea resulted from a dream of whirling snakes.

'I turned my chair to the fire [after having worked on the problem for some time] and dozed. Again the atoms were gambolling before my eyes. This time the smaller groups kept modestly to the background. My mental eye, rendered more acute by repeated visions of this kind, could now distinguish larger structures, of manifold conformation – long rows, sometimes more closely fitted together – all twining and twisting in snakelike motion. But look! What was that? One of the snakes had seized hold of its own tail, and the form whirled mockingly before my eyes. As if by a flash of lighting I awoke.'

However, even this insight left a number of problems:

- A cyclic triene should show addition reactions, which benzene rarely does.

- Kekulé's structure should give rise to two isomeric di-substituted compounds as shown, using skeletal notation:

- The hexagon would not be symmetrical – double bonds are shorter than single bonds (Figure 6).

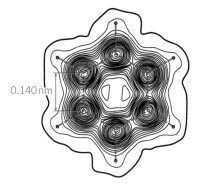

▲ **Figure 6** *A technique called X-ray diffraction shows a contour map of the electron density in an individual benzene molecule. This shows that the benzene molecule is a perfect hexagon and each carbon–carbon bond length is 0.140 nm*

Kekulé himself suggested a solution to the second dilemma by proposing that benzene consisted of structures in rapid equilibrium:

Later this rapid alternation of two structures evolved into the idea of **resonance** between two structures, both of which contribute to the actual structure. The actual structure was thought to be a hybrid (a sort of average) of the two. Such **resonance hybrids** were believed to be more stable than either of the separate structures.

Summary questions

1 What is the empirical formula of benzene?

2 How many molecules of hydrogen, H_2, would need to be added onto a benzene molecule to give a fully saturated product cyclohexane?

3 Explain what is meant by delocalisation of electrons in the benzene ring.

4 Look at the two di-substituted compounds formed with the bromination of Kekulé's proposed structure of benzene. Which of the two hypothetical di-substituted compounds would have the shorter bond between the two carbon atoms bonded to the bromine atoms.

Arenes – physical properties, naming, and reactivity

Physical properties of arenes

Benzene is a colourless liquid at room temperature. It boils at 353 K and freezes at 279 K. Its boiling point is comparable with that of hexane (354 K) but its melting point is much higher than hexane's (178 K). This is because benzene's flat, hexagonal molecules (Figure 1) pack together very well in the solid state. They are therefore harder to separate and this must happen for the solid to melt.

Learning objectives:

→ Describe how substituted arenes are named.

→ State how the arene ring affects reactivity.

Specification reference: 3.3.10

Hint

Freezing point and melting point are exactly the same temperature.

Hint ⚠

Benzene itself is carcinogenic (may cause cancer) so in your practical work in school you are likely to use related compounds that are safer, such as methylbenzene.

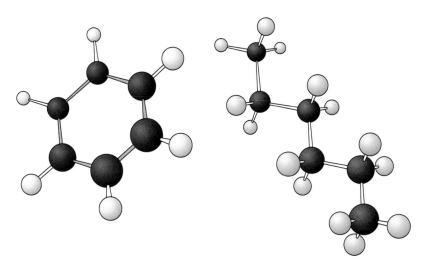

▲ **Figure 1** *Benzene molecules (left) can pack together better than hexane molecules (right) so benzene has a higher melting point than hexane*

Like other hydrocarbons that are non-polar, arenes do not mix with water but mix with other hydrocarbons and non-polar solvents.

Naming aromatic compounds

Substituted arenes are generally named as derivatives of benzene, so benzene forms the root of the name.

$C_6H_5CH_3$, is called methylbenzene.

C_6H_5Cl, is called chlorobenzene, and so on.

If there is more than one substituent, the ring is numbered:

1,2-dichlorobenzene   1,4-dichlorobenzene

Examples

You can test yourself by covering the names or the structures.

ethylbenzene $C_6H_5C_2H_5$

nitrobenzene $C_6H_5NO_2$

1,2-dimethylbenzene $C_6H_4(CH_3)_2$

The reactivity of aromatic compounds

Two factors are important to the reactivity of aromatic compounds:

- The ring is an area of high electron density, because of the delocalised bonding (Topic 26.1) and is therefore attacked by electrophiles.
- The aromatic ring is very stable. It needs energy to be put in to break the ring before the system can be destroyed. This is called the delocalisation energy. It means that the ring almost always remains intact in the reactions of arenes.

The above two points mean that most of the reactions of aromatic systems are **electrophilic substitution** reactions.

Summary questions

1 What intermolecular forces act between non-polar molecules?

2 Name:

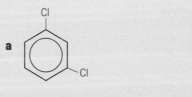

a   b

3 Draw the structure of:

  a   1,4-dimethylbenzene

  b   1,2-dimethylbenzene

4 Which of the following is an electrophile?

    R^+ $:NH_3$ NO_2 Cl^-

Combustion

Arenes burn in air with flames that are noticeably smoky. This is because they have a high carbon : hydrogen ratio compared with alkanes. There is usually unburnt carbon remaining when they burn in air and this produces soot. A smoky flame suggests an aromatic compound.

Electrophilic substitution reactions

Although benzene is unsaturated it does not react like an alkene.

The most typical reaction is an electrophilic substitution that leaves the aromatic system unchanged, rather than addition which would require the input of the delocalisation energy to destroy the aromatic system.

The mechanism of electrophilic substitutions

The delocalised system of the aromatic ring has a high electron density that attracts electrophiles. At the same time the electrons are attracted towards the electrophile, El^+.

A bond forms between one of the carbon atoms and the electrophile. But, to do this, the carbon must use electrons from the delocalised system. This destroys the aromatic system. To get back the stability of the aromatic system, the carbon loses an H^+ ion with the electron in the C—H bond returning to the delocalised system. The sum of these reactions is the substitution of H^+ by El^+.

[benzene electrophilic substitution mechanism diagram showing three structures with arrows, producing $+ H^+$]

The same overall process occurs in, for example, nitration and Friedel–Crafts acylation reactions.

Nitration

Nitration is the substitution of a NO_2 group for one of the hydrogen atoms on an arene ring. The electrophile NO_2^+ is generated in the reaction mixture of concentrated nitric and concentrated sulfuric acids:

$$H_2SO_4 + HNO_3 \rightarrow H_2NO_3^+ + HSO_4^-$$

Sulfuric acid is a stronger acid than nitric acid and donates a proton, H^+, to HNO_3.

$H_2NO_3^+$ then loses a molecule of water to give NO_2^+, which is called the **nitronium ion** or **nitryl cation**.

$$H_2NO_3^+ \rightarrow NO_2^+ + H_2O$$

The overall equation for the generation of the NO_2^+ electrophile is:

$$H_2SO_4 + NHO_3 \rightarrow NO_2^+ + HSO_4^- + H_2O$$

Learning objectives:
→ Explain why arenes react by electrophilic substitution.
→ Describe the mechanism of nitration.
→ Describe the mechanism of acylation.

Specification reference: 3.3.10

▲ **Figure 1** *Arenes burn with a smoky flame*

Hint
Compare the C : H ratio of benzene, C_6H_6, (1:1) to that of cyclohexene, C_6H_{12} (1:2).

Synoptic link
Nitric acid is acting as a Brønsted–Lowry base, see Topic 21.1, Defining an acid.

Synoptic link
You will need to know the electrophilic addtion to alkenes studied in Topic 14.2, Reactions of alkenes.

Study tip

In organic chemistry, curly arrows are used to indicate the movement of a pair of electrons. They run from areas of high electron density to more positively charged areas.

NO_2^+ is an electrophile and the following mechanism occurs:

The overall product of the reaction of the nitronium ion, NO_2^+, with benzene is nitrobenzene:

nitrobenzene

The H^+ then reacts with the HSO_4^- to regenerate H_2SO_4, making sulfuric acid a catalyst. The balanced equation is:

Synoptic link

You will learn about making aromatic amines in Topic 28.1, Introduction to amines.

The uses of nitrated arenes

Nitration is an important step in the production of explosives like TNT. Nitration is the first step in making aromatic amines, and these in turn are used to make industrial dyes.

TNT

▲ **Figure 2** *Filling TNT shells (1940)*

TNT is short for trinitrotoluene. It is made by nitrating methylbenzene, commonly called toluene. TNT is an important high explosive with both military and peaceful applications. It is a solid of low melting point. This property is used both in filling shells (Figure 2) and by bomb disposal teams who can steam TNT out of unexploded bombs.

The explosion of TNT is shown in the following equation:

$$2 \quad \text{(s)} + 10\tfrac{1}{2}\,O_2(g) \longrightarrow 14\,CO_2(g) + 3\,N_2(g) + 5\,H_2O(g)$$

The reaction is strongly exothermic. The rapid formation of a lot of gas as well as heat produces the destructive effect.

Many other compounds with several nitrogen atoms in the molecule are explosive. Another example is 2,4,6-trinitrophenol, which can explode on impact and is therefore useful as a detonator to set off other explosives.

What is the systematic name for for TNT? The methyl group is at position 1.

1-methyl-2,4,6-trinitrobenzene

Further substitution of arenes

An atom or group of atoms already on a benzene ring will affect further substitution reactions in two ways:

1   It may release electrons onto the benzene ring and therefore make it more susceptible to further electrophilic substitution reactions (*i.e.,* these reactions will go faster and there may be more than one substituent). Or it will withdraw electrons from the ring, making it less susceptible to further electrophilic substitution.
2   It will direct further substitution to particular positions on the ring. Electron-releasing groups direct further substitution to the 2, 4, and 6 positions. Electron-withdrawing groups direct further substitution to the 3 and 5 positions.

Electron-releasing groups include $-CH_3$, $-OCH_3$, $-OH$, and $-NH_2$.

Electron withdrawing groups include $-NO_2$, $-COCl$.

Halogens are exceptions to the rule – they withdraw electrons but direct substitution to the 2 and 4 positions.

These rules explain why phenol can be nitrated to 2, 4, 6-trinitrophenol because $-OH$ is electron-releasing.

1   Predict the likely products of single nitration of chlorobenzene.
2   The nitration of methylbenzene can be done as a school practical exercise. Explain why there is no danger of the formation of the explosive 2, 4, 6-trinitromethylbenzene (trinitrotoluene, TNT).

1 2-nitrochlorobenzene 4-nitrochlorobenzene
2 Nitro groups withdraw electrons and make further substitution unlikely.

Friedel–Crafts acylation reactions

These reactions use aluminium chloride as a catalyst. The method of doing this was discovered by Charles Friedel and James Crafts.

The mechanism for acylation is a substitution, with RCO substituting for a hydrogen on the aromatic ring.

Acyl chlorides provide the RCO group. They react with $AlCl_3$ to form $AlCl_4^-$ and RCO^+.

$$RCOCl + AlCl_3 \rightarrow RCO^+ + AlCl_4^-$$

This reaction takes place because the aluminium atom in aluminium chloride has only six electrons in its outer main level and readily accepts a lone pair from the chlorine atom of RCOCl.

RCO^+ is a good electrophile that is attacked by the benzene ring to form substitution products.

The aluminium chloride is a catalyst – it is reformed by reaction of the $AlCl_4^-$ ion with H^+ from the benzene ring:

$$AlCl_4^- + H^+ \rightarrow AlCl_3 + HCl$$

Synoptic link

You will need to know the co-ordinate bonding and bond polarity studied in Topic 3.2, Covalent bonding, and Topic 3.4, Electronegatvity - bond polarity in covalent bonds, and shapes of molecules studied in Topic 3.6, The shapes of molecules and ions.

The mechanism for the reaction is:

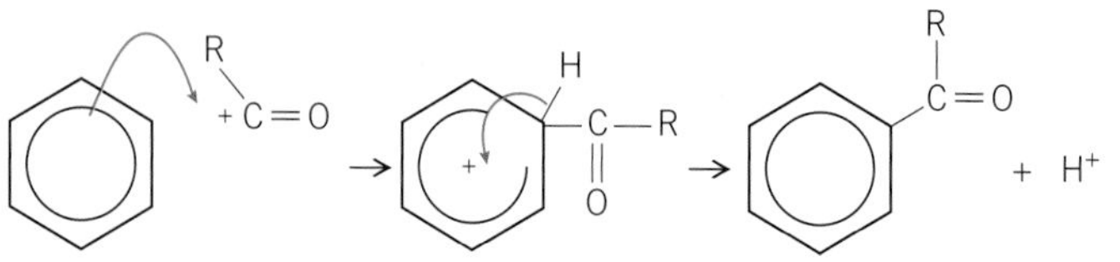

The products are acyl-substituted arenes. The overall reactions are:

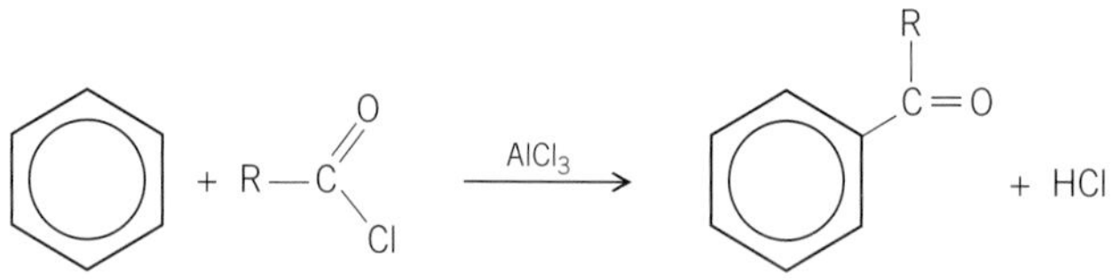

For example, ethanoyl chloride reacts with benzene to form:

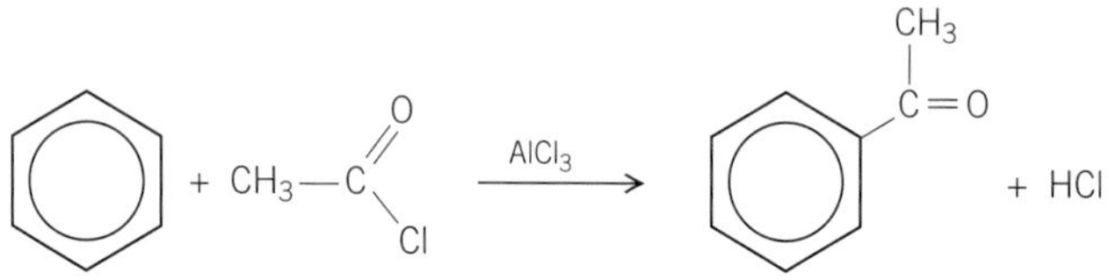

Acylation is a useful step in the synthesis of new substituted aromatic compounds.

Summary questions

1 Classify **a** nitration **b** Friedel– Crafts reactions as:

   A electrophilic substitution

   B nucleophilic substitution

   C electrophilic addition

   D free-radical addition

   E free radical substitution.

2 Name the two isomers of 1,3-dinitrobenzene.

3 Explain why most of the reactions of benzene are substitutions rather than additions.

4 Write the equation for the reaction between propanoyl chloride with benzene. What species attacks the benzene ring?

Practice questions

1. Give reagents and conditions and write equations to show the formation of nitrobenzene from benzene.
   Name and outline a mechanism for this reaction of benzene.

   (8 marks)
   AQA, 2007

2. A possible synthesis of phenylethene (*styrene*) is outlined below.

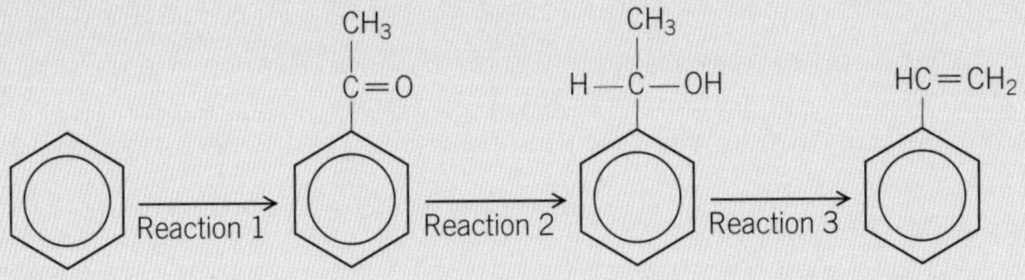

   In Reaction 1, ethanoyl chloride and aluminium chloride are used to form a reactive species which then reacts with benzene.
   Write an equation to show the formation of the reactive species.
   Name and outline the mechanism by which this reactive species reacts with benzene.

   (6 marks)
   AQA, 2006

3. An acylium ion has the structure $R-\overset{+}{C}=O$ where R is any alkyl group.
   In the conversion of benzene into phenylethanone, $C_6H_5COCH_3$, an acylium ion, $CH_3\overset{+}{C}O$, reacts with a benzene molecule.
   Write an equation to show the formation of this acylium ion from ethanoyl chloride and one other substance.
   Name and outline the mechanism of the reaction of this acylium ion with benzene.

   (6 marks)
   AQA, 2007

4. An equation for the formation of phenylethanone is shown below. In this reaction a reactive intermediate is formed from ethanoyl chloride. This intermediate then reacts with benzene.

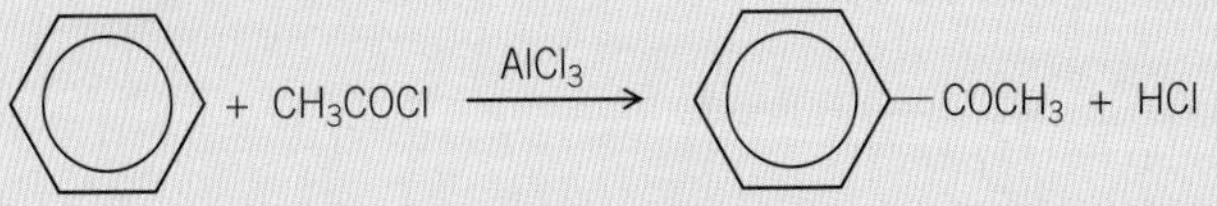

   (a) Give the formula of the reactive intermediate.
   (b) Outline a mechanism for the reaction of this intermediate with benzene to form phenylethanone.

   (4 marks)

5   Consider compound **P** shown below that is formed by the reaction of benzene with an electrophile.

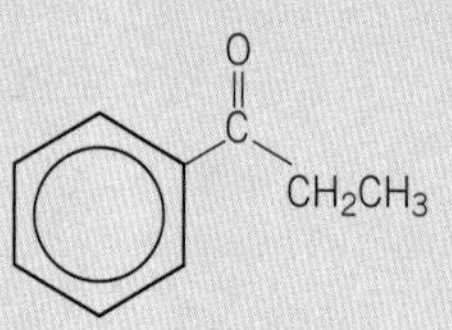

P

(a) Give the **two** substances that react together to form the electrophile and write an equation to show the formation of this electrophile. *(3 marks)*

(b) Outline a mechanism for the reaction of this electrophile with benzene to form P. *(3 marks)*

(c) Compound **Q** is an isomer of **P** that shows optical isomerism. **Q** forms a silver mirror when added to a suitable reagent.
Identify this reagent and suggest a structure for **Q**. *(2 marks)*

AQA, 2010

6   The hydrocarbons benzene and cyclohexene are both unsaturated compounds. Benzene normally undergoes substitution reactions, but cyclohexene normally undergoes addition reactions.

(a) The molecule cyclohexatriene does not exist and is described as hypothetical. Use the following data to state and explain the stability of benzene compared with the hypothetical cyclohexatriene.

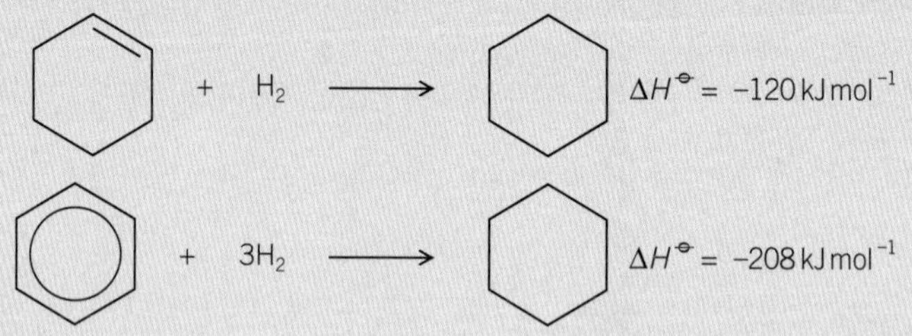

(4 marks)

(b) Benzene can be converted into amine **U** by the two-step synthesis shown below.

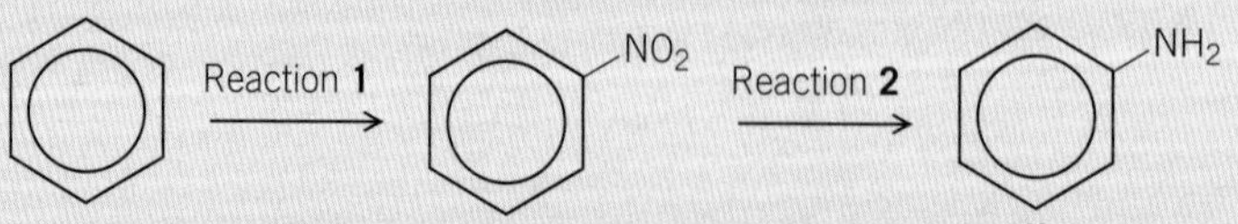

The mechanism of Reaction **1** involves attack by an electrophile.

Give the reagents used to produce the electrophile needed in Reaction **1**.

Write an equation showing the formation of this electrophile.

Outline a mechanism for the reaction of this electrophile with benzene.

(6 marks)

AQA, 2011

7 Propanoyl chloride can be used, together with a catalyst, in the synthesis of
   1-phenylpropene from benzene. The first step in the reaction is shown below.

   The mechanism of this reaction is an electrophilic substitution. Write an equation to show
   the formation of the electrophile from propanoyl chloride. Outline the mechanism of the
   reaction of this electrophile with benzene.

   (5 marks)
   AQA, 2004

8 Use the following data to show the stability of benzene relative to the hypothetical
   cyclohexa-1,2,5-triene.

   Give a reason for this difference in stability.

   $+ H_2 \longrightarrow \quad \Delta H^{\ominus} = -120\,kJ\,mol^{-1}$

   $+ 3H_2 \longrightarrow \quad \Delta H^{\ominus} = -208\,kJ\,mol^{-1}$

   (4 marks)
   AQA, 2004

9 The nitration of benzene is an important industrial reaction.
   (a) State the reagents required for the nitration of benzene.

   (1 mark)

   (b) Name an important material whose manufacture involves the nitration
       of benzene.

   (1 mark)

   (c) **(i)** Write a balanced equation for the nitration of benzene.

   (2 marks)

   (ii) Explain why the NO_2^+ ion is described as an electrophile.

   (1 mark)

   (iii) Name the type of mechanism involved in the nitration of benzene.

   (1 mark)

This chapter is about a group of compounds called amines. Amines can be thought of as derivatives of ammonia in which one or more of the hydrogen atoms in the ammonia molecule have been replaced by alkyl or aryl groups.

$$H—\overset{\cdot\cdot}{N}—H \qquad H—\overset{\cdot\cdot}{N}—H \qquad H—\overset{\cdot\cdot}{N}—R' \qquad R''—\overset{\cdot\cdot}{N}—R'$$
$$\underset{H}{|} \qquad\qquad \underset{R}{|} \qquad\qquad \underset{R}{|} \qquad\qquad \underset{R}{|}$$

ammonia   primary amine   secondary amine   tertiary amine

Amines are very reactive compounds, so they are useful as intermediates in **synthesis** – the making of new molecules.

The terms primary, secondary, and tertiary are used for amines slightly differently from the way they are used with alcohols. In amines, 1°, 2°, and 3° refer to the number of substituents (R-groups) on the *nitrogen* atom. (In alcohols, 1°, 2°, and 3° refer to the number of substituents on the *carbon* atom bonded to the −OH group.)

How to name amines

Primary amines have the general formula RNH_2, where the R can be an alkyl or aryl group. Amines are named using the suffix -amine, for example:

$CH_3—NH_2$ is methylamine

$C_2H_5—NH_2$ is ethylamine.

$C_6H_5NH_2$,   is phenylamine.

▲ **Figure 1** *Phenylamine has almost the same density as water and is not soluble in it. Heat from a bulb at the base of the lava lamp changes the density enough for the phenylamine to float when hot and sink when cool*

Secondary amines have the general formula RR′NH, for example:

$(CH_3)_2NH$, $\overset{CH_3}{\underset{CH_3}{\diagdown N—H}}$ is dimethylamine.

Tertiary amines have the general formula RR′R″N, for example:

$(C_2H_5)_3NH$, $\overset{C_2H_5}{\underset{C_2H_5}{\diagdown N—C_2H_5}}$ is triethylamine.

Different substituents are written in alphabetical order:

$CH_3(C_3H_7)NH$, $\overset{CH_3}{\underset{C_3H_7}{\diagdown N—H}}$ is N−methylpropylamine.

The properties of primary amines

Shape

Ammonia is a pyramidal molecule with bond angles of approximately 107°. The angles of a perfect tetrahedron are 109.5°. The difference is caused by the lone pair, which repels more than the bonding pairs of electrons in the N—H bonds. Amines keep this basic shape (Figure 1).

Boiling points

Amines are polar:

$$R-\underset{\underset{H}{|}}{\overset{\overset{H}{|}}{C}}{}^{\delta+}-\overset{H^{\delta+}}{\underset{H^{\delta+}}{N}}{}^{\delta-}$$

Primary amines can hydrogen bond to one another using their $-NH_2$ groups (in the same way as alcohols with their $-OH$ groups). However, as nitrogen is less electronegative than oxygen (electronegativities – O = 3.5, N = 3.0), the hydrogen bonds are not as strong as those in alcohols. The boiling points of amines are lower than those of comparable alcohols:

methylamine, $M_r = 31$, CH_3-NH_2, boiling point = 267 K

methanol, $M_r = 32$, CH_3-OH, boiling point = 338 K

Shorter chain amines such as methylamine and ethylamine are gases at room temperature, and those with slightly longer chains are volatile liquids. They have fishy smells. Rotting fish and rotting animal flesh smell of di- and triamines, produced when proteins decompose (Figure 3).

Solubility

Primary amines with chain lengths up to about four carbon atoms are very soluble both in water and in alcohols because they form hydrogen bonds with these solvents. Most amines are also soluble in less polar solvents. Phenylamine, $C_6H_5NH_2$, is not very soluble in water due to the benzene ring, which cannot form hydrogen bonds.

The reactivity of amines

Amines have a lone pair of electrons and this is important in the way they react. The lone pair may be used to form a bond with:

- a H^+ ion, when the amine is acting as a **base**
- an electron-deficient carbon atom, when the amine is acting as a **nucleophile**.

Synoptic link

You can remind yourself of the effect of lone pairs on the shapes of molecules in 3.6, The shapes of molecules and ions.

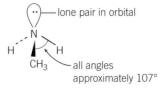

▲ **Figure 2** *The shape of the methylamine molecule*

Hint

A lower boiling point means the molecules are easier to separate.

▲ **Figure 3** *Rotting fish smell of di- and triamines*

Synoptic link

An amine is a proton acceptor so it is a Brønsted–Lowry base, see Chapter 21, Acids, bases, and buffers. An amine is also a Lewis base as it is a lone-pair donor, see Topic 24.1, The acid–base chemistry of aqueous transition metal ions.

Summary questions

1 Classify $C_2H_5-\underset{\overset{|}{H}}{N}-C_3H_7$ as primary, secondary, or tertiary.

2 Name the compound in **1**.

3 Write the structural formula of trimethylamine.

4 Predict whether dimethylamine will be a solid, liquid, or gas at room temperature.

5 Explain your answer to **4**.

28.2 The properties of amines as bases

Learning objectives:

→ Explain why amines behave as Brønsted–Lowry bases.

→ Explain why the base strengths of amines differ from each other and from ammonia.

Specification reference: 3.3.11

Amines as bases

Amines can accept a proton (an H^+ ion) so they are Brønsted–Lowry bases.

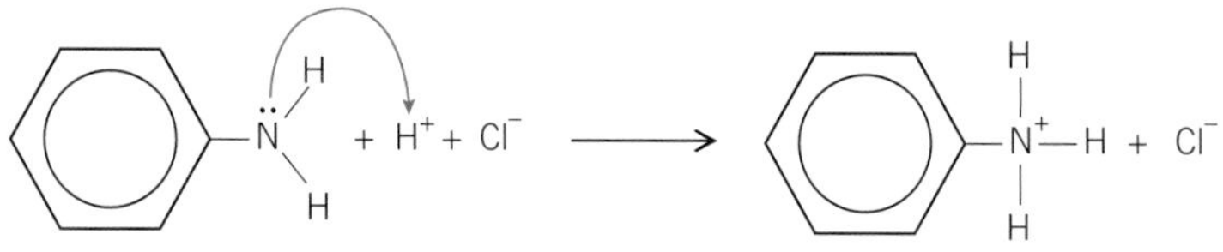

phenylamine

phenylammonium chloride
a water-soluble, ionic salt

> **Hint**
>
> The salts of amines are sometimes named as the hydrochloride of the parent amine.

Reaction as bases

Amines react with acids to form salts. For example, ethylamine, a soluble alkylamine, reacts with dilute hydrochloric acid:

$$C_2H_5\overset{..}{N}H_2 + H^+ + Cl^- \longrightarrow C_2H_5NH_3^+ + Cl^-$$

ethylamine   ethylammonium chloride

The products are ionic compounds that will crystallise as the water evaporates.

Phenylamine, an arylamine, is relatively insoluble, but it will dissolve in excess hydrochloric acid because it forms the soluble ionic salt.

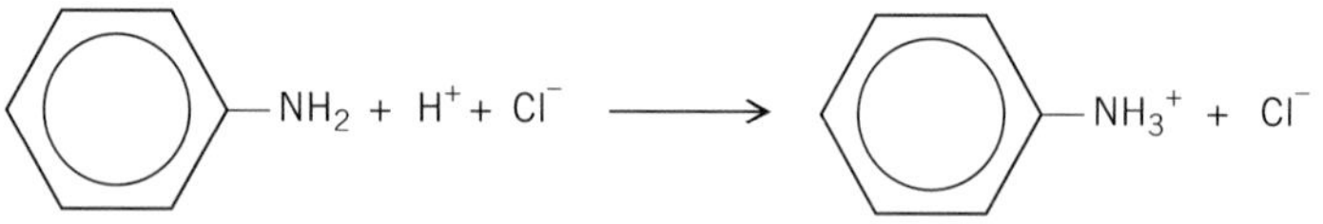

phenylamine

phenylammonium chloride
a water-soluble, ionic salt

If a strong base like sodium hydroxide is added, it removes the proton from the salt and regenerates the insoluble amine.

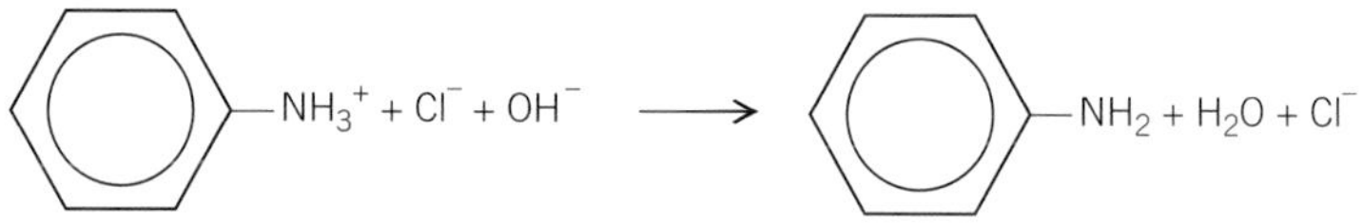

phenylamine

> **Hint**
>
> The smell of a solution of an amine disappears when an acid is added due to the formation of the ionic (and therefore involatile) salt. The smell returns if a strong base is then added.

Comparing base strengths

The strength of a base depends on how readily it will accept a proton, H^+. Both ammonia and amines have a lone pair of electrons that attract a proton.

Alkyl groups *release* electrons away from the alkyl group and towards the nitrogen atom. This is called the **inductive effect** and is shown by an arrow (Figure 1).

$$R \rightarrow \overset{..}{N}H_2$$

▲ **Figure 1** *A primary amine. The arrow shows that R releases electrons. This is called the inductive effect*

The inductive effect of the alkyl group increases the electron density on the nitrogen atom and therefore makes it a better electron pair donor (i.e., more attractive to protons). So, primary alkylamines are stronger bases than ammonia.

Solubility of drugs

A number of medicinal drugs are amines, for example, the nasal decongestant Sudafed, active ingredient pseudoephedrine. Longer chain amines are relatively insoluble in water so when they are used in medicines they are often supplied as hydrochlorides to make them more soluble in the bloodstream.

Pseudoephedrine has the formula:

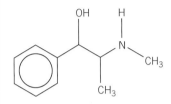

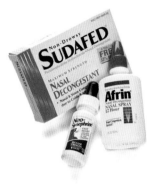

▲ **Figure 2** *Nasal decongestant sprays reduce swelling in the blood vessels inside your nose, helping to relieve breathing issues caused by colds or hayfever*

1 Is pseudoephedrine a primary, secondary, or tertiary amine?
2 Draw the formula of pseudoephedrine hydrochloride.
3 Explain why pseudoephedrine hydrochloride is more soluble in water than pseudoephedrine.
4 Is pseudoephedrine hydrochloride likely to be a solid, liquid, or gas? Explain your answer.
5 As well as having an amine group, pseudoephedrine has two other functional groups. Name them.
6 Pseudoephedrine has two chiral centres in its molecule. Mark them with a * on your formula of pseudoephedrine hydrochloride. Hint, it may help if you draw in the hydrogen atoms that are not marked on the skeletal formula.
7 What problems might this chirality have in pseudoephedrine's use as a drug?

Secondary alkylamines have two inductive effects and are therefore stronger bases than primary alkylamines. However, tertiary alkylamines are not stronger bases than secondary ones because they are less soluble in water.

Aryl groups *withdraw* electrons from the nitrogen atom because the lone pair of electrons overlaps with the delocalised system on the benzene ring, as shown for phenylamine.

The nitrogen is a weaker electron pair donor and therefore less attractive to protons, so arylamines are weaker bases than ammonia.

ethylamine > ammonia > phenylamine

strongest ⟶ weakest

Summary questions

1 **a** Write the equation for dimethylamine reacting with hydrochloric acid.

   b Name the product.

2 Phenylamine is not very soluble in water. It forms oily drops that float in the water. Predict what you would see if you:

   a add concentrated hydrochloric acid to a mixture of phenylamine and water.

   b then add sodium hydroxide solution to the resulting solution.

3 Suggest whether dimethylamine will be a weaker or stronger base than ethylamine. Explain your answer.

28.3 Amines as nucleophiles and their synthesis

Learning objectives:

→ Explain why ammonia and amines act as nucleophiles.

→ State how halogenoalkanes react with ammonia and amines.

→ State how amines are prepared from nitriles.

→ State how aromatic amines are synthesised from benzene.

Specification reference: 3.3.11

The lone pair of electrons from an amine will attack positively charged carbon atoms. So amines, like ammonia, will act as nucleophiles.

Reactions of ammonia with halogenoalkanes

Primary aliphatic amines are produced when halogenoalkanes are reacted with ammonia. There is nucleophilic substitution of the halide by NH_2.

$$NH_3 + RX \rightarrow [RNH_3]^+ X^-$$
$$[RNH_3]^+X^- + NH_3 \rightarrow RNH_2 + [NH_4]^+ X^-$$
primary
amine

However, the primary amine produced is also a nucleophile and this will react with the halogenoalkane to produce a secondary amine:

$$RNH_2 + RX \rightarrow [R_2NH_2]^+X^-$$
$$[R_2NH_2] + X^- + NH_3 \rightarrow R_2NH + [NH_4]^+X^-$$
secondary
amine

The secondary amine will react to give a tertiary amine:

$$R_2NH + RX \rightarrow [R_3NH]^+X^-$$
$$[R_3NH]^+X^- + NH_3 \rightarrow R_3N + [NH_4]^+X^-$$
tertiary
amine

This in turn will react to a produce a quarternary ammonium salt:

$$R_3N + RX \rightarrow [R_4N]^+X^-$$

So a mixture of primary, secondary, and tertiary amines and a quarternary ammonium salt is produced. This means that this is not a very efficient way of preparing an amine, though the products may be separated by fractional distillation. A large excess of ammonia gives a better yield of primary amine.

Synoptic link

You will need to know the nucleophilic substitution reactions in halogenoalkanes studied in Topic 13.2, Nucleophilic substitution in halogenoalkanes, and redox reactions and oxidation states studied in Topic 7.2, Oxidation states, and Topic 7.3, Redox equations.

The mechanism of the reaction

For all the above reactions the mechanism is essentially the same:

Initially ammonia acts as a nucleophile. In the second stage, it acts as a base.

Study tip

Remember that in these reactions a proton is removed from the initial substitution intermediate so two moles of ammonia or amine are required for each mole of halogenoalkane.

Synoptic link

You will cover the formation of nylon in Topic 29.1, Condensation polymers.

Preparation of amines

Primary amines
Reduction of nitriles
Primary aliphatic amines can be prepared from halogenoalkanes in a two-step process:

Step 1: Halogenoalkanes react with the cyanide ion in aqueous ethanol. The cyanide ion replaces the halide ion by nucleophilic substitution to form a nitrile:

$$RBr + CN^- \rightarrow R\text{—}C\equiv N + Br^-$$

Step 2: Nitriles contain the functional group $-C\equiv N$. They can be reduced to primary amines, for example, with a nickel/hydrogen catalyst:

$$R\text{—}C\equiv N + 2H_2 \rightarrow R\text{—}CH_2NH_2$$

This gives a purer product than a bromoalkane and ammonia because only the primary amine can be formed. The carbon chain of the product is one carbon atom longer than in the starting material.

Phenylamine

Phenylamine is the simplest arylamine. It is the starting point for making many other chemicals and is made in industry using benzene produced from crude oil.

Making phenylamine
Phenylamine can be made from benzene.

Step 1: Benzene is reacted with a mixture of concentrated nitric and concentrated sulfuric acid. This produces nitrobenzene:

benzene $+$ HNO$_3$ $\xrightarrow{\text{conc.}\ H_2SO_4}$ nitrobenzene $+$ H$_2$O

Step 2: Nitrobenzene is reduced to phenylamine, using tin and hydrochloric acid as the reducing agent.

The tin and hydrochloric acid react to form hydrogen, which reduces the nitrobenzene by removing oxygen atoms of the NO_2 group and replacing them with hydrogen atoms.

NO$_2$ $+$ 6[H] $\xrightarrow[\text{room temp.}]{\text{Sn/HCl}}$ NH$_2$ $+$ 2H$_2$O

This could also be written:

$$C_6H_5NO_2 + 6[H] \rightarrow C_6H_5NH_2 + 2H_2O$$

Since the reaction is carried out in hydrochloric acid, the salt $C_6H_5NH_3{}^+Cl^-$ is formed and sodium hydroxide is added to liberate the free amine:

$$C_6H_5NH_3{}^+Cl^- + NaOH \rightarrow C_6H_5NH_2 + H_2O + NaCl$$

> **Study tip**
>
> Nitriles cannot be reduced to amines by NaBH$_4$.

> **Hint** 🧪
>
> This preparation would not be attempted in a school laboratory as benzene is a carcinogen.

> ▲ **Figure 1** *Aromatic amines are used in the manufacture of dyestuffs*

> **Synoptic link**
>
> Remember a mixture of nitric and sulfuric acids generates the electrophile $NO_2{}^+$ – see Topic 27.3, Reactions of arenes. So this reaction is an electrophilic substitution.

> **Study tip**
>
> It is acceptable to use [H] when writing equations for reduction, but the equation *must* balance.

The formation of amides

Amines will react with acid chlorides and acid anhydrides. These are nucleophilic substitution reactions (sometimes called addition – elimination reactions) and the products are N-substituted amides.

The mechanism is:

The amine adds on to the acid chloride and then HCl is eliminated.

This reaction is useful in forming polymers such as nylon.

The economic importance of amines

Amines are used in the manufacture of synthetic materials such as nylon and polyurethane, dyes, and drugs.

Quaternary ammonium compounds are used industrially in the manufacture of hair and fabric conditioners. They have a long hydrocarbon chain and a positively charged organic group, so they form cations:

Both wet fabric and wet hair pick up negative charges on their surfaces. So the positive charges of the cations attract them to the wet surface and form a coating that prevents the build-up of static electricity. This keeps the surface of the fabric smooth (in fabric conditioner) and prevents flyaway hair in hair conditioners.

They are called cationic surfactants because in aqueous solution the ions cluster with their charged ends in the water and their hydrocarbon tails on the surface.

▲ **Figure 1** *Hair conditioner*

Sulfa drugs

The story of the antibiotic penicillin is well known. It was the result of a chance observation of mould on a discarded Petri dish by Alexander Fleming, and was developed by Howard Flory and Ernst Chain (and a massive industrial effort) into a drug that saved thousands of lives in the Second World War and since. However, it was not the first anti-bacterial drug. Another class of drugs, the sulfanilamides, were already in use before penicillin and may also have had an effect on the course of the war – by saving the life of Prime Minister Winston Churchill.

Towards the end of the nineteenth century, it had been noticed that some dyes used to stain bacteria to make them visible under the microscope could also kill them. Since these dyes were absorbed by the bacteria rather than their surroundings, they might be expected to kill the bacteria but not their host. Eventually the dye Prontosil Rubrum began to be used in medicine to fight bacterial infections.

By the early 1940s it had been established that Prontosil was converted in the body into the compound sulfanilamide which was the active ingredient.

Prontosil Rubrum   sulfanilamide

The drug worked by preventing the bacteria making folic acid, which they need to synthesise DNA and therefore replicate. Bacteria make folic acid from a compound called *para*-aminobenzoic acid (PABA). The sulfanilamide molecule is of a similar shape to PABA and the bacteria try to use it to make folic acid but without success, as it is the wrong molecule. Humans do not need to synthesise folic acid – they get it from their food – and so sulfanilamide kills bacteria but is harmless to humans.

PABA

Since the 1940s over 5000 variations on the sulfanilamide molecule have been synthesised by chemists in an effort to find molecules that are more effective, have fewer side effects, are absorbed at a different rate, and so on. This is one of the main methods used to discover new drugs – to take a molecule with a known beneficial effect and make variations on it in the hope of maintaining or enhancing its activity but reducing any disadvantages that it might have. Nowadays, this process can be sped up by the technique of combinatorial chemistry.

Although less common than they once were, sulfa drugs are still used today. The one that cured Winston Churchill's pneumonia in 1943 was sulfapyridine, known at the time as M & B 693, after the makers the May & Baker Company. May & Baker is still in business and supplies chemicals for school laboratories as well as making drugs. Look for their labels in your school preparation room.

What is the systematic name of PABA?

4-aminobenzenecarboxylic acid (or 4-aminobenzoic acid)

Hint

Para- is part of an older naming system for locating substituents on aromatic rings. It means opposite the original substituent, (i.e., in the 4 position).
Ortho- means adjacent (the 2 position) and *meta-* the 3 position.

Synoptic link

The use of robots in synthesis is described in Topic 31.1, Synthetic routes.

Summary questions

1   Why is nucleophilic substitution of a halogenolkane not a good method for preparing a primary amine?

2   **a**   Write the equation for the reaction of chloroethane with an excess of ammonia. Give the reaction mechanism.

    b   What are the other possible products of this reaction?

Practice questions

1 **(a)** Name the compound $(CH_3)_2NH$.

(1 mark)

(b) $(CH_3)_2NH$ can be formed by the reaction of an excess of CH_3NH_2 with CH_3Br. Name and outline a mechanism for this reaction.

(5 marks)

(c) Name the type of compound produced when a large excess of CH_3Br reacts with CH_3NH_2 Give a use for this type of compound.

(2 marks)

AQA, 2006

2 Consider the following reaction sequence.

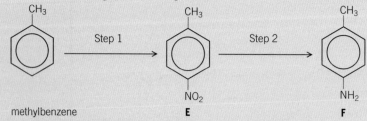

methylbenzene   **E**   **F**

(a) For Step 2, give a reagent or combination of reagents. Write the equation for this reaction using [H] to represent the reducing agent.

(2 marks)

(b) Draw the structure of the species formed by **F** in an excess of hydrochloric acid.

(2 marks)

(c) Compounds **G** and **H** are both monosubstituted benzenes and both are isomers of **F**. **G** is a primary amine and **H** is a secondary amine. Draw the structures of **G** and **H**.

(2 marks)

AQA, 2005

3 **(a)** Name and outline a mechanism for the formation of butylamine, $CH_3CH_2CH_2CH_2NH_2$, by the reaction of ammonia with 1-bromobutane, $CH_3CH_2CH_2CH_2Br$.

(5 marks)

(b) Butylamine can also be prepared in a two-step synthesis starting from 1-bromopropane, $CH_3CH_2CH_2Br$. Write an equation for each of the two steps in this synthesis.

(3 marks)

(c) Explain why butylamine is a stronger base than ammonia.

(2 marks)

(d) Draw the structure of a tertiary amine which is an isomer of butylamine.

(1 mark)

AQA, 2004

4 Propylamine, $CH_3CH_2CH_2NH_2$, can be formed either by nucleophilic substitution or by reduction.

(a) Draw the structure of a compound which can undergo nucleophilic substitution to form propylamine.

(1 mark)

(b) Draw the structure of the nitrile which can be reduced to form propylamine.

(1 mark)

(c) State and explain which of the two routes to propylamine, by nucleophilic substitution or by reduction, gives the less pure product. Draw the structure of a compound formed as an impurity.

(3 marks)

AQA, 2006

5   This question is about the primary amine $CH_3CH_2CH_2NH_2$
    (a) The amine $CH_3CH_2CH_2NH_2$ reacts with CH_3COCl
        Name and outline a mechanism for this reaction.
        Give the IUPAC name of the organic product.
                                                                      (6 marks)

    (b) Isomers of $CH_3CH_2CH_2NH_2$ include another primary amine, a secondary amine, and a
        tertiary amine.
        (i)   Draw the structures of these **three** isomers.
              Label each structure as primary, secondary, or tertiary.
                                                                      (3 marks)

6   (a) Name and outline a mechanism for the reaction of $CH_3CH_2NH_2$ with CH_3CH_2COCl
        Name the amide formed.
                                                                      (6 marks)

    (b) Halogenoalkanes such as CH_3Cl are used in organic synthesis.
        Outline a three-step synthesis of $CH_3CH_2NH_2$ starting from methane. Your first step
        should involve the formation of CH_3Cl
        In your answer, identify the product of the second step and give the reagents and
        conditions for each step.
        Equations and mechanisms are **not** required.
                                                                      (6 marks)
                                                                      AQA, 2013

7   Consider the reaction shown below.

    In this reaction phenylamine reacts with hydrochloric acid to form phenylammonium
    chloride.
    (a) Explain how this reaction shows that phenylamine is a Brønsted-Lowry base.
                                                                      (1 mark)

    (b) Explain why phenylammonium chloride is soluble in water.
                                                                      (1 mark)

Learning objectives:

→ Describe a condensation polymer.

→ Explain what sorts of molecules react to form condensation polymers.

Specification reference: 3.3.12

▲ **Figure 1** *Examples of polyesters and polyamides (nylon)*

A condensation reaction occurs when two molecules react together and a small molecule, often water or hydrogen chloride, is eliminated. For example, esters are formed when carboxylic acids and alcohols react together. This is a condensation reaction because water, H_2O, is eliminated – hydrogen from the alcohol and an –OH group from the carboxylic acid.

$$R-C\overset{O}{\underset{OH}{}} \;+\; HO-R' \longrightarrow R-C\overset{O}{\underset{O-R'}{}} \;+\; H_2O$$

carboxylic acid   alcohol   ester   water

Condensation polymers are normally made from two different monomers, each of which has *two* functional groups. Both functional groups can react, forming long-chain polymers.

Polyesters, polyamides, and polypeptides are all examples of condensation polymers (Figure 1).

Polyesters

A *poly*ester has the ester linkage –COO– repeated over and over again.

To make a polyester diols are used, which have two –OH groups, and dicarboxylic acids, which have two carboxylic acid, –COOH, groups:

$$HO-A-OH \qquad\qquad HO\overset{O}{\underset{}{}}C-B-C\overset{O}{\underset{}{}}OH$$

diol   dicarboxylic acid

A and B represent unspecified organic groups, often $-(CH_2)_n-$. The functional groups on the ends of each molecule react to form a chain. For example, diols and dicarboxylic acids react together to give a polyester by eliminating molecules of water (Figure 2).

$$HO-A-OH \quad HO-\overset{O}{\overset{\|}{C}}-B-\overset{O}{\overset{\|}{C}}-OH \quad HO-A-OH \; HO-\overset{O}{\overset{\|}{C}}-B-\overset{O}{\overset{\|}{C}}-OH$$

diol   dicarboxylic acid   diol

$H_2O \longleftarrow \qquad\qquad \longrightarrow H_2O \qquad \longrightarrow H_2O$

$$-O-A-O-\overset{O}{\overset{\|}{C}}-B-\overset{O}{\overset{\|}{C}}-O-A-O-\overset{O}{\overset{\|}{C}}-B-\overset{O}{\overset{\|}{C}}-$$

polyester

▲ **Figure 2** *Making a polyester*

The fibre Terylene is a polyester made from benzene-1,4-dicarboxylic acid and ethane-1,2-diol (Figure 3).

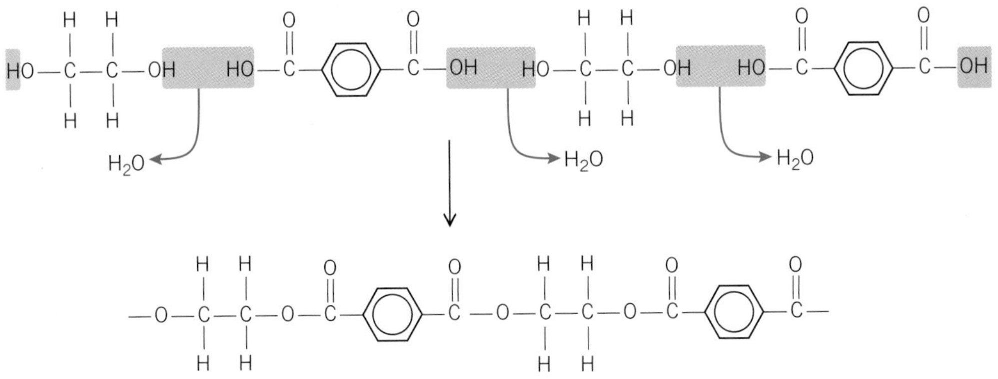

▲ **Figure 3** *Terylene is a polyester. Notice how the C—O is alternately to the left and to the right of the C═O*

Polyamides

An amide is formed when a carboxylic acid and an amine react together:

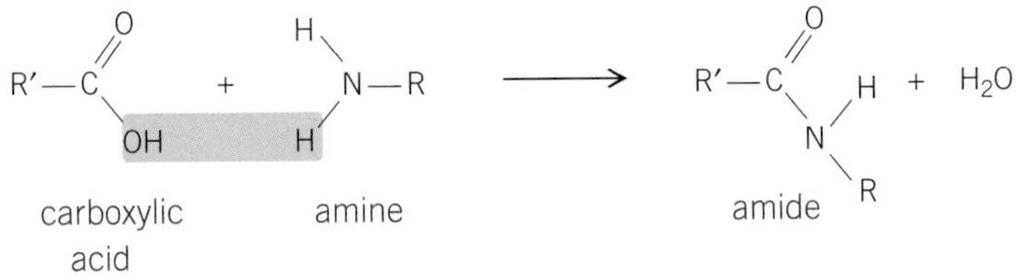

*Poly*amides have the amide linkage –CONH– repeated over and over again. To make polyamides from two different monomers, a diaminoalkane (which has two amine groups) reacts with a dicarboxylic acid (which has two carboxylic acid groups) (Figure 4).

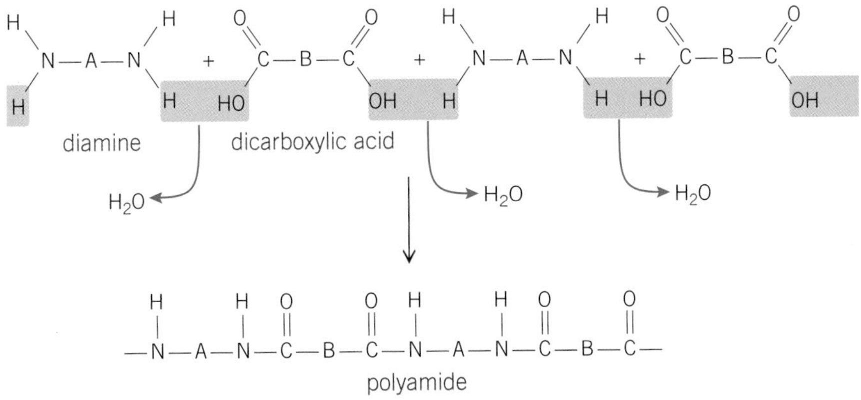

▲ **Figure 4** *The general equation for making a polyamide, such as Nylon-6,6 or Kevlar*

Both Nylon and Kevlar are condensation polymers.

Study tip

It is important you understand that, unlike polyalkenes, polyester and polyamides can be hydrolysed and are biodegradable.

Nylon

Industrially, Nylon-6,6 is made from 1, 6-diaminohexane and hexane-1,6-dicarboxylic acid:

▲ **Figure 6** *When 1-6-diaminohexane and hexane-1,6-dioyldichloride meet, Nylon-6,6 is formed at the interface. This demonstration was first performed by Stephanie Kwolek who developed Kevlar*

1,6-diaminohexane   hexane-1,6-dicarboxylic acid

Nylon-6,6

In the laboratory, the reaction goes faster if a diacid chloride is used rather than the dicarboxylic acid, and in this case hydrogen chloride is eliminated. Nylon-6,10 is made from 1,6-diaminohexane and decane-1,10-dicarboxylic acid. Many other Nylons are made each with slightly different properties.

Kevlar

Kevlar is made from benzene-1,4-diamine and benzene-1,4-dicarboxylic acid (Figure 6).

benzene-1,4-diamine   benzene-1,4-dicarboxylic acid

Kevlar

▲ **Figure 6** *Kevlar is a polyamide. Because the amide groups are linking rigid benzene rings, Kevlar has very different properties to Nylon*

▲ **Figure 7** *Formula 1 drivers' helmets need to be lightweight as the less weight it adds to a drivers head, the smaller the risk of whiplash injuries under the extreme G-forces experienced in accelerating and braking. Helmets worn by racing drivers contain Kevlar which is five times stronger than steel, weight for weight*

Kevlar's strength is due to the rigid chains and the ability of the flat aromatic rings to pack together held by strong intermolecular forces. The polymer, developed in the 1960s by Stephanie Kwolek of the DuPont company, is credited with saving some 3000 lives because of its use in bullet proof vests and anti-stab clothing as worn by the police. You may have Kevlar oven gloves at home.

Polypeptides and proteins

Polypeptides are also polyamides. They may be made from a single amino acid monomer, or many different ones.

In a polypeptide, *each* amino acid has both an amine group and a carboxylic acid group. So the amine group of one amino acid can react with the carboxylic acid group of another. A molecule of water is eliminated and a condensation polymer can begin to form:

Hint

Once a dipeptide is formed, tri-, tetra, and polypeptides can form by further reaction at each end of the molecule.

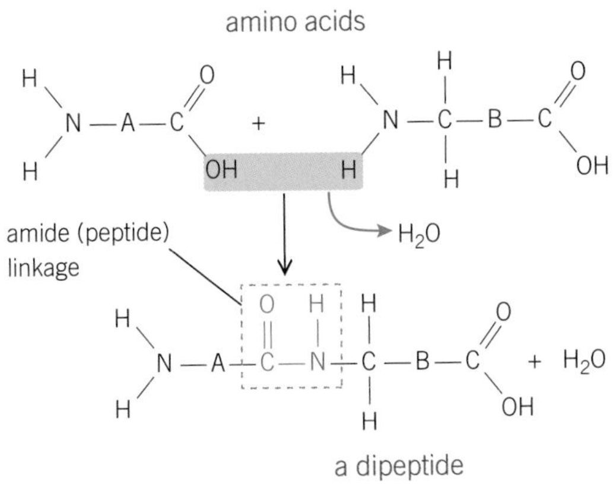

amino acids

amide (peptide) linkage

a dipeptide

Hint

When an amide linkage is formed between amino acids, it is often called a peptide link.

There is a difference between a polymer like Nylon-6,6 (where there are two monomers (one a diamine, H_2N—X—NH_2 and one a dicarboxylic acid, HOOC—Y—COOH)) and a polypeptide (where each amino acid monomer has one –NH_2 group and one –COOH group, $H_2NCHRCOOH$. There are 20 naturally occurring varieties of amino acids.

Synoptic link

Amino acids are covered in Topic 30.1, Introduction to amino acids.

Identifying the repeat unit of a condensation polymer

The repeat unit of a condensation polymer is found by starting at any point in the polymer and stopping when the same pattern of atoms begins again (Figure 8).

Study tip

Check that you can deduce the repeating unit in each type of polymer.

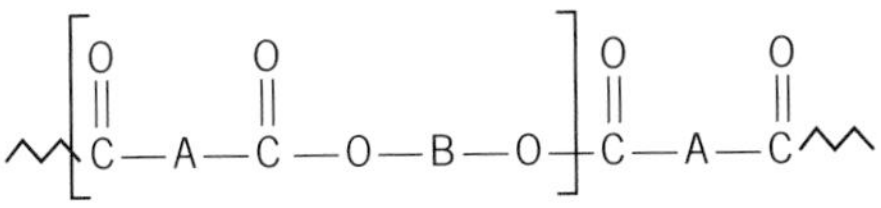

Identifying the monomer(s) of a condensation polymer

The best way to work out the monomer(s) in a condensation polymer is to try and recognise the links formed by familiar functional groups (Table 1).

▲ **Figure 8** *The repeat unit is in brackets*

1 Start with the repeat unit.

▼ **Table 1** *Condensation polymers – the repeat unit is inside the bracket*

Study tip

It is important that you can deduce the monomers from which a polymer is formed.

Monomer 1	Monomer 2	Polymer
HO—C—A—C—OH (with O double bonds) dicarboxylic acid	HO—B—OH diol	polymer chain
HO—C—A—C—OH (with O double bonds) dicarboxylic acid	H—N—B—N—H diamine	polymer chain
HO—C—C—N—H (with O double bond and R, H) amino acid		polymer chain

2 Break the linkage (at the C—O for a polyester or C—N for a polyamide).
3 Add back the components of water for each ester or amide link.

For example:

▲ **Figure 9** *Undecomposed poly(ethene) and poly(propene) cause problems for wildlife*

 Recycling plastics

Many polyester materials are now recycled. They are being collected, sorted, and then melted and reformed. Fleece garments may well be made from recycled soft drink bottles. With all recycling, the costs and benefits have to be balanced. Melting and reforming of plastics can only be done a limited number of times as during the process the polymer chains tend to break and shorten, thus degrading the properties of the polymer.

Advantages of recycling

Almost all plastics are derived from crude oil. Recycling saves this expensive and ever diminishing resource, as well as the energy used in refining it.

If plastics are not recycled they mostly end up in landfill sites.

Disadvantages of recycling

The plastics need to be collected, transported, and sorted, which uses energy and manpower and is therefore expensive.

This is exactly the same process that occurs when condensation polymers are hydrolysed.

Disposal of polymers

Poly(ethene) and poly(propene) are not **biodegradable** because they are basically long-chain alkane molecules. Alkanes are unreactive because they have only strong, non-polar C—H and C—C bonds. There is nothing in the natural environment that will easily break them down and they persist for many years. They are usually disposed of in landfill sites, along with other rubbish, or by incineration. Some may be melted down and remoulded.

Poly(alkenes) can be burnt to carbon dioxide and water to produce energy, although poisonous carbon monoxide may be released into the atmosphere if combustion is incomplete (when there is a shortage of oxygen).

Burning poly(alkenes) does add to the problem of increasing the level of carbon dioxide in the atmosphere:

$$\text{-CH}_2\text{-}_n + 1\tfrac{1}{2}n\text{O}_2 \rightarrow n\text{CO}_2 + n\text{H}_2\text{O}$$

Other addition polymers, such as polystyrene, may release toxic products on burning. Complete combustion of polystyrene (a hydrocarbon) would produce carbon dioxide and water only. However, under certain conditions the polymer may depolymerise to produce toxic styrene vapour. Incomplete combustion produces carbon monoxide and unburnt carbon particles – black smoke.

Condensation polymers like polyesters and polyamides can be broken down by hydrolysis and are potentially biodegradable by the reverse of the polymerisation reaction by which they were formed.

The reaction below shows the hydrolysis of a polyamide such as nylon. However, this reaction is so slow under everyday conditions that you do not need to worry about your nylon umbrella depolymerising in the rain.

Throughout the polymer, the N—C bond is broken.

Hermann Staudinger

Hermann Staudinger is considered to be the father of polymer chemistry and he received the 1953 Nobel Prize for chemistry for his discoveries in this field – work which started in the 1920s.

Today, the idea of giant molecules (macromolecules) made up of chains of smaller ones is universally accepted. However, in the 1920s this idea was at odds with the established theory, and molecules with relative molecular masses (then called molecular weights) of over 5000 or so – such as rubber, starch, proteins, and so on – were considered to be made up of small molecules held together by some unknown force. Staudinger was already an established academic chemist (he had a reaction named after him) and put his reputation on the line by taking up the study of rubber. One distinguished colleague, with ill-disguised contempt, advised him to: 'Drop the idea of large molecules – organic molecules with a molecular weight higher than 5000 do not exist. Purify your rubber, then it will crystallise.'

Staudinger proved that polymers were indeed giant molecules made up of monomers by linking together molecules of methanal (formaldehyde, HCHO) one at a time to make successively bigger molecules, CH_2O, $(CH_2O)_2$, $(CH_2O)_3$, and so on until he produced the high molecular weight substance paraldehyde. He showed that the properties of these molecules gradually changed from those typical of small molecules to those of very large ones. So the very large molecule was simply a chain of small molecules held together by normal covalent bonds

– no unknown force was required. A few years later, X-ray diffraction was able to confirm the structures of polymers.

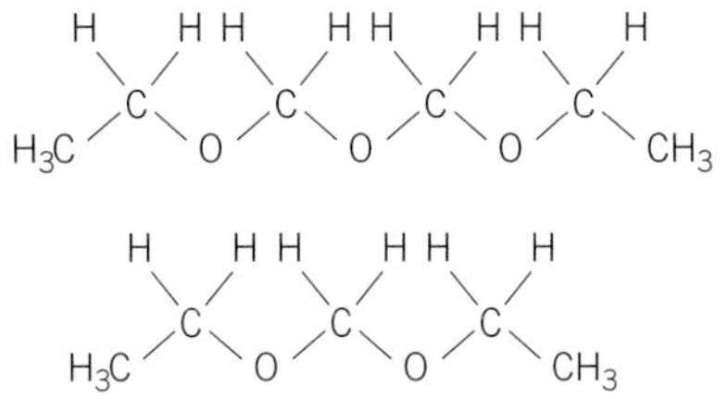

▲ **Figure 10** *Two of Staudinger's molecules – the CH_3 groups are the ends of the chains*

Staudinger's work led to modern synthetic polymers such as polythene (poly(ethene)) and nylon (a polyamide) and to an understanding of the structures of natural ones such as proteins, starch, and of course, rubber – which is poly(isoprene). Staudinger actually predicted artificial fibres – nylon was produced by Wallace Carothers in the late 1930s.

1 This is the structural formula of isoprene, the monomer from which rubber is made. What is its systematic (IUPAC) name?
2 Is isoprene likely to form an addition or a condensation polymer? Explain your answer.

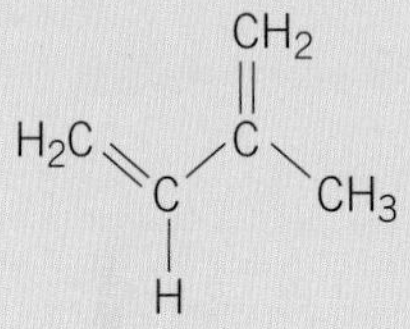

Summary questions

1 There are a number of different types of nylon made from two monomers – a dicarboxylic acid and a diamine.

   a The one made from hexane-1,6-dicarboxylic acid and 1,6-diaminohexane is called Nylon-6,6. Suggest where the numbers come from.

   b Nylon-6,10 is made from the same dicarboxylic acid as Nylon-6,6. What is the other monomer? Give its name and its formula.

2 Nylons are polyamides. Explain why proteins and peptides are also called polyamides.

3 Terylene is a polyester made from benzene-1,4-dicarboxylic acid and ethane-1,2-diol. Suggest another diol that would react with this acid to make a different polyester.

4 What are the linkages called in the the two polymers below?

   a   —A—C—N—B—N—C—A—
           ‖ | | ‖
           O H   H O

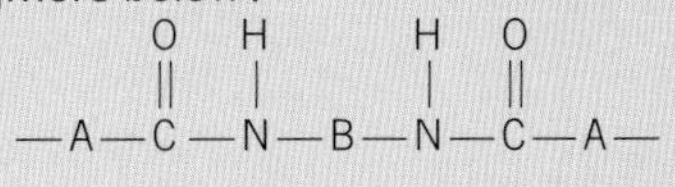

   b   —A—C—O—B—O—C—A—
           ‖   ‖
           O   O

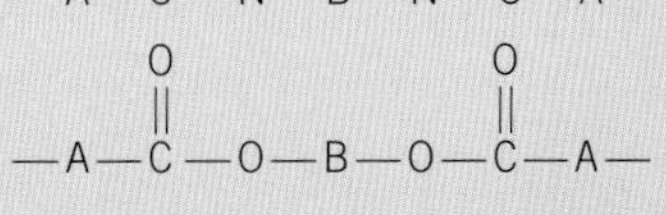

5 Write an equation for the hydrolysis of a polyester.

Practice questions

1   The repeating units of two polymers, **P** and **Q**, are shown below.

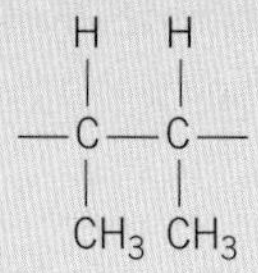

P 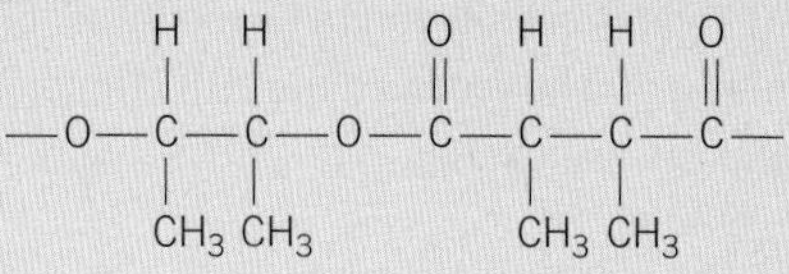   **Q**

   (a) Draw the structure of the monomer used to form polymer **P**. Name the type of
       polymerisation involved.

   (2 marks)

   (b) Draw the structures of **two** compounds which react together to form polymer **Q**.
       Name these **two** compounds and name the type of polymerisation involved.

   (5 marks)

   (c) Identify a compound which, in aqueous solution, will break down polymer **Q**
       but not polymer **P**.

   (1 mark)
   AQA, 2006

2   The structure below shows the repeating unit of a polymer.

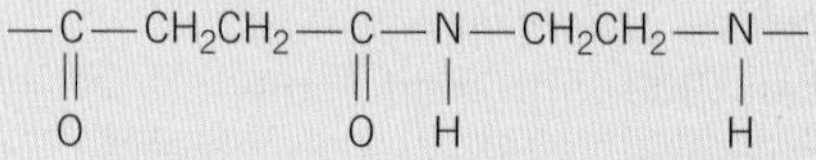

   By considering the functional group formed during polymerisation, name this type of
   polymer and the type of polymerisation involved in its formation.

   (2 marks)
   AQA, 2006

3   **(a)** The compound $H_2C{=}CHCN$ is used in the formation of acrylic polymers.
       (i) Draw the repeating unit of the polymer formed from this compound.
       (ii) Name the type of polymerisation involved in the formation of this polymer.

   (2 marks)

   (b) The repeating unit of a polyester is shown below.

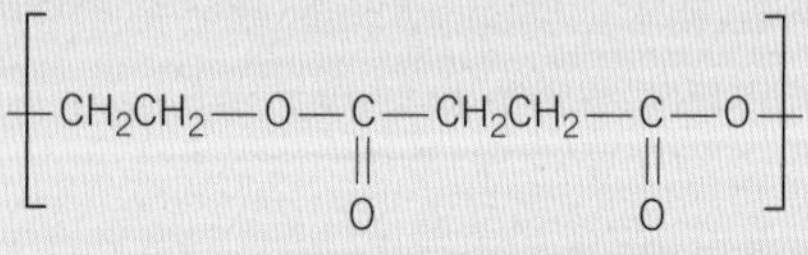

       (i) Deduce the empirical formula of the repeating unit of this polyester.
       (ii) Draw the structure of the acid which could be used in the preparation of this
           polyester and give the name of this acid.
       (iii) Give **one** reason why the polyester is biodegradable.

   (4 marks)
   AQA, 2004

4   Consider the hydrocarbon **G**, $(CH_3)_2C{=}CHCH_3$, which can be polymerised.
   (a) Name the type of polymerisation involved and draw the repeating unit of the
       polymer.

   (2 marks)

   (b) Draw the structure of an isomer of **G** which shows geometrical isomerism.

   (1 mark)

   (c) Draw the structure of an isomer of **G** which does not react with bromine water.

   (1 mark)
   AQA, 2004

5 **(a)** The hydrocarbon **M** has the structure shown below.

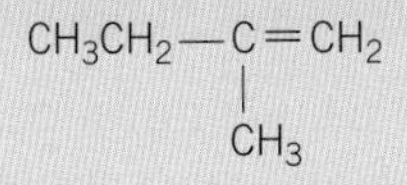

$$CH_3CH_2—C=CH_2$$
$$|$$
$$CH_3$$

  (i)   Name hydrocarbon **M**.
  (ii) Draw the repeating unit of the polymer which can be formed from **M**.
        State the type of polymerisation occurring in this reaction.

                                                                        (3 marks)

  (b) Draw the repeating unit of the polymer formed by the reaction between
      butanedioic acid and hexane-1,6-diamine. State the type of polymerisation
      occurring in this reaction and give a name for the linkage between the
      monomer units in this polymer.

                                                                        (4 marks)
                                                                        AQA, 2003

6 **(a)** Synthetic polyamides are produced by the reaction of dicarboxylic acids with
      compounds such as $H_2N(CH_2)_6NH_2$
      (i)   Name the compound $H_2N(CH_2)_6NH_2$
      (ii) Give the repeating unit in the polyamide nylon 6,6.

                                                                        (2 marks)

  (b) Synthetic polyamides have structures similar to those found in proteins.
      (i)   Draw the structure of 2-aminopropanoic acid.
      (ii) Draw the organic product formed by the condensation of two molecules of
            2-aminopropanoic acid.

                                                                        (2 marks)
                                                                        AQA, 2002

7 **(a)** Explain why polyalkenes are chemically inert.

                                                                        (2 marks)

  (b) Explain why polyesters and polyamides are biodegradeable.

                                                                        (2 marks)

  (c) Discuss the advantages of recycling polymers.

                                                                        (2 marks)

8 The displayed formula of two organic compounds are shown below.

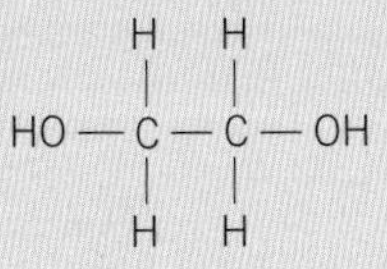

                    H   H   O   H   H   H   O
                    | | ‖ | | | ‖
            HO — C — C — OH   C — C — C — C — C
                    | | HO | | | OH
                    H   H   H   H   H
              monomer A 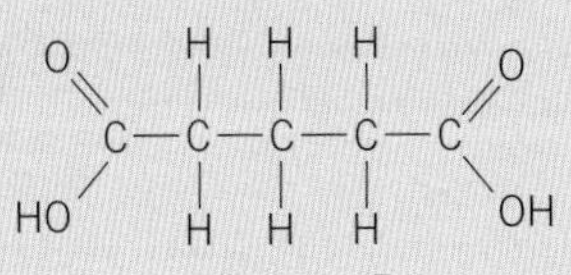   monomer B

  (a) **(i)**   Monomer A is diol. Name compound A.

                                                                        (1 mark)

      (ii) What type of compound is monomer B?

                                                                        (1 mark)

  (b) Monomer A and monomer B can react together to form a useful new substance
      named compound C.
      (i)   Draw a repeat unit of the new substance compound C.

                                                                        (1 mark)

      (ii) Circle the ester linkage in compound C.

                                                                        (1 mark)

      (iii) Name the non-organic product of this reaction.

                                                                        (1 mark)

      (iv) State the type of reaction that has taken place.

                                                                        (1 mark)

      (v) Suggest why a lab coat made from compound C may be damaged if
            concentrated sodium hydroxide was accidentally spilt on it.

                                                                        (1 mark)

30 Amino acids, proteins, and DNA
30.1 Introduction to amino acids

Learning objectives:

→ State what amino acids are.

→ Describe why they have both acidic and basic properties.

Specification reference: 3.3.13

Amino acids are the building blocks of proteins, which in turn are a vital component of all living systems.

Amino acids have two functional groups – a carboxylic acid and a primary amine. There are 20 important naturally occurring amino acids and they are all α-amino acids (also called 2-amino acids), which means that the amine group is on the carbon next to the $-CO_2H$ group (Figure 1).

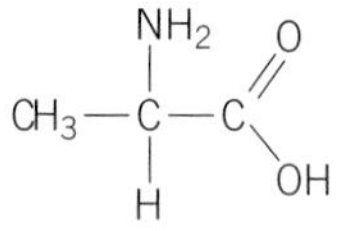

▲ **Figure 1** *α-aminopropanoic acid, also called alanine, written in shorthand as $CH_3CH(NH_2)COOH$*

α-amino acids have the general formula:

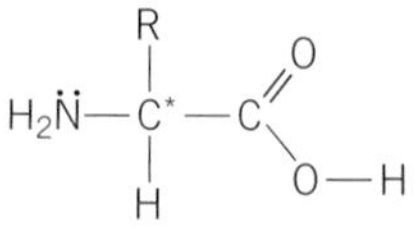

This structure has a carbon bonded to four different groups. The molecule is therefore chiral. Almost all naturally occurring amino acids exist as the (–) enantiomer.

Synoptic link

Look back at topic 25.2, Optical isomerism.

Acid and base properties

Amino acids have both an acidic and a basic functional group.

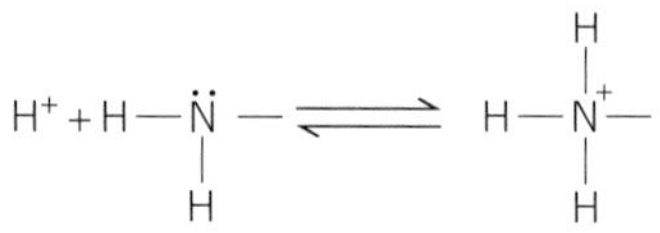

Hint

Compounds with two functional groups are called **bifunctional compounds**.

- The carboxylic acid group has a tendency to lose a proton (act as an acid):

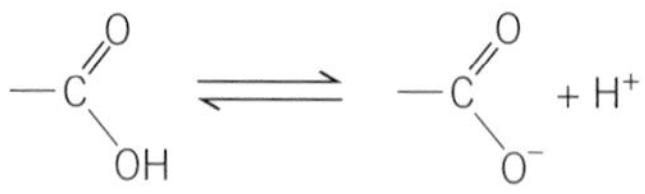

- The amine group has a tendency to accept a proton (act as a base):

Synoptic link

You will need to know the nature of ionic bonding and states of matter studied in Topic 3.7, Bonding and physical properties.

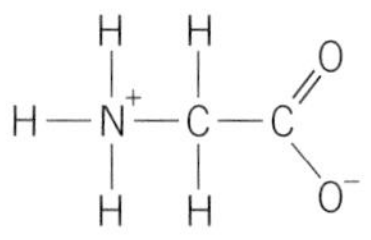

Amino acids exist as **zwitterions**. Ions like these have both a permanent positive charge and a permanent negative charge, though the compound is neutral overall (Figure 2).

Because they are ionic, amino acids have high melting points and dissolve well in water but poorly in non-polar solvents. A typical amino acid is a white solid at room temperature and behaves very much like an ionic salt.

In strongly acidic conditions the lone pair of the H_2N-group accepts a proton to form the positive ion (Figure 3)

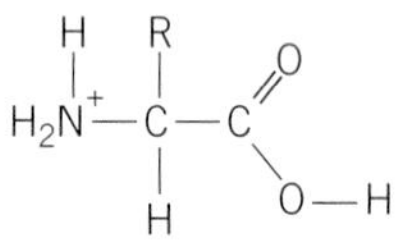

The amino group has gained a hydrogen ion – it is protonated.

In strongly alkaline solutions, the –OH group loses a proton to form the negative ion (Figure 4)

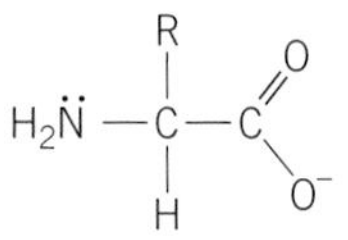

The carboxylic acid group has lost a hydrogen ion – it is **deprotonated**.

▲ **Figure 2** *A zwitterion*

▲ **Figure 3** *A protonated amino acid*

▲ **Figure 4** *A deprotonated amino acid*

Table 1 shows some naturally occurring amino acids. Each of these is usually referred to by its non-systematic name (the IUPAC names can be complex) and also by a three-letter abbreviation, which is useful when describing the sequences of amino acids in proteins, see Topic 30.2.

Formula	Name and abbreviation	Formula	Name and abbreviation
H_2NCHCO_2H $\|$ H	glycine (Gly)	H_2NCHCO_2H $\|$ CHOH $\|$ CH_3	threonine (Thr)
H_2NCHCO_2H $\|$ CH_3	alanine (Ala)	H_2NCHCO_2H $\|$ CH_2SH	cysteine (Cys)
H_2NCHCO_2H $\|$ $CHCH_3$ $\|$ CH_3	valine (Val)	H_2NCHCO_2H $\|$ CH_2 $\|$ $CONH_2$	asparagine (Asn)
H_2NCHCO_2H $\|$ CH_2 $\|$ $CH_3(CH_3)_2$	leucine (Leu)	H_2NCHCO_2H $\|$ CH_2 $\|$ CH_2CONH_2	glutamine (Gln)
H_2NCHCO_2H $\|$ CHC_2H_5 $\|$ CH_3	isoleucine (Ile)	H_2NCHCO_2H $\|$ CH_2—⬡—OH	tyrosine (Tyr)
HN—$CHCO_2H$ CH_2 CH_2 CH_2	proline (Pro) (proline is a secondary amine)	H_2NCHCO_2H $\|$ CH_2—C=CH HN N CH	histidine (His)
H_2NCHCO_2H $\|$ CH_2 ⬡⬡ C CH NH	tryptophan (Try)	H_2NCHCO_2H $\|$ $(CH_2)_3$ $\|$ NH $\|$ NH=C—NH_2	arginine (Arg)
H_2NCHCO_2H $\|$ CH_2 $\|$ CH_2SCH_3	methionine (Met)	H_2NCHCO_2H $\|$ $(CH_2)_3$ $\|$ CH_2NH_2	lysine (Lys)
H_2NCHCO_2H $\|$ CH_2—⬡	phenylalanine (Phe)	H_2NCHCO_2H $\|$ CH_2CO_2H	aspartic acid (Asp)
H_2NCHCO_2H $\|$ CH_2OH	serine (Ser)	H_2NCHCO_2H $\|$ CH_2 $\|$ CH_2CO_2H	glutamic acid (Glu)

Summary questions

1 The systematic name of glycine is 2-aminoethanoic acid. What is the systematic name of alanine (Table 1)?

2 Explain why alanine is chiral whereas glycine is not.

30.2 Peptides, polypeptides, and proteins

Learning objectives:

→ State what peptides are.

→ Describe how amino acids form proteins.

→ Describe the primary, secondary, and tertiary structures of proteins.

→ State what bonds hold protein molecules in their particular shapes.

→ Describe how proteins can be broken down.

Specification reference: 3.3.13

Amino acids link together to form peptides. Molecules containing up to about 50 amino acids are referred to as polypeptides. When there are more than 50 amino acids they are called proteins. Naturally occurring proteins are everywhere – enzymes, wool, hair, and muscles are all examples.

Amino acids and the peptide link

An amide has the functional group $-CONH_2$ or $-C\overset{O}{\underset{NH_2}{\big\backslash}}$

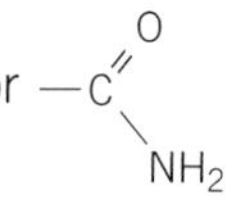

The amine group of one amino acid can react with the carboxylic acid group of another to form an **amide linkage** $-CONH-$.

This linkage is shown by shading in Figure 1.

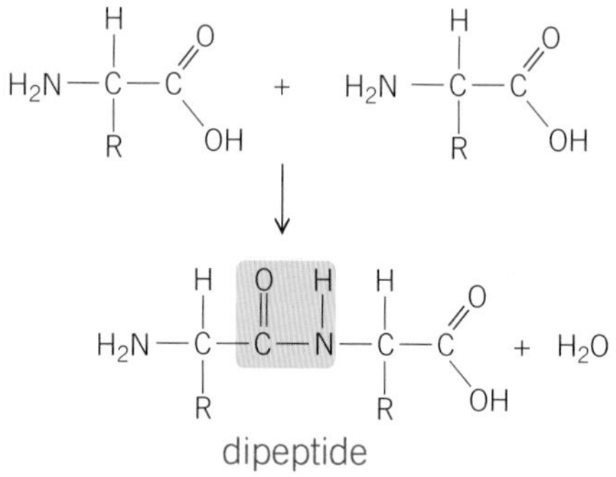

▲ **Figure 1** *Formation of a dipeptide*

Compounds formed by the linkage of amino acids are called **peptides**, and the amide linkage is called a peptide linkage in this context. A peptide with two amino acids is called a **dipeptide**. The dipeptide still retains $-NH_2$ and $-CO_2H$ groups and so can react further to give tri- and tetra-peptides, and so on (Figure 2).

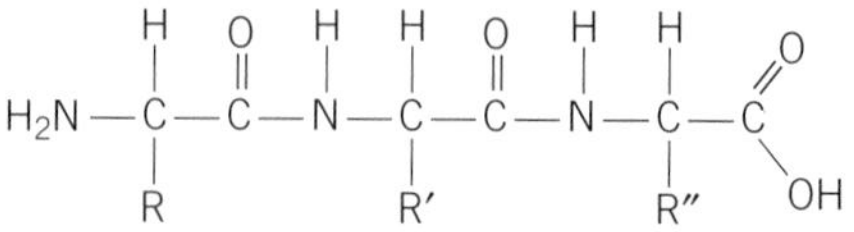

▲ **Figure 2** *A tripeptide – R, R', and R'' may be the same or different*

A particular protein will have a fixed sequence of amino acids in its chain. This is called the **primary structure** of the protein. For example, just one short sequence of the protein insulin (the hormone controlling sugar metabolism) runs:

-ala-glu-ala-leu-tyr-

Polypeptides and proteins are condensation polymers because a small molecule (in this case water) is eliminated as each link of the chain forms.

Hint

Table 1 in Topic 29.1 gives the names, formula, and the three letter abbreviations of twenty naturally occurring amino acids.

Synoptic link

You covered condensation polymers in Topic 29.1, Condensation polymers.

Study tip

Hydrolysis is a reaction with water (often boiling) that may be catalysed by an acid, an alkali, or an enzyme. As they can be hydrolysed, proteins are biodegradable.

Hydrolysis

When a protein or a peptide is boiled with hydrochloric acid of concentration $6 \, mol \, dm^{-3}$ for about 24 hours, it breaks down to a mixture of all the amino acids that made up the original protein or peptide. All the peptide linkages are hydrolysed by the acid (Figure 3).

▲ **Figure 3** *The hydrolysis of the peptide link*

The structure of proteins

Proteins have complex shapes that are held in position by hydrogen bonds and other intermolecular forces as well as sulfur–sulfur bonds. The shapes of proteins are vital to their functions, for example, as enzymes and structural materials in living things. Many proteins are helical (spiral). Hydrogen bonding holds the helix in shape (Figures 4 and 5).

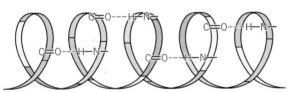

▲ **Figure 4** *The helical structure of a protein. Hydrogen bonds are shown as dotted lines*

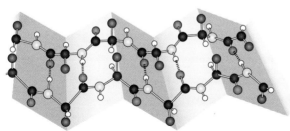

▲ **Figure 5** *The helix of a protein is held together by hydrogen bonding. The coloured strips represent amino acids (there are 18 to every 5 turns of the helix)*

Another arrangement of a protein is called pleating and the protein ends up as a pleated sheet. The hydrogen bonding is shown in Figure 6.

▲ **Figure 6** *A pleated sheet protein showing the hydrogen bonds as dotted lines*

> **Synoptic link**
>
> Bond polarity and intermolecular forces were covered in Topic 3.4, Electronegatvity – bond polarity in covalent bonds, and Topic 3.5, Forces acting between molecules.

➕ Hydrolysis by enzymes

Certain enzymes will partially hydrolyse specific proteins. For example, the enzyme trypsin will only break the peptide bonds formed by lysine and arginine. Detective work, based on this and other techniques, enables chemists to find the sequence of amino acids in different proteins. The first protein to be fully sequenced was insulin. Fred Sanger won the 1958 Nobel Prize for chemistry for this achievement. He also won the Prize in 1980 for sequencing of DNA (see Topic 29.4). He is the only person ever to win two Nobel Prizes for chemistry.

The stretchiness of wool

Wool is a protein fibre with a helix which is, as usual, held together by hydrogen bonds (Figure 7). When wool is gently stretched, the hydrogen bonds stretch (Figure 8) and the fibre extends. Releasing the tension allows the hydrogen bonds to return to their normal length and the fibre returns to its original shape. However, washing at high temperatures can permanently break the hydrogen bonds and a garment may permanently lose its shape.

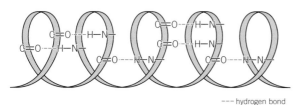

▲ **Figure 7** *Hydrogen bonds in wool*

--- hydrogen bond

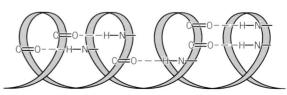

▲ **Figure 8** *The wool is gently stretched*

Bonding between amino acids

The amino acids in a protein chain can bond together in a number of ways.

- Hydrogen bonding between, for example, C=O groups and –N—H groups as shown C=O$\cdots$H—N.

- Ionic attractions between groups on the side chains of amino acids such as –COO$^-$ (for example, on glutamic acid) and –NH$_3^+$ (for example, on lysine).

- Sulfur–sulfur bonds. The amino acid cysteine has a side chain with an –CH$_2$SH group. Under suitable oxidising conditions, two cysteine molecules may react together to form sulfur–sulfur bond that forms a bridge between the two molecules and creates a double amino acid called cystine. This is called sulfur–sulfur bridging or a disulfide bridge.

$$-CH_2SH + HSCH_2- + [O] \rightarrow -CH_2S-SCH_2- + H_2O$$

> ### Hint
> Remember the use of [O] in equations to represent an oxidising agent.

Levels of protein structure

All proteins have three (and sometimes more) levels of structure – primary, secondary, and tertiary.

Primary structure

The sequence of amino acids along a protein chain is called its primary structure. It can be represented simply by the sequence of three-letter names of the relevant amino acids, for example, gly-ala-ala-val-leu, and so on. This structure is held together by covalent bonding. Therefore it is relatively stable – it requires harsh conditions such as boiling with 6 mol dm^{-3} hydrochloric acid to break the amino acids apart.

Secondary structure

A protein chain may form a helix (the α-helix) or a folded sheet (called a β-pleated sheet). This is called the secondary structure and is held in place by hydrogen bonding between, for example, C=O groups and –N–H groups. Hydrogen bonds are much weaker than covalent bonds and this level of structure can relatively easily be disrupted by gentle heating or changes in pH.

Tertiary structure

The α-helix or β-pleated sheet can itself be folded into a three dimensional shape – this is called the tertiary structure and is held in place by a mixture of hydrogen bonding, ionic interactions, and sulfur–sulfur bonds (as well as van der Waals forces which exist between all molecules). Figure 9 shows an example of part of the tertiary structure and the bonds that hold it in place.

Synoptic link

Make sure that you are confident about the different types of intermolecular forces which are described in Topic 3.5, Forces acting between molecules.

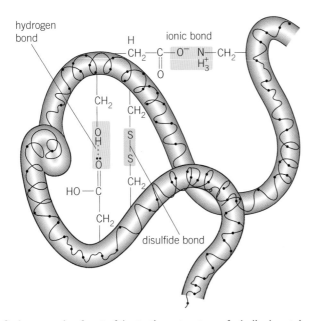

▲ **Figure 9** *An example of part of the tertiary structure of a helical protein*

Many proteins fold into globular shapes. The shapes of protein molecules are vital to their function – especially as enzymes (see Topic 30.3).

Finding the structure of proteins

The shapes of proteins are of great importance. Many techniques are used to determine the secondary and tertiary structures of proteins including X-ray diffraction, which can locate the actual positions of atoms in space. However, these are beyond the scope of this book. The first step in determining the *primary* structure is to find out the number of each type of amino acid present in the protein. To begin this process, the protein is refluxed with $6 \, mol \, dm^{-3}$ hydrochloric acid. This process is called hydrolysis. It breaks the amide bonds between the amino acids and results in a mixture containing all the individual amino acids in the original protein.

Hint

TLC plates may use glass or aluminium rather than plastic sheet. There are alternative materials to silica for the stationary phase. Plastic sheets are convenient because they can be cut to the required size with scissors.

Study tip

The R_f value has no units since both measurements are in cm and therefore cancel. All R_f values must be less than 1 since the spot cannot move further than the solvent.

Synoptic link

You will learn more about other chromatography techniques in Topic 33.1, Chromatography.

 ## Thin-layer chromatography

After hydrolysis, the amino acids can then be separated and identified by a technique called thin-layer chromatography (TLC). TLC is similar to paper chromatography but the paper is replaced by a chromatography plate consisting of a thin, flexible plastic sheet coated with a thin layer of silica (Figure 10). The IUPAC name of silica is silicon dioxide, SiO_2. This white powder is called the stationary phase.

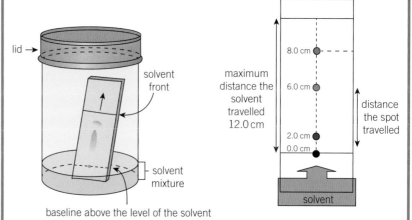

▲ **Figure 10** *A thin-layer chromatography experiment and the chromatogram that results*

1   A small spot containing the mixture of amino acids to be separated is placed on a line about 1 cm up the plate and the plate is placed in a tank containing a suitable solvent to a depth of about $\frac{1}{2}$ cm. The starting line must be above the initial level of the solvent. The solvent (or mixture of solvents) is called the mobile phase (or eluent).

2   A lid is placed on the tank so that the inside of it is saturated with solvent vapour and the solvent is allowed to rise up the plate (Figure 10). As it does so, it carries the amino acids with it. Each amino acid lags behind the solvent front to an extent that depends on its affinity for the solvent compared with its affinity for the stationary phase. This depends on the intermolecular forces that act between the amino acid and the solvent – the stronger they are, the closer the amino acid is to the solvent front.

3   When the solvent has almost reached the top of the plate, the plate is removed from the tank and the position to which the solvent front has moved is marked. Amino acids are colourless, so the positions they have reached have to be made visible. This is done by spraying the plate with a developing agent such as ninhydrin, which reacts with amino acids to form a purple compound, or by shining ultra-violet light on the plate. If the solvent is suitable, the amino acids will be completely separated.

R_f values are then calculated for each amino acid spot.

$$R_f = \frac{\text{distance moved by the spot}}{\text{distance moved by the solvent}}$$

So the R_f value for the red spot (8.0cm) in Figure 10 is $\frac{8.0\text{ cm}}{12.0\text{ cm}} = 0.67$

This allows each amino acid in the mixture to be identified by comparing the R_f value of each spot with the values obtained by known pure amino acids run in the same solvent mixture.

2-dimensional TLC

Often, two amino acids have very similar R_f values in a particular solvent. This makes it hard to distinguish them. One solution to this problem is to use 2-dimensional TLC. Here a square piece of TLC film is used. The plate is spotted in one corner and a chromatogram is run in the usual way so that the spots are separated along one side of the plate. The plate is then turned through 90° and the chromatogram is run again with a different solvent (Figure 11). This makes it easier to see the separation between the spots and gives two R_f values (one for each solvent). If both these values match those for a known amino acid, you can be more confident in your identification.

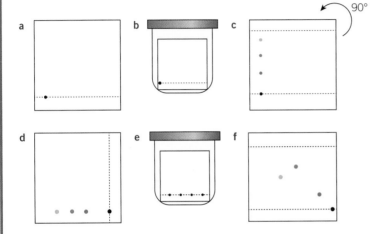

▲ **Figure 11** *The principle of 2-D TLC*

Calculate the R_f values of each of the amino acids represented by the orange and blue spots in Figure 11(f). You will need to use a ruler to measure the distances travelled.

Summary questions

1 **a** What are the functional groups in an amino acid?

   b Which group is acidic and which basic?

2 How many amide (peptide) linkages are there in a tripeptide?

3 In what form will amino acid residues exist after a protein has been hydrolysed with $6 \ mol \ dm^{-3}$ hydrochloric acid. Draw the structural formula of an alanine residue.

4 Draw the formulae of the three amino acids that would be formed by the hydrolysis of the tripeptide shown below.

$$H_2N-\underset{\underset{H}{|}}{\overset{\overset{H}{|}}{C}}-\overset{\overset{O}{||}}{C}-\underset{}{\overset{\overset{H}{|}}{N}}-\underset{\underset{CH_3}{|}}{\overset{\overset{H}{|}}{C}}-\overset{\overset{O}{||}}{C}-\underset{}{\overset{\overset{H}{|}}{N}}-\underset{\underset{CH(CH_3)_2}{|}}{\overset{\overset{H}{|}}{C}}-C\overset{\nearrow O}{\underset{\searrow OH}{}}$$

     gly   ala   val

30.3 Enzymes

Learning objectives:
→ Describe what an enzyme is.
→ Describe enzyme action.
Specification reference 3.3.13

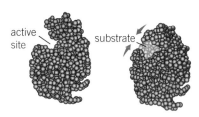

▲ **Figure 1** *The substrate of an enzyme fits its active site*

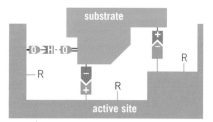

▲ **Figure 2** *Substrate bonding to the active site of an enzyme*

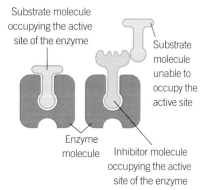

▲ **Figure 3** *Enzyme inhibition by a molecule of a similar structure to the substrate*

Summary question

1 State the two types of bonding between enzyme and substrate that are shown schematically in Figure 2.

2 State two other types of bonding that will also take place.

Enzymes are protein-based catalysts found in living things. They are enormously effective, speeding up reactions by factors of up to 10^{10}. This means that a reaction that takes place in one second when catalysed by an enzyme would take 300 years without it. A single enzyme molecule may catalyse the reaction of up to 500 000 molecules per second. Enzymes are also extremely specific – each enzyme is optimised for just one reaction.

Enzymes are usually globular proteins. Their shape has a cleft or crevice in it called the active site. This is where the reaction takes place. The reacting molecule (or molecules) fit precisely in the active site and are held in just the right orientation to react (Figure 1). Only molecules whose shape fits the active site can react. This is often called the lock and key hypothesis. The actual situation is a little more complex. The reacting molecule (called the substrate) must not only fit the shape of the active site, it must bond to it temporarily by intermolecular forces (Figure 2). Whilst bonded, these forces promote the movement of electrons within the substrate that lower the activation energy for the reaction.

Stereospecificity
The active site of an enzyme can be so selective of the shape of a substrate that many enzymes only catalyse reactions of one or other of a pair of enantiomers. These are called stereoisomers, so the enzyme is said to be **stereospecific**.

Enzyme inhibition
Enzymes control most of the chemical reactions in all living things, from the human body to microorganisms such as bacteria. Drugs can be designed which will affect their action. If you can block an enzyme that catalyses a harmful reaction, you can effectively stop the reaction.

Enzymes can easily be denatured (have their shapes changed) by changes in temperature and pH, but the conditions in the human body cannot be changed enough for this to happen.

A different way of destroying an enzyme's ability to catalyse a reaction is to devise a molecule of similar shape to its substrate. This molecule will bind to its active site and block the active site of the enzyme to the substrate (Figure 3). This is called enzyme inhibition and is the mode of action of some drugs. For example, penicillin inhibits the enzymes that control the building of cell walls in bacteria.

Computer modelling
Increasingly, chemists are beginning to understand the factors that influence the shapes of even extremely complex molecules such as proteins. They are now able to use sophisticated computer modelling techniques to predict the shapes of proteins even before they have been synthesised, and therefore predict their properties. This helps them to design drugs that may be used to treat a range of medical conditions.

DNA, short for deoxyribonucleic acid, is present in all cells and contains the blueprint from which living organisms are made. Everything living has its own particular DNA molecules so there is an infinite variety of DNA.

A single strand of DNA is a polymer made up from just four different monomers, but it is the way these are arranged that leads to this infinite variety of different end products.

The monomers – nucleotides

The monomers from which DNA is made are called nucleotides.

A nucleotide molecule is made up of three parts – a phosphate, a sugar, and a base. These are bonded together as shown.

Learning objectives:

→ State what a nucleotide is.

→ Describe how nucleotides bond together to form a single strand of DNA.

→ Describe how complementary strands of DNA are formed.

Specification reference 3.3.13

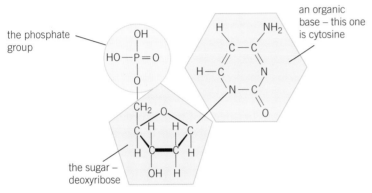

▲ **Figure 1** *The different parts of a nucleotide molecule*

> **Synoptic link**
>
> C, T, A, and G are bases because their nitrogen atoms have lone pairs of electrons which can accept a proton (see Topic 20.1, Defining an acid).

There are four different bases shown below called cytosine, thymine, adenine, and guanine. These are usually identified by their initials, C, T, A, and G (Figure 2).

Sometimes it is simpler just to show the basic skeleton of a nucleotide as:

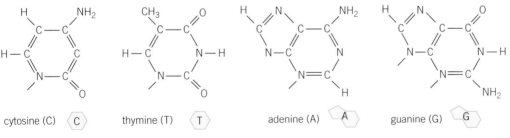

cytosine (C)   C   thymine (T)   T   adenine (A)   A   guanine (G)   G

▲ **Figure 2** *The four bases of DNA*

Polymerisation

Two nucleotides can link together when an –OH group of a phosphate on one nucleotide reacts with an –OH group on a sugar molecule on another nucleotide to eliminate a molecule of water as shown.

More nucleotide molecules can add on in the same way to form a polymer chain, which has a backbone of phosphate and sugar molecules with the bases attached (Figure 3).

The arrangement of the bases along this chain may be in any order and it is this that leads to the variety of DNA molecules.

Hint

This is an example of condensation polymerisation.

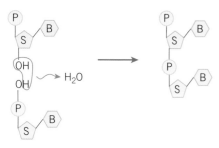

▲ **Figure 4** *A simplified model of the polymerisation of DNA bases*

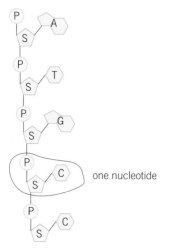

▲ **Figure 5** *Part of a single strand of DNA*

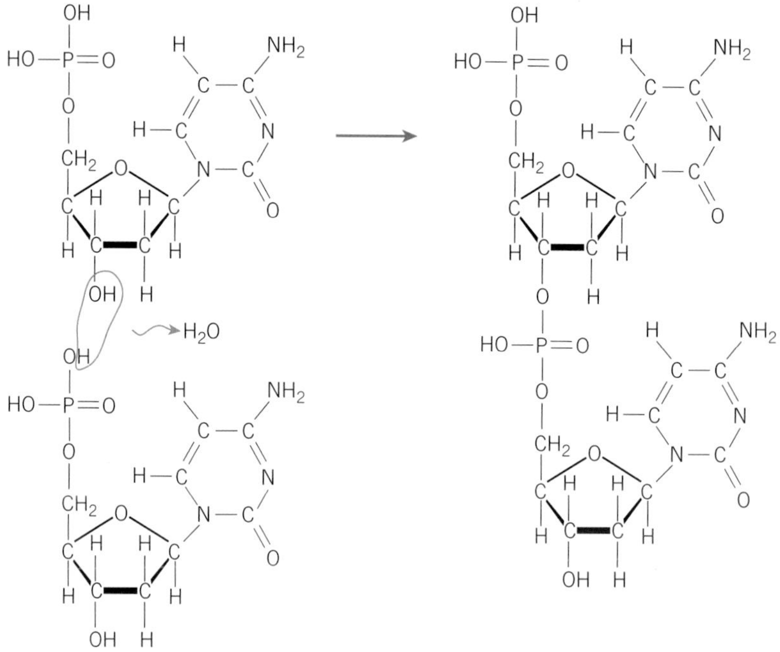

▲ **Figure 3** *Polymerisation of DNA bases*

So, DNA molecules can be specified by the order of the bases, for example, CCAGTTCAGGCTT, and so on. This is rather like a four-letter code.

Cells in the bodies of living things can read this alphabet, which holds the instructions for making a living thing. Chemists can now read the sequence of bases in material taken from living things, including humans. This is called DNA sequencing.

The double helix

DNA exists as two strands held together by hydrogen bonding (Figure 6).

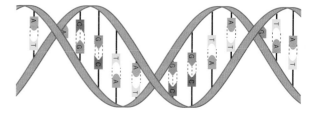

▲ **Figure 6** *A strand of the double helix of DNA held together by hydrogen bonding between A and T, and C and G*

The bases adenine and thymine can hydrogen bond with each other as can guanine and cytosine. Only these pairs can hydrogen bond together effectively (Figure 7).

This means that two DNA molecules can bond together forming a pair but only if their bases are in a complementary order – an A on one strand matching a T on the other and a G on one strand matching a C on the other. So if a strand of DNA had the base sequence CCAGTTCAGGCTT, the other, complementary strand would have the sequence GGTCAAGTCCGAA.

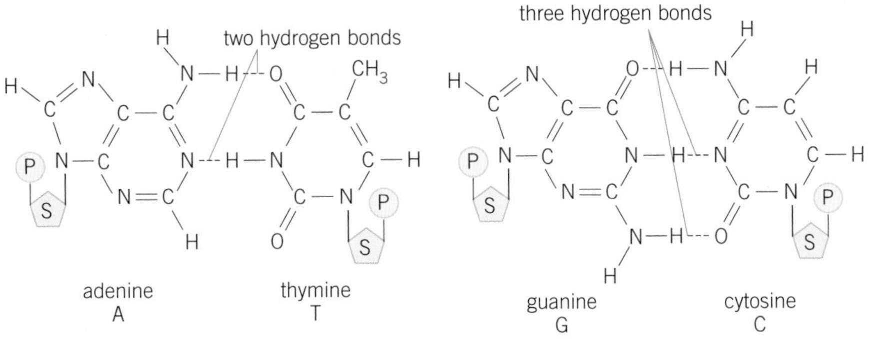

▲ **Figure 7** *Hydrogen bonding of adenine and thymine, and of guanine and cytosine*

The two strands would hydrogen bond together as shown:

C–C–A–G–T–T–C–A–G–G–C–T–T

G–G–T–C–A–A–G–T–C–C–G–A–A

The shapes of the molecules are such that, in order to fit neatly together, the two strands wind around each other in the shape of a double helix (Figure 6) .

Crick, Watson, and the Double Helix

Francis Crick and James Watson worked in Cambridge in the 1950s with molecular models to understand the idea of bases pairing by hydrogen bonding. The models had to be specially made – geometrically correct plastic modelling kits were not available at this time. They relied on data obtained by X-ray diffraction that showed that DNA molecules were spiral (helical) in shape. This data was obtained by Rosalind Franklin and Maurice Wilkins working in London. The relations between the two groups were often stormy. Franklin in particular had difficulties being accepted in a male-dominated world that was less -accepting of women than would be the case today.

Crick, Watson, and Wilkins won the Nobel Prize in 1962 for their work. Sadly, Rosalind Franklin died of cancer before the Prize was awarded – Nobel Prizes are not awarded posthumously. Most people believe that she would have been honoured had she lived.

DNA, amino acids, and proteins

The sequence of bases along a DNA chain acts as a template for arranging amino acids into protein chains. Three-base sections of DNA, called codons, each represent a particular amino acid in a protein sequence. For example, the codon GGA codes for the amino acid glycine.

Draw the displayed formula of glycine.

Cell division

DNA holds the instructions for making an organism. It is contained in every cell, and for the organism to function correctly, must be copied exactly when cells divide.

Hydrogen bonds are only about 10% of the strength of covalent bonds. This means that they can break under conditions that leave the covalent bonds of the DNA chain unaffected.

During cell division, the hydrogen bonds of the DNA double helix break and the strands start to unravel but the covalently bonded chain remains intact retaining the sequence of bases.

The cell contains a mixture of separate nucleotide molecules. These move in and pair up with the newly-exposed bases – T to A and C to G. The new bases then link together by phosphate-sugar bonding. This results in two double helix molecules each exactly the same as the original one. This process is called **replication** (Figure 8).

How is the information used?

The information contained by DNA molecules is used in the cell as a 'recipe' for making proteins (poly-amino acids), which are the basis of living things (see Topic 30.2). They form structural material such as flesh and also the enzymes that control the reactions by which other parts of the organism are made.

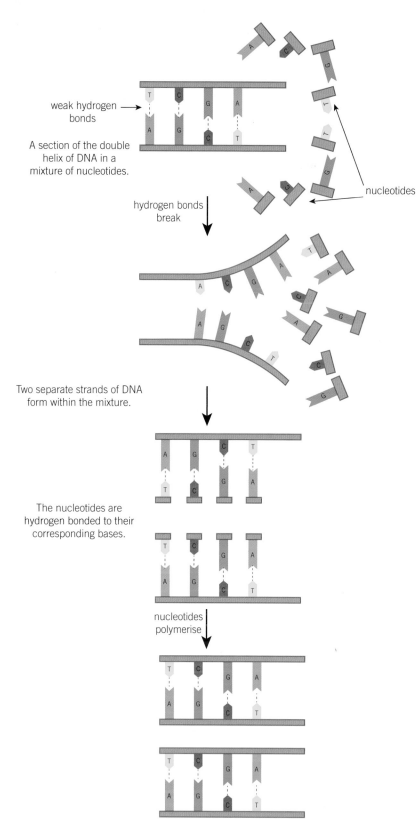

weak hydrogen bonds

A section of the double helix of DNA in a mixture of nucleotides.

hydrogen bonds break

nucleotides

Two separate strands of DNA form within the mixture.

The nucleotides are hydrogen bonded to their corresponding bases.

nucleotides polymerise

▲ **Figure 8** *Replication of DNA*

Summary questions

1 State the complementary sequence of bases to the following stretch of DNA
CCAGTTGACC

2 Explain why DNA (deoxyribonucleic acid) is acidic.

Cisplatin, an anti-cancer drug

Cancer is not one disease but many. What cancers have in common is 'rogue' cells which have lost control over their growth and replication and grow much faster than normal cells. Cisplatin was discovered in 1965 and is one of the most successful cancer treatments, for example, giving survival rates of up to 90% in testicular cancer.

Cisplatin is square planar and has the formula:

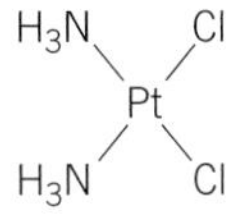

It works by bonding to strands of DNA, distorting their shape and preventing replication of the cells. The molecule bonds to nitrogen atoms on two adjacent guanine bases on a strand of DNA.

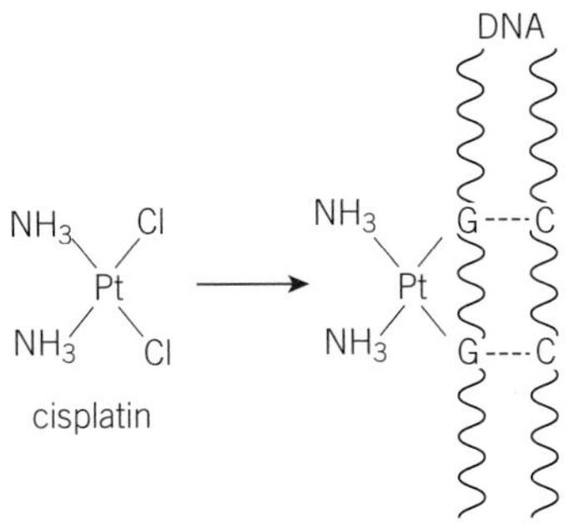

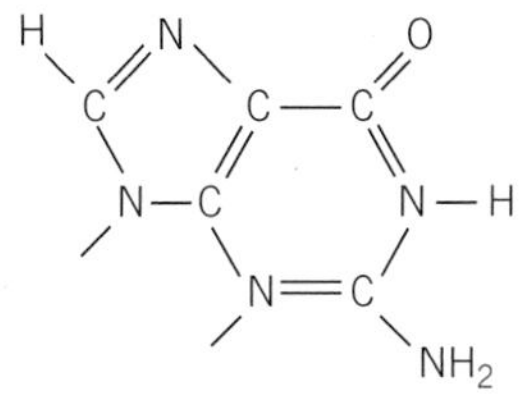

Learning objective:

→ Describe how the anti-cancer drug cisplatin works.

Specification reference 3.3.13

▲ **Figure 1** *The structure of the guanine base*

This works because the nitrogen atoms of the guanine molecules have lone pairs of electrons which form dative covalent bonds with the platinum. They displace the chloride ions because they are better ligands. This is an example of a ligand substitution reaction.

Like all drugs, cisplatin has side effects – it will bond to DNA in healthy cells as well as in cancerous ones but cancer cells are replicating faster than healthy cells, and so the effect of the drug is greater on cancer cell, than on normal cells. However, healthy cells that replicate quickly, such as hair follicles, are significantly affected and this is why patients undergoing chemotherapy (drug treatment for cancer) often lose their hair. Work is underway to find drugs and delivery systems that can better discriminate between healthy and cancerous cells.

Synoptic link

Ligand substitution reactions where covered in Topic 24.2, Ligand substitution reactions.

Summary questions

1 Draw the formula of transplatin and suggest why it is not an effective anti-cancer drug.

1 **(a)** The structure of the amino acid alanine is shown below.

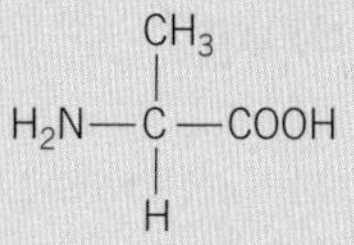

CH$_3$
|
H$_2$N—C—COOH
|
H

     (i)   Draw the structure of the zwitterion formed by alanine.
     (ii) Draw the structure of the organic product formed in each case from alanine
          when it reacts with:
         •   CH$_3$OH, in the presence of a small amount of concentrated sulfuric acid.
         •   Na$_2$CO$_3$
         •   CH$_3$Cl in a 1 : 1 mole ratio.

(4 marks)

(b) The amino acid lysine is shown below.

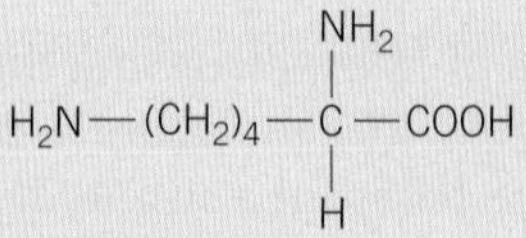

NH$_2$
|
H$_2$N—(CH$_2$)$_4$—C—COOH
|
H

Draw the structure of the lysine species present in a solution at low pH.

(1 mark)

(c) The amino acid proline is shown below.

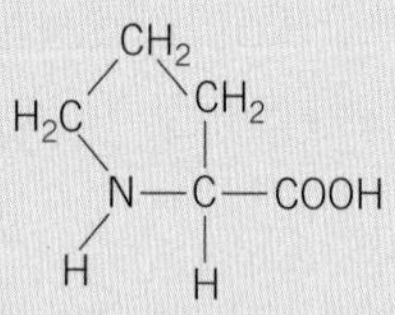

CH$_2$
H$_2$C   CH$_2$
    N—C—COOH
    | |
    H   H

Draw the structure of the dipeptide formed from two proline molecules.

(1 mark)

AQA, 2007

2 Draw the structures of the **two** dipeptides which can form when one of the amino acids
shown below reacts with the other.

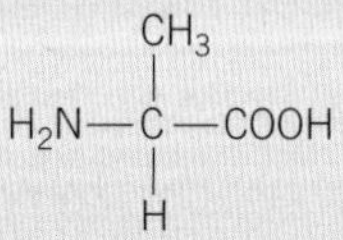

CH$_3$ CH$_2$OH
| |
H$_2$N—C—COOH   H$_2$N—C—COOH
| |
H   H

structure 1 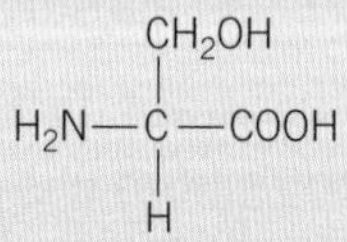   structure 2

(2 marks)

AQA, 2006

3 Consider the following amino acid.

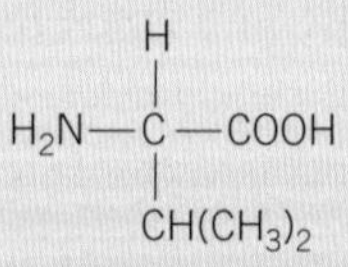

H
|
H$_2$N—C—COOH
|
CH(CH$_3$)$_2$

(a) Draw the structure of the amino acid present in the solution at pH 12.
(b) Draw the structure of the dipeptide formed from two molecules of this amino acid.
(c) Protein chains are often arranged in the shape of a helix. Name the type of
interaction that is responsible for holding the protein chain in this shape.

(3 marks)

AQA, 2004

4 The structures of the amino acids alanine and glycine are shown below.

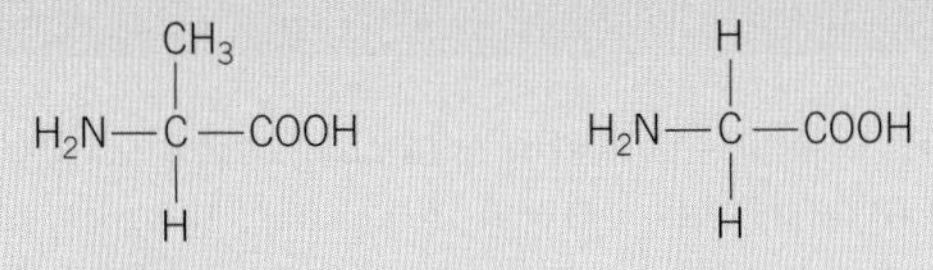

alanine   glycine

Alanine exists as a pair of stereoisomers.
(a) Explain the meaning of the term stereoisomers.
(b) State how you could distinguish between the stereoisomers.

(4 marks)
AQA, 2003

5 The anticancer drug cisplatin operates by reacting with the guanine in DNA.
Figure 1 shows a small part of a single strand of DNA. Some lone pairs are shown.

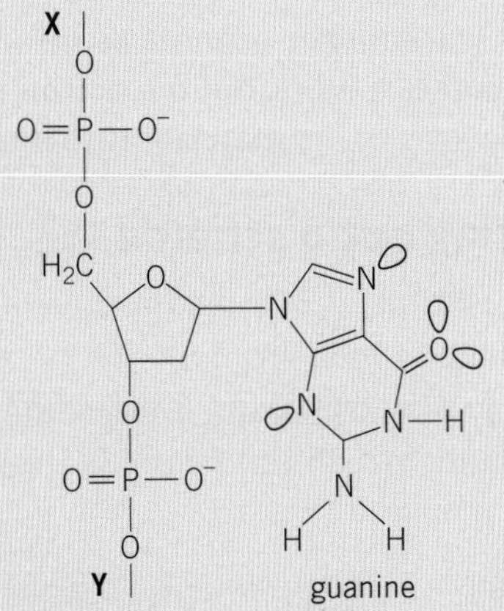

guanine

▲ **Figure 1**

The DNA chain continues with bonds at **X** and **Y**.
State the name of the molecule that is attached to the bond at **X**.

(1 mark)

Figure 2 shows two more bases found in DNA.

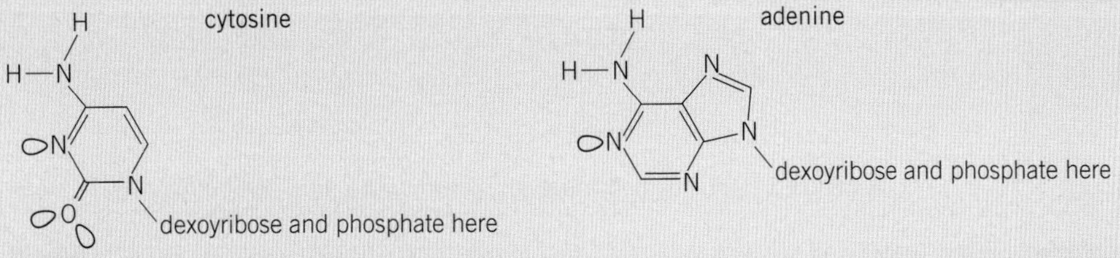

cytosine

adenine

dexoyribose and phosphate here

dexoyribose and phosphate here

▲ **Figure 2**

State which of these two bases, cytosine or adenine, pairs with the guanine in
Figure 1 when two separate strands of DNA form a double helix.

(1 mark)

Explain how the base that you have chosen forms a base pair with guanine.

(3 marks)

Cisplatin works because one of the atoms on guanine can form a co-ordinate bond with
platinum, replacing one of the ammonia or chloride ligands. Another atom on another
guanine can also form a co-ordinate bond with the same platinum by replacing another
ligand.
Explain how the action of cisplatin is able to stop the growth of cancer cells.

(3 marks)
AQA Specimen

This chapter is about working out a series of reactions for making (synthesising) a given molecule, usually called the **target molecule**.

Synthesis of a target molecule is a common problem in industries like drug or pesticide manufacture. Suppose a molecule is found to have a particular effect, for example, as an antibiotic. Drug companies may synthesise, on a small scale, a number of compounds of similar structures. These will be screened for possible antibiotic properties. Any promising compounds may then be made in larger quantities for thorough investigation of their effectiveness, safety, side effects, and so on, before the final step goes ahead – producing them commercially.

> Using the organic reactions you have already met, you can work out a reaction scheme to convert a starting material into a target molecule.

Working out a scheme

Start by writing down the formula of the starting molecule, A, and that of the target molecule, X.

One way of working out what route to take is to write down all the compounds which can be made from A and all the ways in which X can be prepared (Figure 1).

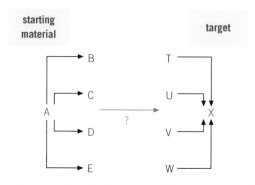

▲ **Figure 1** *Devising a synthesis of compound X from compound A*

You may then see how B, C, D, or E can be converted, in one or more steps to T, U, V, W, or direct to X. It is important to keep the number of steps as small as possible to maximise the yield of the target.

Sometimes you will be able to see straight away that a particular reaction will be needed. For example, if the target molecule has one more carbon atom than the starting material, it is probable that the reaction of cyanide ions with a halogenoalkane will be needed at some stage, as this reaction increases the length of the carbon chain by one, for example:

$$CH_3Br + CN^- \rightarrow CH_3C{\equiv}N + Br^-$$

$\quad\quad$ bromomethane $\quad\quad\quad$ ethanenitrile

How the functional groups are connected

The inter-relationships between the functional groups you should know are shown in Figure 2. Make sure you can recall the reagents and conditions for each conversion.

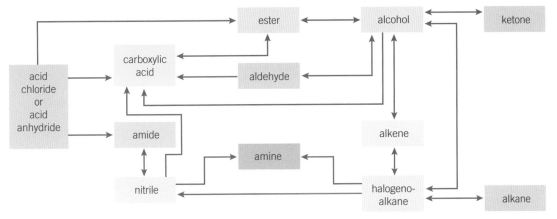

▲ **Figure 2** *Inter-relationships between functional groups. You can use this chart to revise your knowledge of organic reactions*

Synthetic robots

Routine chemical synthesis in the pharmaceutical industry is now often done by robots — not androids but arrays of reaction tubes along with computer-controlled syringes to measure out and mix the reactants. This produces a 'library' of related compounds. For example, you could oxidise several alcohols of different chain lengths to produce a library of aldehydes. The target compound can then be tested to see if any of them have any potential for use as medicines. A chemist is still needed to work out the reaction and program the computer.

Reagents used in organic chemistry

Oxidising agents

Potassium dichromate(VI), $K_2Cr_2O_7$, acidified with dilute sulfuric acid will oxidise primary alcohols to aldehydes, and aldehydes to carboxylic acids. Secondary alcohols are oxidised to ketones.

Reducing agents

Different reducing reagents have different capabilities:

- Sodium tetrahydridoborate(III), $NaBH_4$, will reduce C=O but not C=C. It can be used in aqueous solution. This reducing agent will reduce polar unsaturated groups, such as $C^{\delta+}=O^{\delta-}$, but not non-polar ones, such as C=C. This is because it generates the nucleophile :H^- which attacks $C^{\delta+}$ but is repelled by the electron-rich C=C.

- Hydrogen with a nickel catalyst, H_2/Ni, is used to reduce C=C but not C=O.

- Tin and hydrochloric acid, Sn/H^+, may be used to reduce R—NO_2 to R—NH_2.

Dehydrating agents

Alcohols can be converted to alkenes by passing their vapours over heated aluminium oxide or by acid-catalysed elimination reactions.

Examples of reaction schemes

1 How can propanoic acid be synthesised from 1-bromopropane?

Both the starting material and the target have the same number of carbon atoms, so no alteration to the carbon skeleton is needed.

Write down all the compounds which can be made in one step from 1-bromopropane and all those from which propanoic acid can be made in one step as shown in Figure 3. You may use Figure 2 to help you.

In this case two of the compounds are the same – the ones in red.

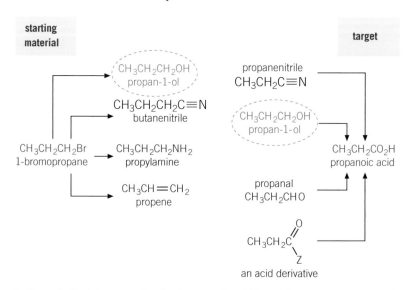

▲ Figure 3 Devising a synthesis of propanoic acid from 1-bromopropane

So, 1-bromopropane can be converted into propan-1-ol which can be converted into propanoic acid. The conversion required can be done in two steps:

Step 1 $CH_3CH_2CH_2Br \xrightarrow{\text{reflux with NaOH(aq)}} CH_3CH_2CH_2OH$

1-bromopropane   propan-1-ol

Step 2 $CH_3CH_2CH_2OH \xrightarrow{\text{reflux with K}_2\text{Cr}_2\text{O}_7\text{/H}^+} CH_3CH_2CO_2H$

propan-1-ol   propanoic acid

Both these reactions have a good yield.

2 How can propylamine by synthesised from ethene?

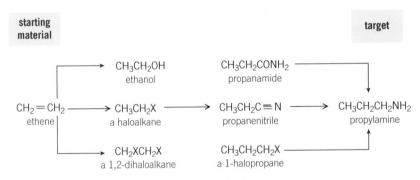

Propylamine has one more carbon atom than ethene. This suggests that the formation of a nitrile is involved at some stage.

Write down all the compounds that can be made from ethene and all the compounds from which the propylamine can be made (Figure 4).

starting material   target

CH_3CH_2OH — ethanol

$CH_3CH_2CONH_2$ — propanamide

$CH_2=CH_2$ — ethene

CH_3CH_2X — a haloalkane

$CH_3CH_2C\equiv N$ — propanenitrile

$CH_3CH_2CH_2NH_2$ — propylamine

CH_2XCH_2X — a 1,2-dihaloalkane

$CH_3CH_2CH_2X$ — a 1-halopropane

▲ **Figure 4** *Devising a synthesis of propylamine from ethene*

In this instance, no compound that can be made in one step from the starting material, can be converted into the product, so more than two steps must be required. You already know that the formation of a nitrile is required. A halogenoethane can be converted into propanenitrile so the synthesis can be completed in three steps:

Step 1 $CH_2CH_2 \xrightarrow{\text{HBr}} CH_3CH_2Br$

ethene   bromoethane

Step 2 $CH_3CH_2Br \xrightarrow{\text{KCN/dil. H}_2\text{SO}_4} CH_3CH_2C\equiv N$

bromoethane   propanenitrile

Step 3 $CH_3CH_2C\equiv N \xrightarrow{\text{Ni/H}_2} CH_3CH_2CH_2NH_2$

propanenitrile   propylamine

Chloroethane or iodoethane could have been chosen instead of bromoethane.

Aromatic reactions

Figure 5 summarises some of the important reactions of aromatic compounds using benzene as the starting material.

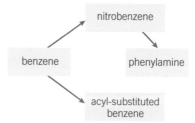

▲ **Figure 5** *Some inter-relationships between functional groups in aromatic compounds. Make sure you can recall the reagents and conditions for each conversion*

Summary questions

1   Give a one step reaction to convert:

   a   1-bromobutane to pentanenitrile

   b   ethanoic acid to methyl ethanoate

   c   but-1-ene to butan-2-ol

   d   cyclohexanol to cyclohexene.

2   Give a two step reaction to convert:

   a   ethene to ethanoic acid

   b   propanone to 2-bromopropane.

3   For each step, name the type of reaction taking place and the reagents required.

COC_2H_5 COC_2H_5 COC_2H_5

Step 1 → Step 2 → Step 3 → NO_2 NH_2

When identifying an organic compound, you need to know the functional groups present.

Chemical reactions

Some tests are very straightforward:

- Is the compound acidic (suggests carboxylic acid)?
- Is the compound solid (suggests long carbon chain or ionic bonding), liquid (suggests medium length carbon chain or polar or hydrogen bonding), or gas (suggests short carbon chain, little or no polarity)?
- Does the compound dissolve in water (suggests polar groups) or not (suggests no polar groups)?
- Does the compound burn with a smoky flame (suggests high $C:H$ ratio, possibly aromatic) or non-smoky flame (suggests low $C:H$ ratio, probably non-aromatic)?

Some specific chemical tests are listed in Table 1.

▼ **Table 1** *Chemical tests for functional groups*

Functional group	Test	Result
alkene $-C{=}C-$	shake with bromine water	red-brown colour disappears
halogenoalkane R—X	1. add NaOH(aq) and warm 2. acidify with HNO_3 3. add $AgNO_3$(aq)	precipitate of AgX
alcohol R—OH	add acidified $K_2Cr_2O_7$	orange colour turns green with primary or secondary alcohols (also with aldehydes)
aldehyde R—CHO	warm with Fehling's solution or warm with Tollens' solution or add acidified $K_2Cr_2O_7$	blue colour turns to red precipitate silver mirror forms orange colour turns green
carboxylic acid R—COOH	add $NaHCO_3$(aq)	bubbles observed as carbon dioxide given off

Learning objective:

→ Describe how organic groups can be identified.

Specification reference: 3.3.6

Synoptic link

You will need to know all the organic chemistry studied in your A Level course.

Synoptic link

Look back at Topic 10.3, Reactions of halide ions, for more detail on how the silver precipitate, AgX, can be used to identify the halogen.

Hint ⚗

You cannot use this test to identify a fluoroalkane as silver(I) fluoride, AgF, is soluble in water.

Summary questions

1 How could you tell if R—X was a chloroalkane, a bromoalkane, or an iodoalkane?

2 ⚗ In the test for a halogenoalkane:

   a Explain why it is necessary to acidify with dilute acid before adding silver nitrate.

   b Why would acidifying with hydrochloric acid not be suitable?

3 A compound decolourises bromine solution and fizzes when sodium hydrogencarbonate solution is added:

   a What two functional groups does it have?

   b Its relative molecular mass is 72. What is its structural formula?

   c Give equations for the two reactions.

1   Describe how you could distinguish between the compounds in the following pairs using **one** simple test-tube reaction in each case.
    For each pair, identify a reagent and state what you would observe when both compounds are tested separately with this reagent.

(a)
$$H_3C-\underset{\underset{CH_3}{|}}{\overset{\overset{CH_3}{|}}{C}}-CH_2OH \qquad H_3C-\underset{\underset{OH}{|}}{\overset{\overset{CH_3}{|}}{C}}-CH_2CH_3$$

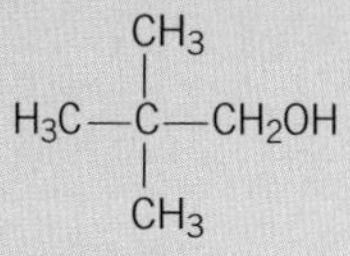

         R 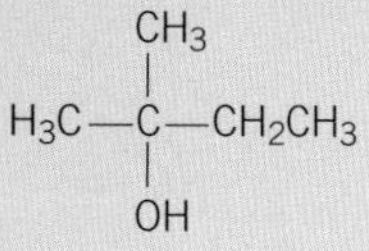   S

(3 marks)

(b)
$$O=C\overset{CH_3}{\underset{OCH_2CH_3}{\big\langle}} \qquad O=C\overset{OH}{\underset{CH_2CH_3}{\big\langle}}$$

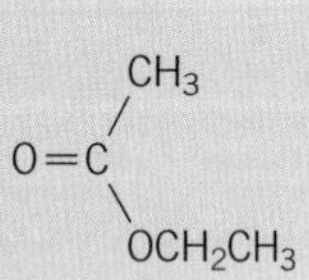

         T 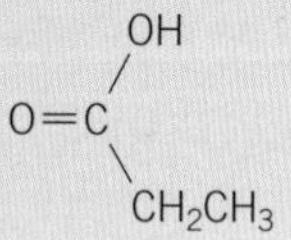   U

(3 marks)

(c) $H_3C-\underset{\underset{O}{\|}}{C}-CH_2-\underset{\underset{O}{\|}}{C}-CH_3 \qquad H-\underset{\underset{O}{\|}}{C}-CH_2-\underset{\underset{O}{\|}}{C}-H$

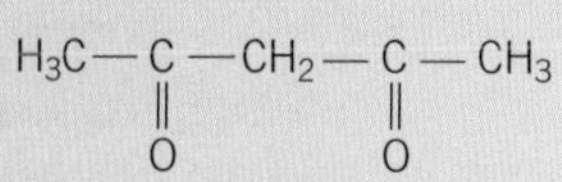

         V 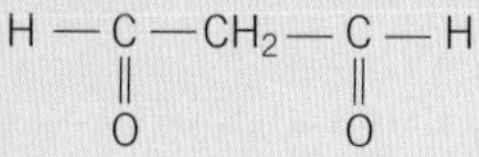   W

(3 marks)
AQA, 2013

2   (a)   A chemist discovered four unlabelled bottles of liquid, each of which contained a different pure organic compound. The compounds were known to be propan-1-ol, propanal, propanoic acid, and 1-chloropropane.
        Describe four **different** test-tube reactions, one for each compound that could be used to identify the four organic compounds.
        Your answer should include the name of the organic compound, the reagent(s) used, and the expected observation for each test.

(8 marks)
AQA, 2012

3   Chemists have to design synthetic routes to convert one organic compound into another.
    Propanone can be converted into 2-bromopropane by a three-step synthesis.
    Step 1: propanone is reduced to compound **L**.
    Step 2: compound **L** is converted into compound **M**.
    Step 3: compound **M** reacts to form 2-bromopropane.
    Deduce the structure of compounds **L** and **M**.
    For each of the three steps, suggest a reagent that could be used and name the mechanism.
    Equations and curly arrow mechanisms are **not** required.

(8 marks)
AQA, 2012

4 **(a)** Complete the diagram by giving the structural formula of the product in each of the boxes provided.

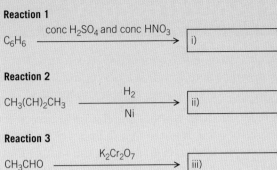

Reaction 1

C_6H_6 $\xrightarrow{\text{conc } H_2SO_4 \text{ and conc } HNO_3}$ i)

Reaction 2

$CH_3(CH)_2CH_3$ $\xrightarrow[\text{Ni}]{H_2}$ ii)

Reaction 3

CH_3CHO $\xrightarrow[\text{dilute}]{K_2Cr_2O_7}$ iii)

(3 marks)

(b)

(i) State the role of the concentrated sulfuric acid in **Reaction 1**.

(1 mark)

(ii) State the role of the nickel in **Reaction 2**.

(1 mark)

(iii) Why is potassium dichromate(VI) used in **Reaction 3**?

(1 mark)

5 A chemist is given a sample of a halogenoalkane labelled compound A. Explain how the chemist could test to see if compound A was a chloroalkane. Describe the test the chemist could carry out and how they could use the results of the test to confirm whether or not compound A is a chloroalkane.

(4 marks)

6 One mole of compound X has a mass of 58.0 g. A chemist tests the compound by warming a sample of X with Fehling's solution. The chemist observes that the Fehling's solution turns from a blue solution to a red precipitate.

(a) What type of substance is compound X.

(1 mark)

(b) Name compound X

(1 mark)

7 Describe how a chemist could test for the presence of the alkene functional group. Describe the how to carry out the test and how to interpret the results of the test.

(2 marks)

32.1 Nuclear magnetic resonance (NMR) spectroscopy

Learning objectives:

→ Explain the principles of NMR.

→ Describe the ^{13}C NMR spectrum.

→ Explain the chemical shift.

→ Describe what information a ^{13}C NMR spectrum gives.

Specification reference: 3.3.15

Nuclear magnetic resonance spectroscopy (NMR) is used particularly in organic chemistry. It is a powerful technique that can help find the structures of even quite complex molecules.

A magnetic field is applied to a sample, which is surrounded by a source of radio waves and a radio receiver. This generates an energy change in the nuclei of atoms in the sample that can be detected. Electromagnetic energy is emitted, which can then be interpreted by a computer.

A brief theory of NMR

Although you will only be examined on *interpreting* NMR spectra, this background reading may help you to understand how NMR works, although in some respects it is an oversimplification.

Many nuclei with odd mass numbers, such as ^{1}H, ^{13}C, ^{15}N, ^{19}F, and ^{31}P, have the property of *spin* (as do electrons). This gives them a magnetic field like that of a bar magnet.

If bar magnets are placed in an external magnetic field, they will line up parallel to the field (Figure 1a).

It is also possible that the bar magnets could line up anti-parallel to the field, as in (Figure 1b) but this orientation has a higher energy as the bar magnets have to be forced into position against the repulsion of the external magnetic field. The stronger the external magnetic field and the stronger the bar magnets, the larger the energy gap between the parallel and anti-parallel states.

Something similar applies to nuclei with spin, such as ^{1}H and ^{13}C. There will be some of the nuclei in each energy state but more of them will be in the lower (parallel) one. If electromagnetic energy just equal in energy to the difference between the two positions (ΔE in Figure 2) is supplied, some nuclei will flip between the parallel and anti-parallel positions. This is called **resonance**. The energy required to cause this is in the radio region of the electromagnetic spectrum. It is supplied by a radio frequency source, and the resonances are detected by a radio receiver (Figure 3). The frequency of the radio waves required to cause flipping for a particular magnetic field is called the **resonant frequency** of that atomic nucleus. A higher frequency corresponds to a larger energy gap between the two states. If the magnetic field is kept constant and the radio frequency gradually increased, different atomic nuclei will come into resonance at different frequencies depending on the strength of their atomic magnets.

In fact, modern instruments use pulses of radio waves of a range of frequencies all at once and analyse the response by a computer technique called Fourier transformation, but the principle remains of finding the frequencies at which different nuclei resonate.

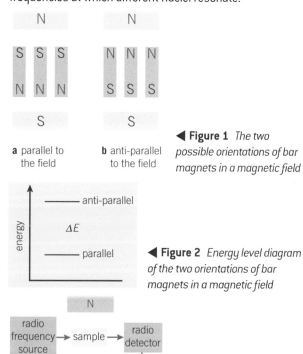

◀ **Figure 1** *The two possible orientations of bar magnets in a magnetic field*

a parallel to the field   **b** anti-parallel to the field

◀ **Figure 2** *Energy level diagram of the two orientations of bar magnets in a magnetic field*

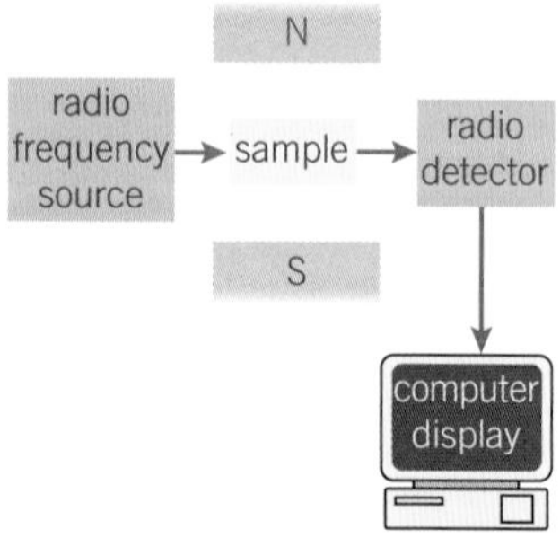

◀ **Figure 3** *Schematic diagram of an NMR spectrometer*

What do you notice about the relative atomic masses of the nuclei with spin?

Carbon-13, ^{13}C, NMR

NMR is most often used with organic compounds. Although carbon-12, ^{12}C, has no nuclear spin carbon-13, ^{13}C, does have one. Whilst only 1% of carbon atoms are carbon-13, modern instruments are sensitive enough to obtain a carbon-13 spectrum.

Not all the carbon-13 atoms in a molecule resonate at exactly the same magnetic field strength. Carbon atoms in different functional groups feel the magnetic field differently. This is because all nuclei are **shielded** from the external magnetic field by the electrons that surround them. Nuclei with more electrons around them are better shielded. The greater the electron density around a carbon-13 atom, the smaller the magnetic field felt by the nucleus and the lower the frequency at which it resonates. The NMR instrument produces a graph of energy absorbed (from the radio signal) vertically against a quantity called **chemical shift** (which is related to the resonant frequency) horizontally.

The chemical shift

Chemical shift δ is measured in units called parts per million (ppm) from a defined zero related to a compound called tetramethylsilane, TMS (see Topic 31.2). Chemical shift is related to the difference in frequency between the resonating nucleus and that of TMS. In ^{13}C NMR values of δ range from 0 to around 200 ppm.

The main point about ^{13}C NMR is that carbon atoms in different environments will give different chemical shift values. Figure 5 shows the ^{13}C NMR spectrum of ethanol. It has two peaks, one for each carbon, because the carbon atoms are in different environments – one is further from the oxygen atom than the other. The oxygen atom, being electronegative, draws electrons away from the carbon atom to which it is directly bonded.

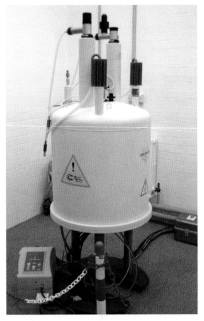

▲ **Figure 4** *Modern NMR instruments use electromagnets with superconducting coils to produce the strong magnetic fields required. The large white tank holds a jacket of liquid nitrogen surrounding an inner jacket of liquid helium which cools the magnet coils to 4 K*

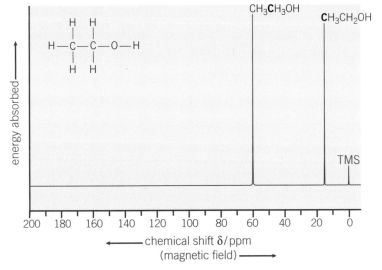

▲ **Figure 5** *Carbon-13 spectrum of ethanol*

Hint

On NMR spectra the chemical shift *increases* from right to left.

Table 1 shows values of ^{13}C chemical shifts for carbon atoms in a variety of environments. The carbon atom at $\delta = 60$ ppm in the ethanol spectrum is the carbon bonded to the oxygen ($CH_3\mathbf{C}H_2OH$), whilst that at $\delta = 15$ ppm is the other carbon ($\mathbf{C}H_3CH_2OH$).

This is because the electronegative oxygen atom draws electrons away from the carbon bonded to it $CH_3\mathbf{C}H_2OH$. It is deshielded and feels a greater magnetic field and so resonates at a higher frequency and therefore has a *greater* δ value than the other carbon. The other carbon $\mathbf{C}H_3CH_2OH$ is surrounded by more electrons and therefore shielded and has a *smaller* δ value.

More examples of ^{13}C NMR spectra

Figures 6 and 7 show the ^{13}C NMR spectra of the isomers propanone, CH_3COCH_3, and propanal, CH_3CH_2CHO. In propanone, there are just two different environments for the carbon atoms – the two CH_3 groups and the C=O. The spectrum shows two peaks:

- At δ = 205 ppm due to the C=O.
- At δ = 30 ppm due to the CH_3 groups.

Propanal has three different carbon environments and so shows three peaks:

- The CH_3 group at δ = 5 ppm.
- The CH_2 at δ = 37.
- The CHO group at δ = 205 ppm.

▼ **Table 1** ^{13}C chemical shift values

Type of carbon	δ / ppm
—C—C—	5–40
R—C—Cl or Br	10–70
R—C—C— (‖O)	20–50
R—C—N	25–60
—C—O— alcohols, ethers, or esters	50–90
C=C	90–150
R—C≡N	110–125
(benzene ring)	110–160
R—C— esters or acids (‖O)	160–185
R—C— aldehydes or ketones (‖O)	190–220

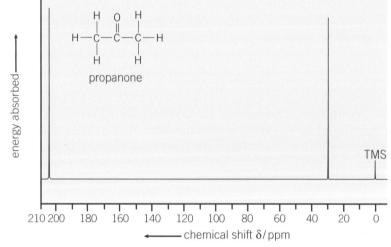

▲ **Figure 6** ^{13}C NMR spectrum of propanone

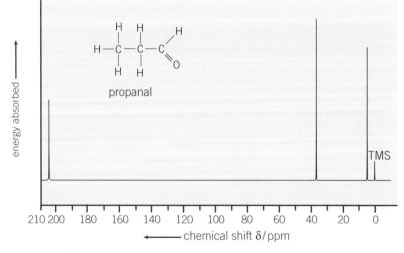

▲ **Figure 7** ^{13}C NMR spectrum of propanal

Summary questions

1 The ^{13}C NMR spectrum of ethanol is discussed above and has two peaks. Methoxymethane is an isomer of ethanol:

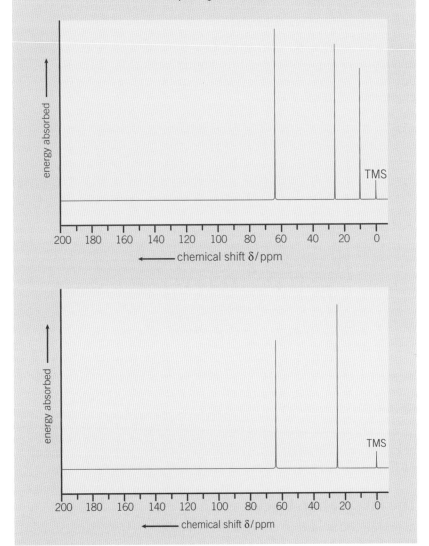

methoxymethane

   a How many peaks would you expect to find in its ^{13}C NMR spectrum?

   b Explain your answer.

2 The ^{13}C NMR spectra of propan-1-ol and propan-2-ol are given below. State which is which and explain your answer.

energy absorbed →

TMS

200 180 160 140 120 100 80 60 40 20 0
← chemical shift δ/ppm

energy absorbed →

TMS

200 180 160 140 120 100 80 60 40 20 0
← chemical shift δ/ppm

Hint

It will help to draw the displayed or structural formula of the two isomers.

In proton NMR, it is the ^1H nucleus that is being examined. Nearly all hydrogen atoms are ^1H so it is easier to get an NMR spectrum for ^1H than for ^{13}C.

Here it is hydrogen atoms attached to different functional groups that feel the magnetic field differently, because all nuclei are shielded from the external magnetic field by the electrons that surround them. Nuclei with more electrons around them are better shielded. The greater the electron density around a hydrogen atom, the smaller the chemical shift δ. The values of chemical shift in proton NMR are smaller than those for ^{13}C NMR – most are between 0 and 10 ppm.

If all the hydrogen nuclei in an organic compound are in identical environments, you get only one chemical shift value. For example, all the hydrogen atoms in methane, CH_4, are in the same environment and have the same chemical shift:

But, in a molecule like methanol, there are hydrogen atoms in two different environments – the three on the carbon atom, and the one on the oxygen atom. The NMR spectrum will show the two environments (Figure 1).

▼ **Table 1** *Chemical shift values for proton, ^1H, NMR*

Type of proton	δ / ppm
RO**H**	0.5–5.0
RC**H**$_3$	0.7–1.2
RN**H**$_2$	1.0–4.5
R$_2$C**H**$_2$	1.2–1.4
R$_3$C**H**	1.4–1.6
R—C—C— (O, **H**)	2.1–2.6
R—O—C— (**H**)	3.1–3.9
RC**H**$_2$Cl or Br	3.1–4.2
R—C—O—C— (O, **H**)	3.7–4.1
R, **H** C=C	4.5–6.0
R—C (=O, **H**)	9.0–10.0
R—C (=O, O—**H**)	10.0–12.0

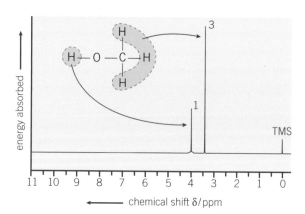

▲ **Figure 1** *The NMR spectrum of methanol – the peak areas are in the ratio 1 : 3*

In general, the further away a hydrogen atom is from an electronegative atom (such as oxygen) the smaller its chemical shift. In ethanol, CH_3CH_2OH, there are three values of δ.

In ^1H NMR the areas under the peaks (shown here by the numbers next to them) are proportional to the number of hydrogen atoms of each type – in this case three and one.

The integration trace

In proton NMR spectra, the area of each peak is related to the number of hydrogen atoms producing it. So, in the spectrum of methanol, CH_3OH, the CH_3 peak is three times the area of the OH peak. This can be difficult to evaluate by eye, so the instrument produces a line called the integration trace, shown in red in Figure 2. The relative heights of the steps of this trace give the relative number of each type of hydrogen – 3 : 1 in this case.

The chemical shift value at which the peak representing each type of proton appears tells you about its environment – the type of functional group of which it is a part.

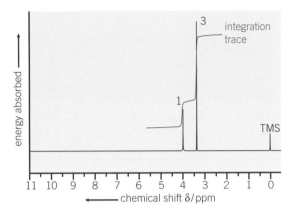

▲ **Figure 2** *The NMR spectrum of methanol showing the integration trace in red*

Chemical shift values

Hydrogen atom(s) in any functional group have a particular chemical shift value (Table 1).

Tetramethylsilane

The δ values of chemical shifts are measured by reference to a standard – the chemical shift of the hydrogen atoms in the compound tetramethylsilane, $Si(CH_4)_4$, TMS (Figure 3).

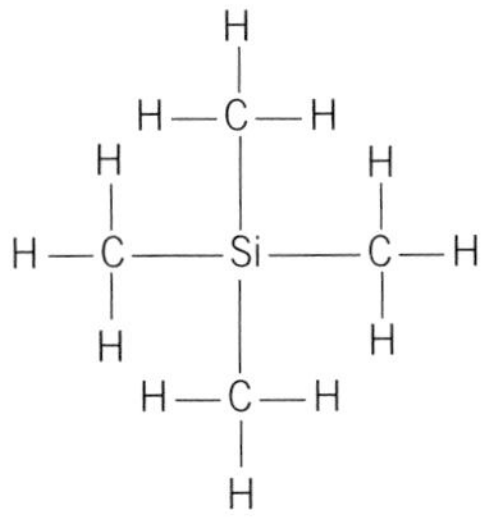

▲ **Figure 3** *Tetramethylsilane (TMS) – all 12 hydrogen atoms are in exactly the same environment, so they produce a single ^{1}H NMR signal*

The chemical shift value of these hydrogen atoms is zero by definition. A little TMS, which is a liquid, may be added to samples before their NMR spectra are run, and gives a peak at a δ value of exactly zero ppm to calibrate the spectrum (although modern techniques do not require this). All the spectra in this book show a TMS peak at δ = 0.

Other reasons for using TMS are that it is inert, non-toxic, and easy to remove from the sample.

Hint

For simplicity, the integrated trace has been omitted from NMR spectra in this book, and the relative number of hydrogen atoms that each peak represents is given.

Summary questions

1 This question is about the isomers propan-1-ol and propan-2-ol.

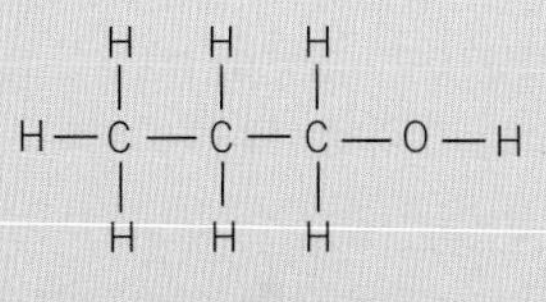

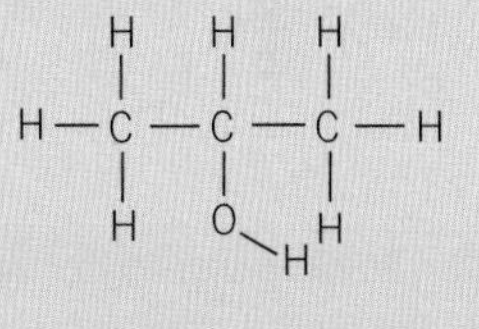

a What is meant by the term isomer?

b Write down the formulae of propan-1-ol and propan- 2-ol and mark each of the hydrogen atoms A, B, and so on, to show which are in different environments.

c How many different environments are there for the hydrogen atoms in:

i propan-1-ol
ii propan-2-ol?

d How many hydrogen atoms in each of the different environments, A, B, and so on, are there in:

i propan-1-ol
ii propan-2-ol?

e Predict the order of the chemical shift for each atom in:

i propan-1-ol
ii propan-2-ol.

Learning objectives:

→ Explain what causes spin–spin coupling.

→ Describe the $n + 1$ rule.

→ Explain how ^{1}H NMR spectra can be interpreted.

Specification reference: 3.3.15

If you are presented with a spectrum of an organic compound, such as in Figure 1, you can find out a lot about its structure.

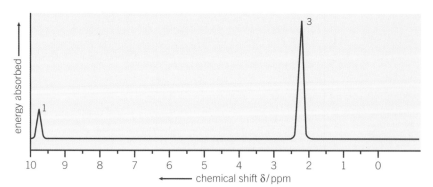

▲ **Figure 1** *The NMR spectrum of an organic compound*

The chemical shift values in Table 1 in Topic 32.2 tell you that the single hydrogen at δ 9.7 is the hydrogen from a –CHO (aldehyde) group and the three hydrogens at δ 2.2 are those of a –COCH$_3$ group. (This peak could also be caused by –COCH$_2$R, but since there are three hydrogens it must be –COCH$_3$.)

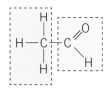

▲ **Figure 2** *The two groups that make up ethanal*

So the compound is likely to be ethanal, CH$_3$CHO (Figure 2).

Spin–spin coupling

If you zoom in on most NMR peaks, they are split into particular patterns – this is called **spin–spin coupling** (also called spin–spin splitting). It happens because the applied magnetic field felt by any hydrogen atom is affected by the magnetic field of the hydrogen atoms on the neighbouring carbon atoms. This spin–spin splitting gives information about the neighbouring hydrogen atoms, which can be very helpful when working out structure.

Figure 3 shows the spin–spin splitting patterns.

The $n + 1$ rule

If there is one hydrogen atom on an adjacent carbon, this will split the NMR signal of a particular hydrogen into two peaks each of the same height.

If there are two hydrogen atoms on an adjacent carbon, this will split the NMR signal of a particular hydrogen into three peaks with the height ratio $1:2:1$.

Three adjacent hydrogen atoms will split the NMR signal of a particular hydrogen into four peaks with the height ratio $1:3:3:1$.

This is called the $n + 1$ rule:

**n hydrogens on an adjacent carbon atom
will split a peak into $n + 1$ smaller peaks.**

Study tip

Spin–spin coupling is not usually seen in ^{13}C NMR spectra due to the low abundance of ^{13}C. This is one reason why ^{13}C spectra are simpler than ^{1}H spectra.

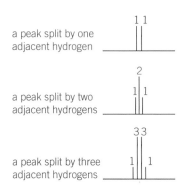

a peak split by one adjacent hydrogen

a peak split by two adjacent hydrogens

a peak split by three adjacent hydrogens

▲ **Figure 3** *NMR splitting patterns*

Some examples of interpreting ^{1}H NMR spectra

Ethanal

If you zoom in on the peaks in the spectrum of ethanal shown in Figure 1, you will see spin–spin splitting (Figure 4).

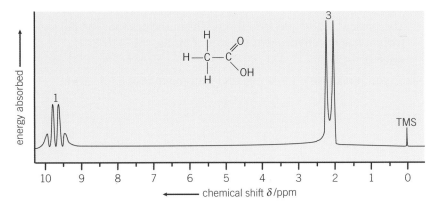

▲ **Figure 4** *The NMR spectrum of ethanal, CH$_3$CHO*

There are two types of hydrogen environments:

- A single peak of δ 9.7. This is the hydrogen of a –CHO group. This peak is split into four (height ratios 1 : 3 : 3 : 1) by the three hydrogens of the adjacent –CH$_3$ group.

- The peak with δ 2.2 is caused by three hydrogens of a –CH$_3$ group. This peak is split into two (height ratios 1 : 1) by the one hydrogen of the adjacent –CHO group.

Propanoic acid

Figure 5 shows the NMR spectrum of propanoic acid.

It is useful to make a table (Table 1) of the chemical shift of the peaks and what group they could correspond to by reference to Table 1 in Topic 32.2.

From the chemical shift value alone, the peak at 2.4 could be caused by either –COCH$_2$R or –COCH$_3$. However the fact that there are just two hydrogens means that it must correspond to –COCH$_2$R.

▼ **Table 1** *Chemical shift of the peaks of the ^{1}H NMR spectrum of propanoic acid, and what groups they could correspond to*

Chemical shift δ	Type of hydrogen	Number of hydrogens
11.7	–COOH	1
2.4	–COCH$_2$R or –COCH$_3$	2
1.1	RCH$_3$	3

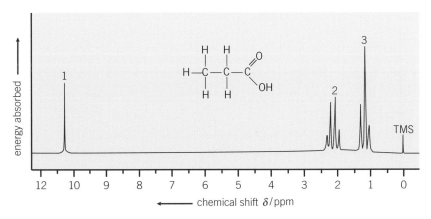

▲ **Figure 5** *The NMR spectrum of propanoic acid, CH$_3$CH$_2$COOH*

Looking at the spin–spin splitting:

- The peak at 11.7 is not split. This is because the adjacent carbon has no hydrogens bonded to it, –COOH.
- The peak at 2.4 is split into four. This indicates that the adjacent carbon has three hydrogens bonded to it. So, the R in –$COCH_2R$ must be –CH_3.
- The peak at 1.1 is split into three. This indicates that the adjacent carbon has two hydrogens bonded to it. So, the R in RCH_3 must be –CH_2.

So, if you put these groups together you make propanoic acid:

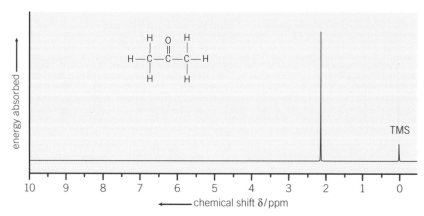

Solvents for ^{1}H NMR

NMR spectra are normally run in solution. The solvent must not contain any hydrogen atoms, otherwise the signal from the hydrogen atoms in the solution would swamp the signals from hydrogen atoms in the sample, because there are vastly more of them.

One solvent commonly used is tetrachloromethane, CCl_4, which has no hydrogen atoms. Other solvents contain deuterium, which is an isotope of hydrogen and has the symbol D. Deuterium does not produce an NMR signal in the same range as hydrogen, though it has the same chemical properties. Some examples of deuterium-based solvents are deuterotrichloromethane, $CDCl_3$, deuterium oxide, D_2O, and perdeuterobenzene, C_6D_6.

More examples of interpreting and predicting NMR spectra

Propanone

The NMR spectrum of propanone (Figure 6) has just one peak. This means that all the hydrogen atoms in the molecule are in identical environments. The chemical shift value of 2.1 indicates that this corresponds to –$COCH_3$ or –$COCH_2R$.

▲ **Figure 6** *The NMR spectrum of propanone*

Predicting NMR spectra

Chemists making new compounds may predict the spectrum of a compound they are making and compare their prediction with that of the compound they actually produce, to check that their reaction has gone as intended.

Ethyl ethanoate

There are three sets of hydrogen atoms in different environments. The values of chemical shift are predicted using Table 1 in Topic 31.2.

You can predict the spectrum shown in Figure 7 by dividing up the molecule as shown:

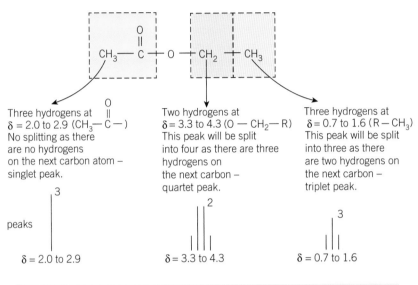

Three hydrogens at
δ = 2.0 to 2.9 (CH$_3$—C—)
No splitting as there
are no hydrogens
on the next carbon atom –
singlet peak.

Two hydrogens at
δ = 3.3 to 4.3 (O — CH$_2$— R)
This peak will be split
into four as there are three
hydrogens on
the next carbon –
quartet peak.

Three hydrogens at
δ = 0.7 to 1.6 (R — CH$_3$)
This peak will be split
into three as there
are two hydrogens on
the next carbon –
triplet peak.

peaks

δ = 2.0 to 2.9 δ = 3.3 to 4.3 δ = 0.7 to 1.6

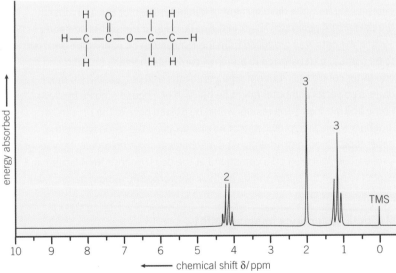

▲ **Figure 7** *The NMR spectrum of ethyl ethanoate*

The birth of NMR

NMR is probably the most important analytical technique used by organic chemists today. Indeed, one Nobel Prize-winning chemist has been quoted as saying 'when the NMR goes down, the organic chemists go home'.

However, the chemical usefulness of the technique was discovered almost by accident. The effect began to be investigated by physicists just before and after the Second World War, and it appears that the researchers were helped in building their apparatus by the availability of cheap electronic components from surplus wartime radar equipment. The aim of the experiment was to measure the magnetic properties of atomic nuclei (their magnetic moments to be precise). They succeeded in their measurements, but were frustrated to find that the same atomic nucleus in different chemicals gave different results. For example, the two nitrogen atoms in ammonium nitrate, NH_4NO_3, gave different values. They realised that this was because the nitrogen nuclei were being shielded from the magnetic field by the electrons that surrounded them, and that as the two nitrogen atoms were in different chemical environments they were shielded to different extents.

What was a frustration to the physicists trying to investigate the nucleus was a gift to chemists whose prime interest was what was happening to the electrons. NMR could tell chemists about the degree to which electrons were surrounding atoms, so it could distinguish between the hydrogen atoms in the CH_3, CH_2, and OH groups in ethanol, CH_3CH_2OH, for example.

Manipulating the data

NMR is a technique that generates a lot of information and it has benefited enormously from the development of computers to process and present the data that it generates. Back in the early days of the 1950s and early 1960s, the data was produced from the instrument on paper tape and had to be manually transferred to punched cards which had to be *posted* to a computer centre to be put onto magnetic tape and processed. (In those days, a powerful computer might be the size of a house — no PC in every home and lab then.) The results would be posted back to the researchers, maybe a week later, provided that no one dropped the cards or tore the paper tape. Later, instruments used mechanical chart recorders. Nowadays, a researcher will drop off a compound at the department's NMR facility and expect to have the spectrum up on their networked PC almost before they are back at the lab.

Magnetic resonance imaging (MRI)

NMR can be used to investigate the human body — this was first realised by Felix Bloch, who found he got a strong signal by placing his finger in an NMR spectrometer. This signal was coming from protons in the water molecules that make up a large proportion of the human body. Water in different parts of the body (e.g., normal cells and cancer cells) gives slightly different NMR signals. MRI scanning of parts of the body, to help diagnose medical conditions, is now routine. The patient passes through a scanner where the magnetic field varies across the body. This, along with

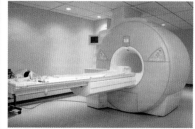

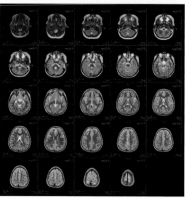

▲ **Figure 9** *MRI scanner and a scan of a female child's brain obtained by using this technique*

sophisticated computer processing of the NMR signal, allows a three-dimensional image of the body to be built up. The technique is harmless as, unlike X-rays, neither the radio waves nor the magnetic field can damage cells. However, the name 'magnetic resonance imaging' is used rather than 'nuclear magnetic resonance' because of the association of the word 'nuclear' with radioactivity in the mind of the public.

Summary questions

1 The ^{1}H NMR spectra shown are those of ethanol and of methoxymethane.

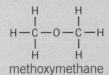

ethanol   methoxymethane

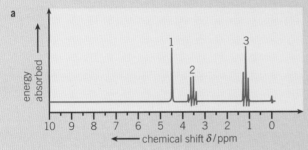

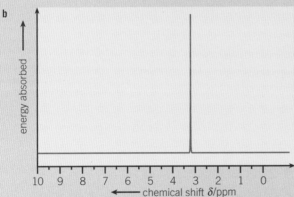

   a Work out which spectrum represents which compound.

   b Say what type of hydrogen each peak represents.

   c How many of each type of hydrogen are there?

2 Predict the NMR spectrum of methyl ethanoate, CH_3COOCH_3, using the same procedure as for ethyl ethanoate above.

1 NMR spectroscopy can be used to study the structures of organic compounds.
  (a) Compound **J** was studied using ¹H NMR spectroscopy.

$$Cl-CH_2-\underset{\underset{CH_3}{|}}{\overset{\overset{CH_3}{|}}{C}}-\overset{a}{CH_2}-CH_2-Cl$$

J

  (i) Identify a solvent in which J can be dissolved before obtaining its ¹H NMR spectrum.

(1 mark)

  (ii) Give the number of peaks in the ¹H NMR spectrum of **J**.

(1 mark)

  (iii) Give the splitting pattern of the protons labelled *a*.

(1 mark)

  (iv) Give the IUPAC name of **J**.

(1 mark)

  (b) Compound **K** was studied using ¹³C NMR spectroscopy.

$$CH_3-\overset{b}{\underset{\underset{O}{\|}}{C}}-CH_2-CH_2-\underset{\underset{O}{\|}}{C}-CH_3$$

K

  (i) Give the number of peaks in the ¹³C NMR spectrum of **K**.

(1 mark)

  (ii) Use Table 1 in Topic 32.1 to suggest a δ value of the peak for the carbon labelled *b*.

(1 mark)

  (iii) Give the IUPAC name of **K**.

(1 mark)
AQA, 2013

2 Atenolol is an example of the type of medicine called a beta blocker. These medicines are used to lower blood pressure by slowing the heart rate. The structure of atenolol is shown below.

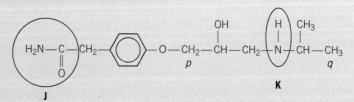

J

K

  (a) Give the name of each of the circled functional groups labelled **J** and **K** on the structure of atenolol shown above.

(2 marks)

  (b) The ¹H NMR spectrum of atenolol was recorded.
      One of the peaks in the ¹H NMR spectrum is produced by the CH₂ group labelled p in the structure of atenolol.
      Use Table 1 in Topic 32.2 to suggest a range of δ values for this peak.
      Name the splitting pattern of this peak.

(2 marks)

  (c) NMR spectra are recorded using samples in solution.
      The ¹H NMR spectrum was recorded using a solution of atenolol in CDCl₃
      (i) Suggest why CDCl₃ and **not** CHCl₃ was used as the solvent.

(1 mark)

      (ii) Suggest why CDCl₃ is a more effective solvent than CCl₄ for polar molecules such as atenolol.

(1 mark)

(d) The ^{13}C NMR spectrum of atenolol was also recorded.
Use the structure of atenolol given to deduce the total number
of peaks in the ^{13}C NMR spectrum of atenolol.

(1 mark)

(e) Part of the ^{13}C NMR spectrum of atenolol is shown below. Use this
spectrum and Table 1 in Topic 32.1 where appropriate, to answer the
questions which follow.

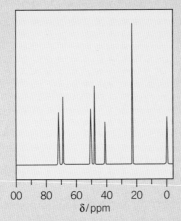

(i) Give the formula of the compound that is used as a standard
and produces the peak at $\delta = 0$ ppm in the spectrum.

(1 mark)

(ii) One of the peaks in the ^{13}C NMR spectrum above is produced by the
CH_3 group labelled q in the structure of atenolol.
Identify this peak in the spectrum by stating its δ value.

(1 mark)

(iii) There are three CH_2 groups in the structure of atenolol.
One of these CH_2 groups produces the peak at $\delta = 71$ in the ^{13}C NMR
spectrum above.
Draw a circle around this CH_2 group in the structure of atenolol
shown below.

$$H_2N-\underset{\underset{O}{\|}}{C}-CH_2-\!\!\!\bigcirc\!\!\!-O-CH_2-\underset{\underset{OH}{|}}{CH}-CH_2-\underset{\underset{H}{|}}{N}-\underset{\underset{CH_3}{|}}{CH}-CH_3$$

(1 mark)

(f) Atenolol is produced industrially as a racemate (an equimolar mixture of
two enantiomers) by reduction of a ketone. Both enantiomers are able to
lower blood pressure. However, recent research has shown that one
enantiomer is preferred in medicines.
(i) Suggest a reducing agent that could reduce a ketone to form atenolol.

(1 mark)

(ii) Draw a circle around the asymmetric carbon atom in the structure
of atenolol shown above.

(1 mark)

(iii) Suggest how you could show that the atenolol produced by reduction
of a ketone was a racemate and **not** a single enantiomer.

(2 marks)

(iv) Suggest **one** advantage and **one** disadvantage of using a racemate
rather than a single enantiomer in medicines.

(2 marks)
AQA, 2011

You will be familiar with paper chromatography, which is often used to separate the dyes in, for example, felt-tip pens.

Chromatography describes a whole family of separation techniques. They all depend on the principle that a mixture can be separated if it is dissolved in a solvent and then the resulting solution (now called the mobile phase) moves over a solid (the stationary phase).

• The moving or **mobile phase** carries the soluble components of the mixture with it. The more soluble the component in the mobile phase, the faster it moves. The solvent in the moving phase is often called the eluent (in column chromatography).

• The **stationary phase** will hold back the components in the mixture that are attracted to it. The more affinity a component in the mixture being separated has for the stationary phase, the slower it moves with the solvent.

So, if suitable moving and stationary phases are chosen, a mixture of similar substances can be separated completely, because every component of the mixture has a unique balance between its affinity for the stationary and for the mobile phase. In fact, chromatography is often the only way that very similar components of a mixture can be separated.

Thin-layer chromatography

Thin-layer chromatography (TLC) is a development of paper chromatography. The filter paper is replaced by a glass, metal, or plastic sheet coated with a thin layer of silica gel (silicon dioxide, SiO_2) or alumina (aluminium oxide, Al_2O_3) which acts as the stationary phase. These are often called plates. Plastic- and metal-backed sheets can be cut to size with scissors.

TLC has several advantages over paper chromatography:

• it runs faster

• smaller amounts of mixtures can be separated

• the spots usually spread out less

• the plates are more robust than paper.

When the chromatogram has run, the position of colourless spots may have to be located by shining ultra-violet light on the plate, or chemically by spraying the plate with a locating agent which reacts with the components of the mixture to give coloured compounds.

After the plate has been run, an R_f value is calculated of each component using:

$$R_f = \frac{\text{distance moved by spot}}{\text{distance moved by solvent}}$$

The R_f values can be used to help identify each component.

▲ **Figure 1** *The cellulose of the paper holds many trapped water molecules (the stationary phase). Here, ethanol is the mobile phase, or eluent*

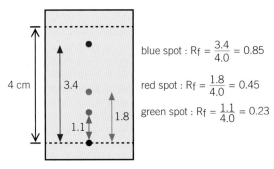

▲ **Figure 2** *Calculating R_f values from thin-layer chromatograms*

blue spot : $R_f = \dfrac{3.4}{4.0} = 0.85$

red spot : $R_f = \dfrac{1.8}{4.0} = 0.45$

green spot : $R_f = \dfrac{1.1}{4.0} = 0.23$

Column chromatography

Column chromatography uses a powder, such as silica, aluminium oxide, or a resin, as the stationary phase. This is packed into a narrow tube – the column – and a solvent (the eluent) is added at the top (Figure 3). As the eluent runs down the column, the components of the mixture move at different rates and can be collected separately in flasks at the bottom. More than one eluent may be used to get a better separation. This method has the advantage that fairly large amounts can be separated and collected. For example, a mixture of amino acids can be separated into its pure components by this method.

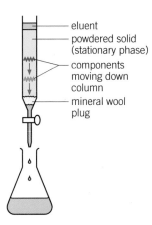

▲ **Figure 3** *Column chromatography*

Gas–liquid chromatography (GC)

This technique is one of the most important modern analytical techniques. The basic apparatus is shown in Figure 4.

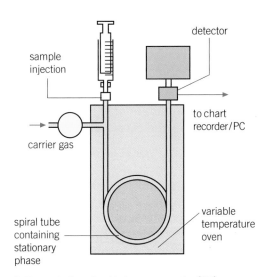

▲ **Figure 4** *Gas–liquid chromatography (GC)*

The stationary phase is a powder, coated with oil. It is either packed into or coated onto the inside of a long capillary tube, up to 100 m long and less than $\frac{1}{2}$ mm in diameter coiled up and placed in an oven whose temperature can be varied. The mobile phase is usually an unreactive gas, such as nitrogen or helium. After injection, the sample is carried along by the gas and the mixture separates as some of the components move along with the gas and some are retained by the oil, each to a

Hint

Gas–liquid chromatography is often simply called gas chromatography.

different degree. This means that the components leave the column at different times after injection – they have different **retention times**.

Various types of detectors are used, including ones that measure the thermal conductivity of the emerging gas. The results may be presented on a graph (Figure 5). The area under each peak is proportional to the amount of that component.

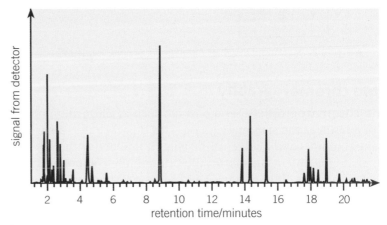

▲ **Figure 5** *Typical GC trace – each peak represents a different component*

In some instruments the components are fed directly into a mass spectrometer, infrared spectrometer, or NMR spectrometer for identification. Today the whole process is automated and computer controlled (Figure 6).

As an analytical method for separating mixtures, GC is extremely sensitive. It can separate minute traces of substances in foodstuffs, and even link crude oil pollution found on beaches with its tanker of origin by comparing oil samples. Perhaps its best-known use is for testing athletes' blood or urine for drug taking.

The identification of a component is done by matching its retention time with that of a known substance under the same conditions. This is then confirmed by comparing the mass spectra of the two substances.

GCMS

GCMS stands for Gas Chromatography–Mass Spectrometry. It is essentially two techniques in one. A mass spectrometer is used as the detector for a gas chromatography system. As each component of a mixture comes out of the gas chromatography column, the time it has taken to pass through the column (its retention time) is noted. Each component is fed automatically into mass spectrometer which enables the compound to be identified either by its fragmentation pattern or by measuring its accurate mass.

▲ **Figure 6** *A gas chromatography machine*

HPLC

HPLC can be taken to stand for High Pressure Liquid Chromatography or High Performance Liquid Chromatography. Both names are appropriate. Here the mixture to be separated is forced through a column containing the stationary phase by a solvent driven by a high pressure pump. It is similar to column chromatography except that the pump drives the solvent (the eluent) rather than gravity. A variety of materials can be used as the stationary phase including chiral ones that can separate optical isomers. A variety of detection methods can be used, for example, the absorption of ultra-violet light.

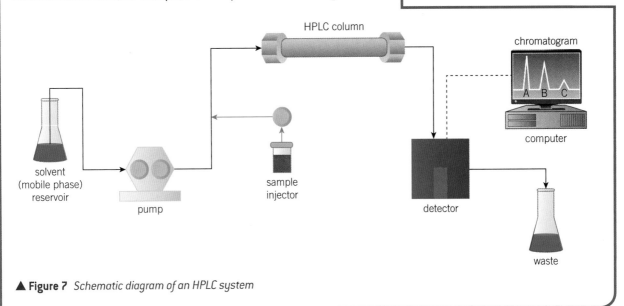

▲ **Figure 7** *Schematic diagram of an HPLC system*

Summary questions

1   What is the difference between column chromatography and gas–liquid chromatography?

2   Why is GC so important in forensic detective work? Give a possible example not in the text.

3   From the GC in Figure 7 above, identify from A, B, and C:

   a   the most abundant component in the mixture

   b   the one with the greatest affinity for the solid phase

   c   the one with the greatest affinity for the gas phase

   d   the one with the greatest retention time.

Synoptic link

Optical isomers was covered in Topic 25.2, Optical isomerism.

1 A peptide is hydrolysed to form a solution containing a mixture of amino acids. This mixture is then analysed by silica gel thin-layer chromatography (TLC) using a developing solvent. The individual amino acids are identified from their R_f values. Part of the practical procedure is given below.

1. **Wearing plastic gloves to hold a TLC plate**, draw a pencil line 1.5 cm from the bottom of the plate.
2. Use a capillary tube to apply a very small drop of the solution of amino acids to the mid-point of the pencil line.
3. Allow the spot to dry completely.
4. In the developing tank, add the developing solvent to **a depth of not more than 1 cm**.
5. Place your TLC plate in the developing tank **and seal the tank with a lid**.
6. Allow the developing solvent to rise up the plate to at least $\frac{3}{4}$ of its height.
7. Remove the plate and quickly mark the position of the solvent front with a pencil.
8. Allow the plate to dry **in a fume cupboard**.

(a) Parts of the procedure are in bold text.
   Explain why these parts of the procedure are essential.

(4 marks)

(b) Outline the steps needed to locate the positions of the amino acids on the TLC plate and to determine their R_f values.

(4 marks)

(c) Explain why different amino acids have different R_f values.

(2 marks)

AQA, Specimen paper 3

2 Figure 1 shows a chromatogram used to separate some amino acids by paper chromatography, using solvent X – a mixture of ethanoic acid, butan-1-ol and water.

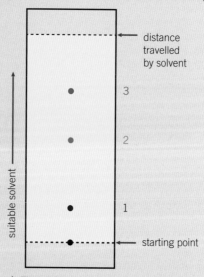

▲ Figure 1

(a) Identify the amino acids using the table below. R_f values of some amino acids using solvent X

(3 marks)

alanine	0.38
arginine	0.16
glycine	0.26
leucine	0.73
tyrosine	0.50
valine	0.60

(b) Why is it essential to know the solvent used in the process?

3   Two-way paper chromatography was used to separate a mixture.
    The results are shown below.

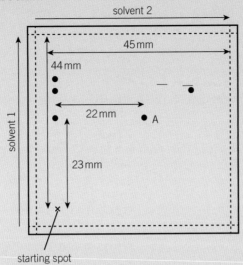

(a) Describe briefly the method of doing this.

(3 marks)

(b) Why does two-way chromatography makes identification of the components of the
    mixture more certain.

(2 marks)

(c) Find the R_f values of:
    (i)   after the first run in solvent 1

(1 mark)

    (ii) after the second run in solvent 2.

(1 mark)

4   A bottle was discovered labelled propan-2-ol. The chemist showed, using infrared
    spectroscopy, that the propan-2-ol was contaminated with propanone.
    The chemist separated the two compounds using column chromatography. The column
    contained silica gel, a polar stationary phase.
    The contaminated propan-2-ol was dissolved in hexane and poured into the column.
    Pure hexane was added slowly to the top of the column. Samples of the eluent
    (the solution leaving the bottom of the column) were collected.
    • Suggest the chemical process that would cause a sample of propan-2-ol to become
      contaminated with propanone.
    • State how the infrared spectrum showed the presence of propanone.
    • Suggest why propanone was present in samples of the eluent collected first
      (those with shorter retention times), whereas samples containing propan-2-ol
      were collected later.

(4 marks)
AQA, 2012

1   Kevlar is a polymer used in protective clothing.
    The repeating unit within the polymer chains of Kevlar is shown.

   (a)   Name the strongest type of interaction between polymer chains of Kevlar.

   (1 mark)

   (b)   One of the monomers used in the synthesis of Kevlar is:

   H_2N—⬡—NH_2

   An industrial synthesis of this monomer uses the following two-stage process starting
   from compound **X**.

   Stage **1**

   Cl—⬡—NO_2 + $2NH_3$ ⟶ H_2N—⬡—NO_2 + NH_4Cl

   X

   Stage **2**

   H_2N—⬡—NO_2 ⟶ H_2N—⬡—NH_2

   (i)   Suggest why the reaction of ammonia with **X** in Stage **1** might be considered
          unexpected.

   (2 marks)

   (ii)   Suggest a combination of reagents for the reaction in Stage **2**.

   (1 mark)

   (iii) Compound **X** can be produced by nitration of chlorobenzene.
          Give the combination of reagents for this nitration of chlorobenzene.
          Write an equation or equations to show the formation of a reactive intermediate
          from these reagents.

   (3 marks)

   (iv)   Name and outline a mechanism for the formation of **X** from chlorobenzene
          and the reactive intermediate in Question **1 (b)** (iii).

   (4 marks)
   AQA, 2014

2   Each of the following conversions involves reduction of the starting material.
   (a)   Consider the following conversion.

   O_2N—⬡—NO_2 ⟶ H_2N—⬡—NH_2

   Identify a reducing agent for this conversion.
   Write a balanced equation for the reaction using molecular formulae for the
   nitrogen-containing compounds and [H] for the reducing agent.
   Draw the repeating unit of the polymer formed by the product of this reaction
   with benzene 1,4-dicarboxylic acid.

   (5 marks)

   (b)   Consider the following conversion.

   ⬡ ⟶ ⬡

   Identify a reducing agent for this conversion.
   State the empirical formula of the product.
   State the bond angle between the carbon atoms in the starting material and
   the bond angle between the carbon atoms in the product.

   (4 marks)

(c) The reducing agent in the following conversion is $NaBH_4$

$$H_3C-\underset{\underset{O}{\|}}{C}-CH_2CH_3 \longrightarrow H_3C-\underset{\underset{OH}{|}}{CH}-CH_2CH_3$$

(i)   Name and outline a mechanism for the reaction.

(5 marks)

(ii)   By considering the mechanism of this reaction, explain why the product formed is optically inactive.

(3 marks)

AQA, 2013

3   Organic chemists use a variety of methods to identify unknown compounds. When the molecular formula of a compound is known, spectroscopic and other analytical techniques are used to distinguish between possible structural isomers. Use your knowledge of such techniques to identify the compounds described below.
Use spectral data where appropriate.
Each part below concerns a different pair of structural isomers.
Draw **one** possible structure for each of the compounds **A** to **J**, described below.

(a)   Compounds **A** and **B** have the molecular formula C_3H_6O
A has an absorption at $1715\,cm^{-1}$ in its infrared spectrum and has only one peak in its ^1H NMR spectrum.
B has absorptions at $3300\,cm^{-1}$ and at $1645\,cm^{-1}$ in its infrared spectrum and does **not** show E–Z isomerism.

(2 marks)

(b)   Compounds **C** and **D** have the molecular formula C_5H_{12}
In their ^1H NMR spectra, **C** has three peaks and **D** has only one.

(2 marks)

(c)   Compounds **E** and **F** are both esters with the molecular formula $C_4H_8O_2$
In their ^1H NMR spectra, **E** has a quartet at $\delta = 2.3$ ppm and **F** has a quartet at $\delta = 4.1$ ppm.

(2 marks)

(d)   Compounds **G** and **H** have the molecular formula $C_6H_{12}O$
Each exists as a pair of optical isomers and each has an absorption at about $1700\,cm^{-1}$ in its infrared spectrum. **G** forms a silver mirror with Tollens' reagent but **H** does not.

(2 marks)

(e)   Compounds **I** and **J** have the molecular formula $C_4H_{11}N$ and both are secondary amines. In their ^{13}C NMR spectra, **I** has two peaks and **J** has three.

(2 marks)

AQA, 2010

4   In 2008, some food products containing pork were withdrawn from sale because tests showed that they contained amounts of compounds called dioxins many times greater than the recommended safe levels.
Dioxins can be formed during the combustion of chlorine-containing compounds in waste incinerators. Dioxins are very unreactive compounds and can therefore remain in the environment and enter the food chain.
Many dioxins are polychlorinated compounds such as tetrachlorodibenzodioxin (TCDD) shown below.

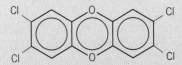

In a study of the properties of dioxins, TCDD and other similar compounds were synthesised. The mixture of chlorinated compounds was then separated before each compound was identified by mass spectrometry.

(a) Fractional distillation is **not** a suitable method to separate the mixture of chlorinated compounds before identification by mass spectrometry.
Suggest how the mixture could be separated.

(1 mark)

(b) The molecular formula of TCDD is $C_{12}H_4O_2Cl_4$
Chlorine exists as two isotopes ^{35}Cl (75%) and ^{37}Cl (25%).
Deduce the number of molecular ion peaks in the mass spectrum of TCDD and calculate the m/z value of the most abundant molecular ion peak.

(2 marks)

(c) Suggest **one** operating condition in an incinerator that would minimise the formation of dioxins.

(1 mark)

(d) TCDD can also be analysed using ^{13}C NMR.
   (i) Give the formula of the compound used as the standard when recording a ^{13}C spectrum.

(1 mark)

   (ii) Deduce the number of peaks in the ^{13}C NMR spectrum of TCDD.

(1 mark)

AQA, 2010

5 (a) A shirt was made from this polyester. A student wearing the shirt accidentally splashed aqueous sodium hydroxide on a sleeve. Holes later appeared in the sleeve where the sodium hydroxide had been.
Name the type of reaction that occurred between the polyester and the aqueous sodium hydroxide. Explain why the aqueous sodium hydroxide reacted with the polyester.

(3 marks)

(b) (i) Complete the following equation for the preparation of aspirin using ethanoic anhydride by writing the structural formula of the missing product.

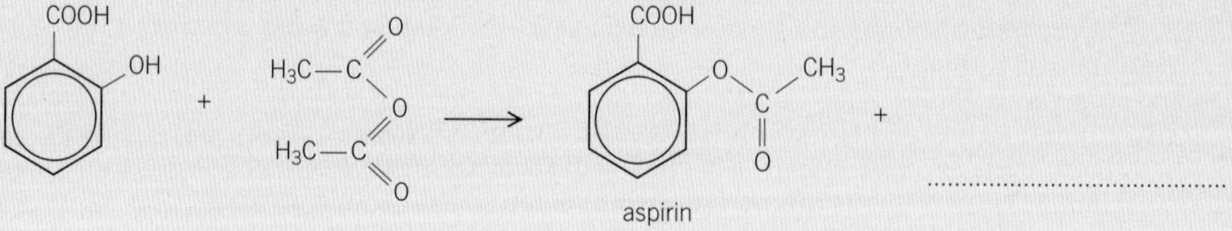

(1 mark)

   (ii) Suggest a name for the mechanism for the reaction in part **(c) (i)**.

(1 mark)

   (iii) Give **two** industrial advantages, other than cost, of using ethanoic anhydride rather than ethanoyl chloride in the production of aspirin.

(2 marks)

(c) Complete the following equation for the reaction of one molecule of benzene 1,2 dicarboxylic anhydride (phthalic anhydride) with one molecule of methanol by drawing the structural formula of the single product.

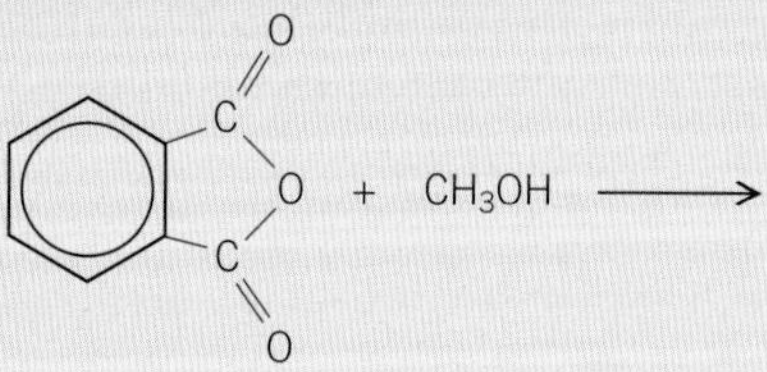

(1 mark)

(d) The indicator phenolphthalein is synthesised by reacting phthalic anhydride with phenol as shown in the following equation.

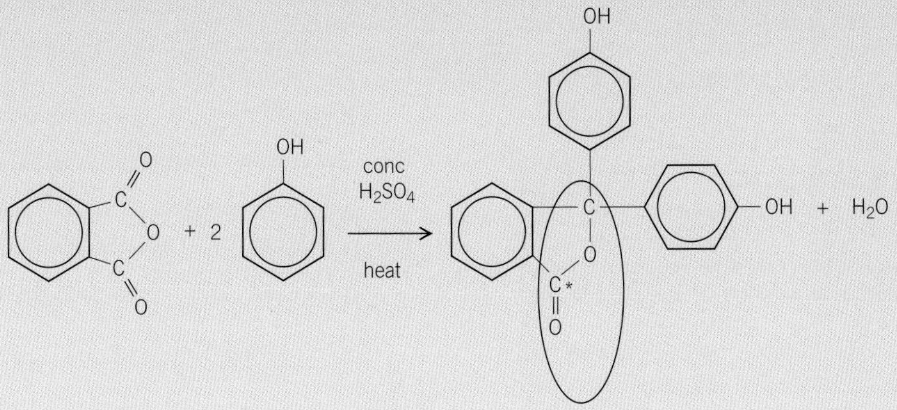

phenol   phenolphthalein

(i)   Name the functional group ringed in the structure of phenolphthalein.   *(1 mark)*
(ii) Deduce the number of peaks in the ^{13}C NMR spectrum of phenolphthalein.
                                                                              (1 mark)
(iii) One of the carbon atoms in the structure of phenolphthalein shown above is labelled with an asterisk (*).
      Use Table 1 in Topic 32.1 to suggest a range of δ values for the peak due to this carbon atom in the ^{13}C NMR spectrum of phenolphthalein.   *(1 mark)*

                                                                              AQA, 2012

6 This question is about the over-the counter painkiller and anti-inflammatory agent ibuprofen, whose skeletal formula is shown below.

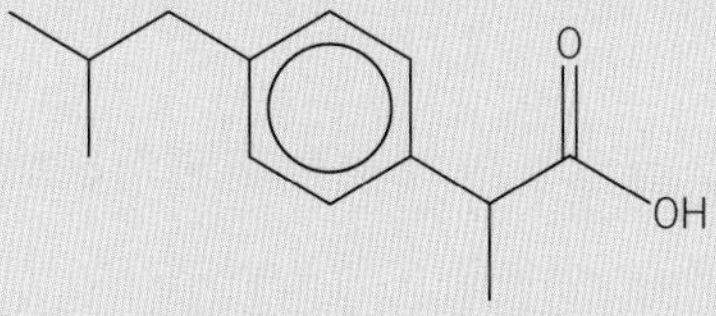

(a) What two functional groups does ibuprofen have?   *(2 marks)*
(b) Ibuprofen is optically active. Identify the asymmetric carbon atom in the structure of ibuprofen and explain your reasoning.   *(3 marks)*
(c) What potential problem does the fact that ibuprofen is optically active have on the use of ibuprofen as a drug?   *(4 marks)*
(d) In order to work quickly, a drug should be soluble in water so that it can quickly get into the bloodstream.
(e) (i)   Explain why ibuprofen is not very soluble in water.   *(1 mark)*
     (ii) Ibuprofen can be made more water-soluble by reacting it with the amino acid lysine whose structure is shown below.

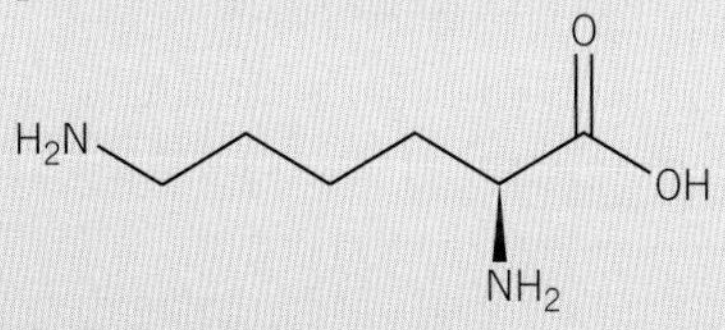

     Part of the reason that lysine is water-soluble is that it can exist as a zwitterion.
     Explain the term zwitterions and show how it makes lysine water-soluble.   *(2 marks)*
     (iii) Ibuprofen and lysine react together to form a soluble salt. Suggest how this happens.   *(2 marks)*
     (iv) Why can we be confident that lysine is non-toxic.   *(1 mark)*
(f) Ibuprofen is synthesised from a 2-methylpropylbenzene which is derived from crude oil.
     (i)   Draw the skeletal formula of 2-methylpropylbenzene.   *(2 marks)*
     (ii) What type of reactions will 2-methylpropylbenzene be most likely to undergo?
           Select the type of reagent from nucleophilic, electrophilic and free radical.
           Select the type of reaction from substitution, addition and elimination.
           Explain your answer.   *(4 marks)*

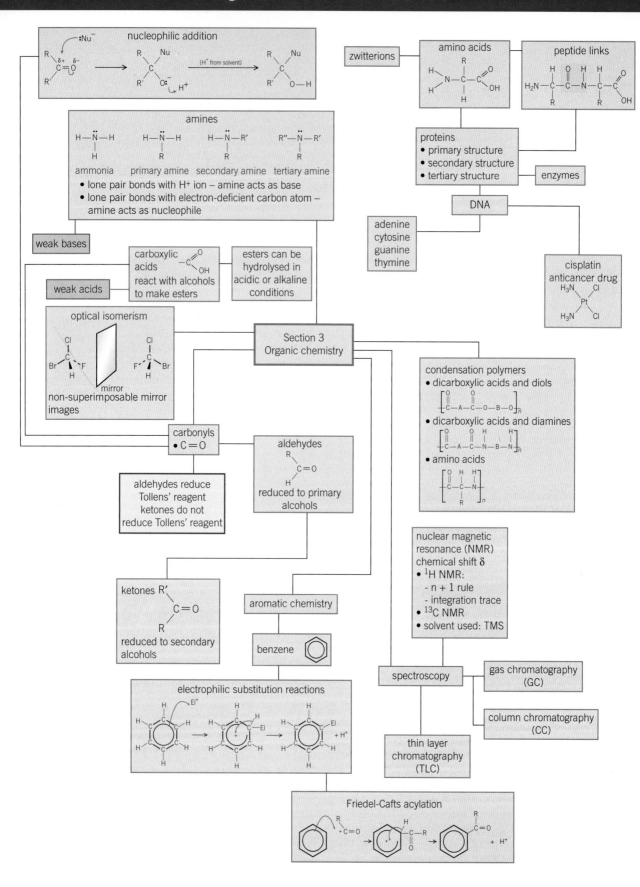

Practical skills

In this section you have met the following ideas:

- Investigating the effect of optical isomers on polarised light.
- Finding out how to distinguish between aldehydes and ketones.
- Finding out how to make and identify esters.
- Finding out how to make soap and biodiesel.
- Investigating the reactions of acid anhydrides and acyl chlorides.
- Finding out how to make aspirin.
- Investigating nitration reactions.
- Using melting points to identify compounds.
- Finding out how to make nylon.

Maths skills

In this section you have met the following maths skills:

- Working out how to draw optical isomers.

Extension

Produce a report explaining the principles of nuclear magnetic resonance and exploring how it can be used.

Suggested resources:

- Hore, P[2015], *Nuclear Magnetic Resonance: Oxford Chemistry Primers*. Oxford University Press, UK. ISBN 978-0-19-870341-9.
- Hore, P[2015], *NMR:THE TOOLKIT: Oxford Chemistry Primers*. Oxford University Press, UK. ISBN 978-0-19-870342-6.

A level additional practice questions

1 (a) Use electron pair repulsion theory to state and explain the shape of an ammonia
        (NH_3), molecule. Draw an NH_3 molecule and include the bond angle. *(5 marks)*
   (b) Chromium is a transition metal. Other than their catalytic activity, state
        three characteristic properties of transition metals. *(3 marks)*
   (c) State the full electron configuration of:
        (i)   Cr
        (ii) Cr^{3+} *(2 marks)*
   (d) $[Cr(H_2O)_4(NH_3)_2]^{2+}$ is a complex ion. Define the terms:
        (i)   complex ion
        (ii) ligand *(2 marks)*
   (e) (i)   Predict the shape of the $[Cr(H_2O)_4(NH_3)_2]^{2+}$ ion.
        (ii) Deduce the co-ordination number of the $[Cr(H_2O)_4(NH_3)_2]^{2+}$ ion *(2 marks)*

2 (a) Describe and explain the trend in atomic radius across Period 3 of the
        Periodic Table. *(3 marks)*
   (b) Describe the bonding and structure in magnesium. Include a diagram in
        your answer. *(4 marks)*
   (c) Explain why magnesium, Mg, has a higher melting point than sodium, Na. *(2 marks)*
   (d) Explain why the melting point of phosphorus, P_4, is greater than the melting
        point of chlorine, Cl_2. *(2 marks)*
   (e) (i)   Magnesium reacts with oxygen to form magnesium oxide. Write the
              equation for the reaction. Include state symbols. *(1 mark)*
        (ii) Magnesium oxide forms an alkaline solution when it reacts with water.
              Explain why and include an equation in your answer. *(2 marks)*

3 Isopropyl alcohol is used as an industrial solvent. The skeletal formula of isopropyl alcohol
   is given below.

   OH

   (a) (i)   State the molecular formula of isopropyl alcohol.
        (ii) Give the IUPAC name for isopropyl alcohol. *(2 marks)*
   (b) A chemist analysed a sample of isopropyl alcohol using ^{13}C NMR. Deduce the number
        of peaks in the NMR spectra of isopropyl alcohol. Choose one answer.
        A 1
        B 2
        C 3
        D 4 *(1 mark)*
   (c) Deduce the relative molecular mass of isopropyl alcohol. Choose one answer.
        A 29
        B 32
        C 36
        D 60 *(1 mark)*
   (d) The sample was then analysed using proton NMR. Deduce the number of peaks in
        the proton NMR of isopropyl alcohol. Choose one answer.
        A 2
        B 3
        C 4
        D 9 *(1 mark)*
   (e) (i)   A student oxidised a sample of isopropyl alcohol using acidified potassium
              dichromate(VI). Using the structural formula $(CH_3)_2CHOH$ to represent isopropyl
              alcohol and [O] to represent the oxidising agent write an equation for this
              reaction.
        (ii) State the IUPAC name for the organic product of the reaction.
        (iii) Describe and explain the colour change that the student would observe during
              the oxidation of isopropyl alcohol by acidified potassium dichromate(VI). *(4 marks)*

(f) The student wanted to use the sample of isopropyl alcohol to produce a carbonyl compound. Explain why the student did not have to distil off the product of the reaction. *(1 mark)*

(g) Draw the displayed formula and give the IUPAC name of the structural isomer of isopropyl alcohol. *(1 mark)*

4 Values for lattice enthalpy can be calculated indirectly using Born-Haber cycles.

Letter	Enthalpy change	Energy/ kJ mol^{-1}
A	Formation of calcium oxide	−635
B	first electron affinity of oxygen	−141
C	second electron affinity of oxygen	+790
D	first ionisation energy of calcium	+590
E	second ionisation energy of calcium	+1145
F	Atomisation of oxygen	+249
G	Atomisation of calcium	+178
H	Lattice enthalpy of calcium oxide	

(a) Give the equation for the formation of one mole of calcium oxide from its constituent elements. Include state symbols. *(2 marks)*

(b) Complete the diagram below by stating the correct letter from the table of enthalpy changes above. *(2 marks)*

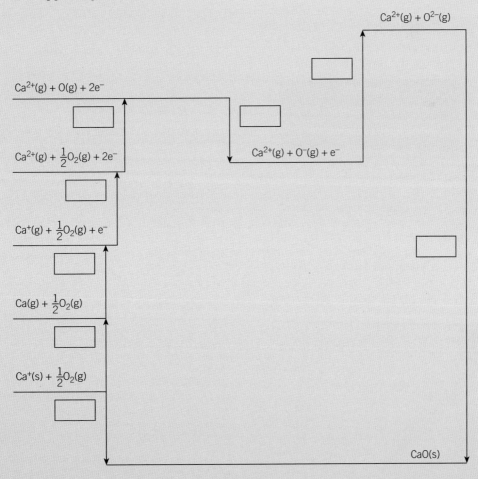

   (c) Calculate the lattice enthalpy of calcium oxide.   (2 marks)
   (d) Predict whether the lattice enthalpy of calcium oxide or barium oxide would be the
        most exothermic. Explain your answer.   (3 marks)

5 Compound B is a secondary alcohol. The display formula of compound B is shown below.

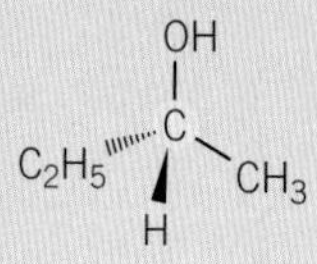

$$C_2H_5 \overset{\displaystyle OH}{\underset{\displaystyle H}{\overset{|}{\underset{|}{C}}}} CH_3$$

   (a) Give the IUPAC name for compound B.   (1 mark)
   (b) Draw the skeletal formula of compound B.   (1 mark)
   (c) Give the IUPAC name of the structural isomer of compound B.   (1 mark)
   (d) Draw the optical isomer of compound B.   (1 mark)

6 Alanine is an α-amino acid

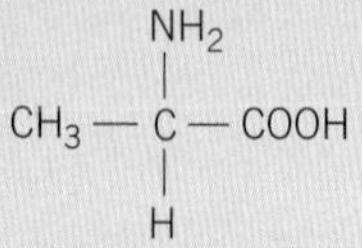

$$CH_3 \overset{\displaystyle NH_2}{\underset{\displaystyle H}{\overset{|}{\underset{|}{C}}}} COOH$$

   (a) Define the term α-amino acid   (1 mark)
   (b) Draw the zwitterion of alanine.   (1 mark)
   (c) Amino acids are crystalline solids which have surprisingly high melting points.
        Explain why zwitterions have relatively high melting points.   (2 marks)
   (d) State the IUPAC name of alanine.   (1 mark)

7 A student carried out a titration between hydrochloric acid and sodium hydroxide and
   recorded the results below.

Experiment	1	2	3
Final burette reading/ cm^3	32.80	32.40	32.70
Initial burette reading/cm^3	50.00	50.00	50.00
Titre/cm^3			

   (a) Complete the table.   (1 mark)
   (b) Name the piece of apparatus used to add sodium hydroxide to the
        hydrochloric acid.   (1 mark)
   (c) Calculate the mean titre. Give your answer to 3 significant figures.   (1 mark)
   (d) The error in the titre for experiment 1 is +/− 0.10 cm^3. Calculate the percentage error
        in experiment 1. Give your answer to an appropriate number of significant figures.
        Choose one answer.   (1 mark)
        A 0.58%
        B 0.6%
        C 5.81%
        D 0.0058%
   (e) The student repeated the experiment using the same volume and concentration of
        hydrochloric acid and the same equipment. Suggest what the student could do to
        reduce the percentage error in their results.   (1 mark)

8 The pH of an acid solution can be calculated from its K_a value.
   (a) Define the term pH.   *(1 mark)*
   (b) Lactic acid has the structural formula $CH_3CHOHCOOH$. Give the IUPAC name of
   lactic acid.   *(1 mark)*
   (c) Lactic acid has a K_a value of 1.4×10^{-4}.
       (i) Give the expression for K_a.
       (ii) Calculate the pH of a $0.10\,mol\,dm^{-3}$ solution of lactic acid. Give your answer to
       two decimal places.   *(4 marks)*

9 A student added sodium hydroxide solution dropwise to a test tube containing a solution
   of iron(II) sulfate. The student then left the test-tube for several hours.
   (a) Describe what the student would see:
       (i) When sodium hydroxide was added to the iron(II) sulfate solution.
       (ii) When the test tube was left for several hours.   *(2 marks)*
   (b) Write ionic equations for the reactions that occur when:
       (i) The sodium hydroxide was added
       (ii) When the test tube was left for several hours. Include state symbols.   *(2 marks)*

10 A chemist investigated the equilibrium system below:
$$2NO(g) + 2CO(g) \rightleftharpoons 2CO_2(g) + N_2(g) \qquad \Delta H = -788\,kJ\,mol^{-1}$$

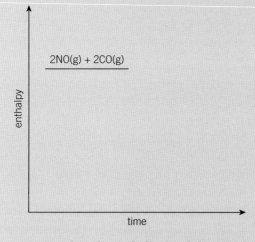

   (a) Complete the enthalpy level diagram for this reaction.   *(2 marks)*
   (b) The chemist mixed 0.76 mol of CO with 0.45 mol of NO. The mixture was left at a
   constant temperature to reach a dynamic equilibrium.
       (i) Define the term dynamic equilibrium
       (ii) The chemist analysed the equilibrium mixture and found that 0.30 moles of
       NO remained. The total volume of the equilibrium mixture was $2.00\,dm^3$.
       Write the expression for K_c for this reaction.   *(2 marks)*
   (c) Deduce the units for K_c. Choose one answer.
       A $mol\,dm^{-3}$
       B $mol^{-1}\,dm^3$
       C $mol^{-2}\,dm^6$
       D $mol^3\,dm^{-9}$ *(1 mark)*
   (d) Calculate the value of K_c for this equilibrium mixture. Show your working. Give your
   answer to 3 significant figures.   *(4 marks)*

11 Ammonia, NH_3, can be produced industrially using the Haber process at a temperature of between 400 and 500 °C.

$$N_2(g) + 3H_2(g) \rightleftharpoons 2NH_3(g) \qquad \Delta H = -92\,kJ\,mol^{-1}$$

(a) State the catalyst used in the Haber process. *(1 mark)*

(b) The diagram below shows a Maxwell-Boltzmann distribution. Explain how the catalyst used in the Haber process will increase the rate of reaction of formation of ammonia.

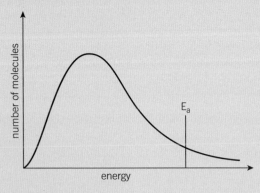

(2 marks)

(c) Ammonia has a higher boiling point than might be expected from other Group 6 hydrides. State the type of intermolecular attraction between ammonia molecules which causes ammonia to have a higher boiling point than expected. Use a diagram to explain your answer. *(5 mark)*

(d) Consider the complex cisplatin, $[Pt(NH_3)_2Cl_2]$, which contains ammonia ligands.

Cl⧵⧵⧵Pt⧸⧸⧸Cl
NH₃ ⧸   ⧵ NH₃

(i) Cisplatin is a neutral complex. Explain why.

(ii) Draw and name the stereoisomer of cisplatin $[Pt(NH_3)_2Cl_2]$. *(3 marks)*

(e) The table below shows the entropy of hydrogen, nitrogen and ammonia.

Substance	$H_2(g)$	$N_2(g)$	$NH_3(g)$
Entropy $S^{\ominus}$ / $JK^{-1}mol^{-1}$	131	191	192

(i) Consider the equations for five processes shown in the table below. For each process predict the sign of the entropy change ΔS.

Process	Sign of ΔS
$H_2O(s) \rightarrow H_2O(l)$	
$NH_3(g) \rightarrow NH_3(l)$	
$NaCl(s) + aq \rightarrow Na^+(aq) + Cl^-(aq)$	
$CaCO_3(s) \rightarrow CaO(s) + CO_2(g)$	
$C_2H_5OH(l) + 3O_2(g) \rightarrow 2CO_2(g) + 3H_2O(g)$	

(5 marks)

(ii) Calculate the entropy change for the formation of ammonia from nitrogen and hydrogen.
$$N_2(g) + 3H_2(g) \rightleftharpoons 2NH_3(g)$$

(iii) Give the expression for the Gibbs free energy change.

(iv) Calculate the Gibbs free energy change for the formation of ammonia from its constituent elements at 25 °C. *(8 marks)*

12 Propanoic acid is a weak acid.

   (a) Select the molecular formula of propanoic acid. Choose one answer.

      A $C_3H_6O_2$

      B CH_3CH_2COOH

      C $CH_3CH_2CH_2COOH$

      D $C_{1.5}H_3O$ *(1 mark)*

   (b) Define the term weak acid.   *(2 marks)*

   (c) A $0.20\,mol\,dm^{-3}$ solution of propanoic acid has a pH of 2.79.
Calculate the K_a value of propanoic acid. Give your answer to two
significant figures.   *(3 marks)*

   (d) A student placed $25.0\,cm^3$ of the $0.100\,mol\,dm^{-3}$ solution of propanoic acid
into a flask.

      (i) Name the piece of apparatus the student should use to measure exactly
$25.0\,cm^3$ of propanoic acid.

      (ii) Calculate the amount, in mol, of propanoic acid in the flask.

      (iii) The propanoic acid was titrated with an aqueous solution of calcium
hydroxide.$22.0\,cm^3$ of calcium hydroxide was required for complete
neutralisation.
Deduce the number of moles of calcium hydroxide required.

      (iv) Calculate the concentration of the calcium hydroxide solution. Give your
answer to three significant figures.

      (v) State the expression for the ionic product of water k_w.

      (vi) Calculate the pH of the calcium hydroxide solution at 25 °C. Give your
answer to two decimal places.   *(8 marks)*

13 A student analysed a multi vitamin and mineral tablet. The tablet contained a small
amount of an iron(II) salt. The tablet was powdered and 0.525 g was dissolved in water.
A small amount of dilute sulfuric acid was added. The solution was then titrated with
potassium manganate(VII).
$15.50\,cm^3$ of $0.002\,00\,mol\,dm^{-3}$ potassium manganate(VII) was required.

   (a) State the colour that the student would observe when exactly the right
amount of potassium manganate(VII) was added.   *(1 mark)*

   (b) Consider the two half equations below.

      $Fe^{2+}(aq) \rightarrow Fe^{3+}(aq) + e^-$

      $MnO_4^-(aq) + 8H^+(aq) + 5e^- \rightarrow Mn^{2+}(aq) + 4H_2O(l)$

      Combine the two half equations to give the overall equation for the reaction. *(1 mark)*

   (c) **(i)** Calculate the amount, in mol, of potassium manganate(VII) used.

      (ii) Calculate the amount, in mol, of Fe^{2+} used.

      (iii) Calculate the percentage by mass of iron in the tablet. Give your answer
to three significant figures.   *(5 marks)*

Section 4
Practical skills

Practical work is firmly at the heart of any chemistry course. It helps you to understand new ideas and difficult concepts, and helps you to appreciate the importance of experiments in testing and developing scientific theories. In addition, practical work develops the skills that scientists use in their everyday work. Such skills involve planning, researching, making and processing measurements, analysing, and evaluating experimental results, as well as the ability to use a variety of apparatus and chemicals safely.

During this course you will be carrying out a number of practicals using a range of apparatus and techniques. Table 1 lists some of the practical activities you will do in your AS/1st year A level course, and questions may be set on these in the written exam. In carrying out these activities, you should become proficient in all the practical skills assessed directly and indirectly in your AS/1st year A-level course. For each activity, the table references the relevant topic or topics in this book. (In addition there will be a teacher assessed pass/fail endorsement of your practical skills.)

▼ **Table 1** *Some practical activities you will cover as part of your course*

	Practical	Topic
1	Making up a volumetric solution and carrying out a titration	2.5, Balanced equations an related calculations
2	Measurement of an enthalpy change	4.3, Measuring enthalpy changes 4.4, Hess's law
3	Investigation of how the rate of a reaction changes with temperature	5.2, The Maxwell-Boltzmann distribution
4	Carrying out test-tube reactions to identify cations and anions in aqueous solutions	9.1, The alkaline earth elements
5	Distillation of a product from a reaction	15.2, Ethanol production
6	Tests for alcohol, aldehyde, alkene, and carboxylic acid	15.3, The reaction of alcohols

Practice questions

The following questions will give you some practice – you may not have done the actual experiment mentioned but you should think about similar practical work that you have done.

Practical 7
Measuring the rate of a reaction

A student was investigating the factors that affect the rate of a reaction between a metal and hydrochloric acid which gives off hydrogen gas. The graphs below show her results for two different concentrations of acid.

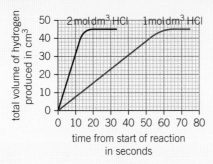

1 Suggest two experimental methods for following this reaction.
2 Use the graphs to calculate the initial rate of the reaction in each experiment.
3 What is important about the *initial* rate of a reaction?
4 What do these results suggest about the order of the reaction with respect to the concentration of hydrochloric acid?
5 What other factors must be kept constant if this conclusion is to be valid?

Practical 8
Measuring E$^{\ominus}$ for an electrochemical cell

A student set up the apparatus below to measure the value of E$^{\ominus}$ for Ni^{2+}(aq) $\rightleftharpoons$ Ni(s)

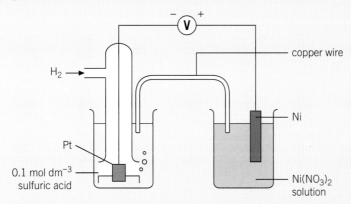

1 Point out three significant errors in the experimental setup.
2 State three experimental conditions not mentioned on diagram the that must be controlled if the potential difference measured is to be E$^{\ominus}$.

Practical 9
Investigating pH change during an acid-base titration

A student obtained the following figures when she added a solution of a base to a solution of an acid of the same concentration.

Volume of base added / cm^3	pH
0	1.00
5	1.13
10	2.24
15	1.31
20	1.41
25	7.01
30	12.00
35	12.11
40	12.24
45	12.34
50	12.41

1 Plot a graph of pH (vertically) against Volume of base added (horizontally)
2 Which value is clearly wrong?
3 The student's teacher suggested that she should have made measurement every 1 cm^3 rather than every 5 cm^3 between 20 cm^3 of base added and 30 cm^3 of base added. Explain why this would be a good idea.
4 Name the instrument used to measure the pH values.
5 Why would universal indicator *not* be suitable for measuring the pH?
6 Which of the following best represents the acid and base used: strong acid/strong base; strong acid/weak base; weak acid/strong base/; weak acid/weak base?

Practical 10
Preparing an organic solid

A student prepared a sample of methyl 3-nitrobenzoate by the nitration of methyl benzoate. Her notes are summarised below.

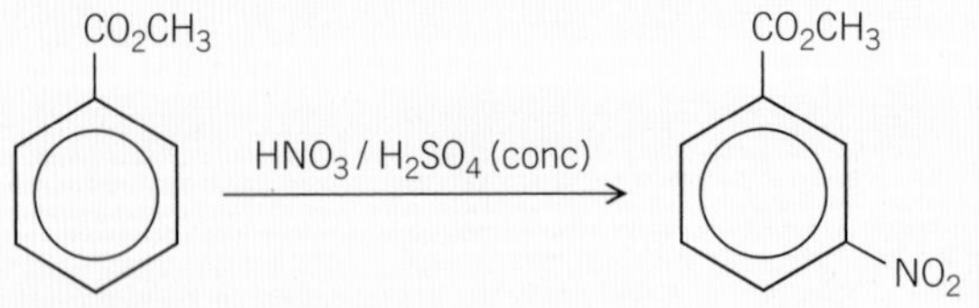

The mixture of acids was slowly added to the methyl benzoate (a liquid) in a conical flask, cooling the flask in an ice-bath to keep the reaction mixture below room temperature. When the reaction was complete, the mixture was poured onto crushed ice to precipitate the solid product. After the ice has melted, the product was separated by suction filtration. It was then purified by **recrystallisation** – the product was dissolved in the minimum volume of boiling ethanol which was then cooled in an ice bath until crystals of the product appeared. These were again separated by suction filtration.

The purity of the product was checked by thin layer chromatography (TLC).

1 Suggest why it was important to keep the reaction mixture cool.
2 What side-product of the reaction is possible?
3 How would the TLC plate appear if the product was a) pure, b) contaminated by a side-product?
4 Suggest another method of checking the purity (or otherwise) of the product.
5 Name and briefly describe the method of purifying a liquid organic compound.

Practical 11
Identifying transition metal ions in solution
You have two very dilute solutions A and B that you want to identify and you know they are solutions of transition metal nitrates.

You have access to sodium hydroxide solution, ammonia solution and sodium carbonate solution.

You first add a little sodium hydroxide solution to a sample of each and this produces a brown gelatinous precipitate in A and a pale blue precipitate in B, so you think A might contain Fe^{3+} ions and B might contain Cu^{2+} ions.

1 What reaction has taken place to produce the results above?
2 Suggest how you could use the ammonia solution and sodium carbonate solution to confirm that A might contain Fe^{3+} ions and B might contain Cu^{2+} ions explaining clearly the results you would expect.

Practical 12
Separation of amino acids by thin layer chromatography (TLC)
A student wrote the following account to explain how to separate a mixture of amino acids using TLC.

A sheet of plastic backed TLC sheet is cut 15 cm long and 5 cm wide. Using a dropping pipette a large spot of the amino acid-containing solution is placed 1 cm up from the bottom of the sheet. The separation solvent is placed to a depth of 2 cm in a beaker 12 cm tall and the TLC sheet is placed so that it rests on the bottom of the beaker. The solvent is then allowed to rise up the TLC plate until it reaches the very top of the plate. The position of the amino acids is then located by shining infra-red light on the plate. The R^f value of each amino acid is then calculated by dividing the length of the plate by the distance moved by the spot.

Point out six errors the student has made.

Section 5
Mathematical skills

Units in calculations

You are expected to use the correct units in calculations.

Units

You still describe the speed of a car in miles per hour. The units miles per hour could be written miles/hour or miles hour^{-1} where the superscript '$^{-1}$' is just a way of expressing per something. In science you use the metric system of units, and speed has the unit metres per second, written m s^{-1}. In each case, you can think of per as meaning divided by.

Units can be surprisingly useful

A mile is a unit of distance and an hour a unit of time, so the unit miles per hour reminds you that speed is distance divided by time.

In the same way, if you know the units of density are grams per cubic centimetre, usually written g cm^{-3}, where cm^{-3} means per cubic centimetre, you can remember that density is mass divided by volume.

Multiplying and dividing units

When you are doing calculations, units cancel and multiply just like numbers. This can be a guide to whether you have used the right method.

For example:

The density of a liquid is 0.8 g cm^{-3}. What is the volume of a mass of 1.6 g of it?

Density = mass / volume

So volume = mass / density

Putting in the values and the units:

volume = 1.6 g / 0.8 g cm^{-3}

Cancelling the gs

volume = 2.0 / cm^{-3}

volume = 2.0 cm^3

If you had started with the wrong equation, such as

volume = density/mass or

volume = mass × density, you would not have ended up with the correct units for volume.

Units to learn

It is a good idea to learn the units of some basic quantities by heart.

	Unit	Comment
volume	dm^3	1 dm^3 is 1 litre (l) which is 1000 cm^3
concentration	mol dm^{-3}	
pressure	Pascals, Pa = N m^{-2}	N m^{-2} are newtons per square metre
enthalpy	kJ mol^{-1}	kJ is kilojoule. Occasionally J mol^{-1} is used.
entropy	J K^{-1} mol^{-1}	joules per kelvin per mole

Standard form

This is a way of writing very large and very small numbers in a way that makes calculations and comparisons easier.

The number is written as number multiplied by ten raised to a power. The decimal point is put to the right of the first digit of the number.

For example:

22 000 is written 2.2 × 10^4.

0.000 002 2 is written 2.2 × 10^{-6}.

How to work out the power to which ten must be raised

Count the number of places you must move the decimal point in order to have one digit before the decimal point.

For example:

0.000 51 = 5.1 × 10^{-4}

51 000 = 5.1 × 10^4

Moving the decimal point to the right gives a negative index (numbers less than 1), and to the left a positive index (numbers greater than one).

(The number 1 itself is 10^0, so the numbers 1–9 are followed by ×10^0 when written in standard form. Can you see why?)

Multiplying and dividing

When *multiplying* numbers expressed in this way, *add* the powers (called indices) and when *dividing*, *subtract* them.

Worked examples

Calculate

a $2 \times 10^5 \times 4 \times 10^6$

b $\dfrac{8 \times 10^3}{4 \times 10^2}$

c $\dfrac{5 \times 10^8}{2 \times 10^{-6}}$

Answer

a $2 \times 10^5 \times 4 \times 10^6$

Multiply $2 \times 4 = 8$. Add the indices to give 10^{11}

Answer $= 8 \times 10^{11}$

b $\dfrac{8 \times 10^3}{4 \times 10^2}$

Divide 8 by 4 = 2. Subtract the indices to give 10^1

Answer $= 2 \times 10^1 = 20$

c $\dfrac{5 \times 10^8}{2 \times 10^{-6}}$

Divide 5 by 2 = 2.5. Subtract the indices $(8 - (-6))$ to give 10^{14}

Answer $= 2.5 \times 10^{14}$

A handy hint for non-mathematicians

Non-mathematicians sometimes lose confidence when using small numbers such as 0.002. If you are not sure whether to multiply or divide, then do a similar calculation with numbers that you are happy with, because the rule will be the same.

Example:

How many moles of water in 0.000 1 g? A mole of water has a mass of 18 g.

Do you divide 18 by 0.000 1 or 0.000 1 by 18?

If you have any doubts about how to do this, then in your head change 0.000 1 g into a more familiar number such as 100 g.

How many moles of water in 100 g? A mole of water has a mass of 18 g.

Now you can see that you must divide 100 by 18. So in the same way you must divide 0.000 1 by 18 in the original problem.

$$\dfrac{0.000\,1}{18} = 5.6 \times 10^{-6}$$

Prefixes and suffixes

In chemistry you will often encounter very large numbers (such as the number of atoms in a mole) or very small numbers (such as the size of an atom).

Prefixes and suffixes are often used with units to help express these numbers. You will come across the following which multiply the number by a factor of 10^n. The red ones are the ones you are most likely to use.

Prefix	Conversion Factor	Symbol
pico	10^{-12}	p
nano	10^{-9}	n
micro	10^{-6}	μ
milli	10^{-3}	m
centi	10^{-2}	c
deci	10^{-1}	d
kilo	10^3	k
mega	10^6	M

So 5400 g $= 5.4 \times 10^3$ g $= 5.4$ kg

Converting to base units

If you want to convert a number expressed with a prefix to one expressed in the base unit, multiply by the conversion factor. If you have a very small or very large number (and have to handle several zeros) the easiest way is to first convert the number to standard form.

Worked example

Convert a) 2 cm and b) 100 000 000 mm to metres

a 2 cm $= 2 \times 10^{-2}$ m $= 0.02$ m

b 100 000 000 mm $= 1 \times 10^8$ mm
$= 1 \times 10^8 \times 10^{-3}$ m $= 1 \times 10^5$ m

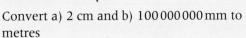 conversion factor

Base units

The SI system is founded on base units. The ones you will meet in chemistry are:

Unit	Symbol	Used for
metre	m	length
kilogram	kg	mass
second	s	time
ampere (amp)	A	electric current
kelvin	K	temperature
mole	mol	amount of substance

Handling data

Sorting out significant figures

Many of the numbers used in chemistry are measurements – for example, the volume of a liquid, the mass of a solid, the temperature of a reaction vessel – and no measurement can be exact. When you make a measurement, you can indicate how uncertain it is by the way you write it. For example a length of 5.0 cm means that you have used a measuring device capable of reading to 0.1 cm, a value of 5.00 cm means that you have measured to the nearest 0.01 cm and so on. So the numbers 5, 5.0 and 5.00 are different, you say they have different numbers of *significant figures*.

What exactly is a significant figure?

In a number that has been found or worked out from a measurement, the significant figures are all the digits known for certain, *plus the first uncertain one* (which may be a zero). The last digit is the uncertain one and is at the limit of the apparatus used for measuring it (Figure 1).

most significant digit   first uncertain digit

32.34

▲ **Figure 1** *A number with four significant figures*

For example if you say a substance has a mass of 4.56 grams it means that you are certain about the 4 and the 5 but not the 6 as you are approaching the limit of accuracy of our measuring device (you will have seen the last figure on a top pan balance fluctuate). The number 4.56 has three significant figures (s.f.).

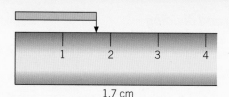

1.7 cm
This ruler gives an answer to two significant figures

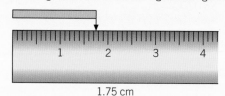

1.75 cm
This ruler gives an answer to three significant figures

▲ **Figure 2** *Rulers with different precision*

When a number contains zeros, the rules for working out the number of significant figures are given below.

- Zeros between digits are significant.
- Zeros to the left of the first non-zero digit are not significant (even when there is a decimal point in the number).
- When a number with a decimal point ends in zeros to the right of the decimal point, these zeros are significant.
- When a number with no decimal point ends in several zeros, these zeros may or may not be significant. The number of significant figures should ideally be stated. For example 20 000 (to 3 s.f.) means that the number has been measured to the nearest 100 but 20 000 (to 4 s.f.) means that the number has been measured to the nearest 10.

The following examples should help you to work out the number of significant figures in your data.

Worked examples

What is the number of significant figures in each of the following?

a   11.23

Answer
4 s.f. all non-zero digits are significant.

b   1100

Answer
2 s.f. (but it could be 2, 3, or 4 significant figures). The number has no decimal point so the zeros may or may not be significant. With numbers with zeros at the end it is best to state the number of significant figures.

c   1100.0

Answer

5 s.f. the decimal point implies a different accuracy of measurement to example (b).

d 0.025

Answer

2 s.f. zeros to the left of the decimal point only fix the position of the decimal point. They are not significant.

Question

1 How many significant figures?
   a 40 000
   b 1.030
   c 0.22
   d 22.00

Using significant figures in answers

When doing a calculation, it is important that you don't just copy down the display of your calculator, because this may have a far greater number of significant figures than the data in the question justifies. Your answer cannot be more certain than the least certain of the information that you used to calculate it. So your answer should contain the same number of significant figures as the measurement that has the smallest number of them.

Worked example

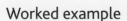

81.0 g (3 s.f.) of iron has a volume of 10.16 cm³ (4 s.f.). What is its density?

Answer

$$\text{Density} = \frac{\text{mass}}{\text{volume}} = \frac{81.0 \text{ g}}{10.16 \text{ cm}^3}$$

= 7.972 440 94 g cm⁻³ (this number has 9 s.f.)
Since our least certain measurement was to 3 s.f., our answer should have 3 s.f., ie 7.97 g cm⁻³

If our answer had been 7.976 440 94, you would have rounded it ip to 7.98 because the fourth significant figure (6) is five or greater.

The other point to be careful about is *when* to round up. This is best left to the very end of the calculation. Don't round up as you go along as this could make a difference to your final answer.

Decimal places and significant figures

The apparatus you use in the laboratory usually reads to a given number of decimal places (for example hundredths or thousandths of a gram). For example, the top pan balances in most schools and colleges usually weigh to 0.01 g which is to two decimal places.

The number of significant figures of a measurement obtained by using the balance depends on the mass you are finding. A mass of 10.38 g has 4 s.f. but a mass of 0.08 has only 1 s.f. Check this with the rules above.

> ### Hint
>
> Calculator displays usually show numbers in standard form in a particular way. For example 2.6 × 10⁻⁴ may appear as 2.6 − 04, a shorthand form which is not acceptable as a way of writing an answer.

Algebra

Equations

You can write an equation if you can show a connection between sets of measurements (variables).

For example, at a *fixed* volume, if you double the temperature (in Kelvin) of a gas, the pressure doubles too.

Mathematically speaking, the pressure p is directly proportional to the temperature T.

$$P \propto T$$

The symbol ∝ means is proportional to.

This is shown in the data in Table 1

▼ Table 1

Temperature/K	Pressure/Pa
100	1000
150	1500
200	2000
250	2500

This also means that the pressure, P, is equal to some constant, k, multiplied by the temperature:

$$P = kT$$

In this case, the constant is 10 and if you multiply the temperature in K by 10, you get the pressure in Pa.

Pressure and volume of a gas also vary. At constant temperature, as the pressure of a gas goes up, its volume goes down. More precisely, if you double the pressure, you halve the volume. Mathematically speaking, volume V, is *inversely* proportional to pressure P.

$$V \propto \frac{1}{P}$$

So $\quad\quad V = k \times \frac{1}{P}$

Or simply $\quad V = \frac{k}{P}$

This is shown by the data in Table 2.

In this case the constant is 24. If you multiply $\frac{1}{P}$ by 24, you get the volume.

▼ **Table 2**

Pressure/Pa	Volume/l	1/P =1/Pa
1	24	1.00
2	12	0.50
3	8	0.33
4	6	0.25

Mathematical symbols	
Symbol	**Meaning**
$\rightleftharpoons$	equilibrium
$<$	less than
$\ll$	much less than
$>$	greater than
$\gg$	much greater than
$\sim$	approximately equal to
$\propto$	proportional to

Question

2a If two variables, x and y, are directly proportional to each other, what happens to one if you quadruple the other?

b Write an expression that means x is inversely proportional to y.

c What happens to the volume of a gas if you triple the pressure at constant temperature?

Handling equations
Changing the subject of an equation

If you can confidently do the next exercise, go straight to the section Substituting into equations and try that. Otherwise work through Rearranging equations.

Question

3 The equation that connects the pressure P, volume V and temperature T of a mole of gas is

$$PV = RT$$

Where P, V, and T are variables and R is a constant called the gas constant.

Rearrange the equation to find:

    a P in terms of V, R and T

    b V in terms of P, R and T

    c T in terms of P, V and R

    d R in terms of P, V and T

Rearranging equations

Start with a simple relation because the rules are the same however complicated the equation.

$$a = \frac{b}{c}$$

where $b = 10$ and $c = 5$,

It is easy to see that substituting these values into the expression

$$a = \frac{10}{5} = 2.$$

But what do you do if you need to find b or c from this equation?

you need to rearrange the equation so that b (or c) appears on its own on the left hand side of the equation like this

$$b = ?$$

$a = \frac{b}{c}$ means $a = b \div c$

Step 1: Multiply both sides of the equation by c, because b is *divided* by c, so to get b on it own you must *multiply* by c.

Remember that to keep an equation valid whatever you do to one side you must do to the other – think of it as a see-saw, with the = sign as the pivot.

So now $c \times a = \frac{b \times c}{c}$

usually written $ca = \frac{bc}{c}$

Now cancel the c's on the right since b is being both multiplied and divided by c

Which leaves $c \times a = b$

Or $b = c \times a$ usually written $b = ca$

You can now rearrange this equation in the same way to find c.

Step 2: $b = c \times a$. Divide both sides by a

$$\frac{b}{a} = \frac{cax}{a}$$

Now cancel the a's on the right

So $\frac{b}{a} = c$

Notice that because c started on the bottom, a two-step process was necessary. You found as expression for b first and then found one for c.

Question

4   Find the variable in brackets in terms of the others.

  a $p = \dfrac{q}{r}$ (q)

  b $n = mt$ (m)

  c $g = \dfrac{fe}{h}$ (h)

  d $\dfrac{pr}{e} = s$ (r)

Substituting into equations

When you are asked to substitute numerical values into an equation, it is essential that you carry out the mathematical operations in the right order.

A useful aid to remembering the order is the word **BIDMAS**:

Brackets

Indices

Division

Multiplication

Addition

Subtraction

Graphs

The graph in Figures 3 and 4 show the rate of reaction between hydrochloric acid and magnesium to produce hydrogen gas.

The gradient of a straight line section of a graph is found by dividing the length of the line A (vertical) by the length of line B (horizontal).

In the case of experiment 1 above this tells us the rate at which hydrogen is produced, between 0 ands 10 seconds. It will have units:

Rate $= 20\,\text{cm}^3/10\,\text{s} = 2\,\text{cm}^3/\text{s}$

or $2\,\text{cm}^3\,\text{s}^{-1}$

The steeper the line is, the greater the rate.

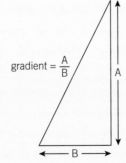

gradient $= \dfrac{A}{B}$

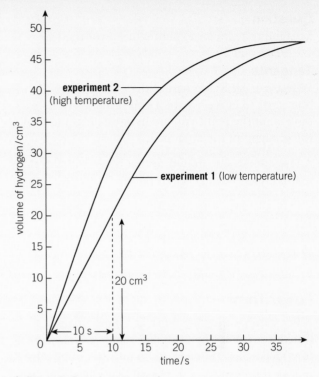

▲ **Figure 3** *Reaction rate graph*

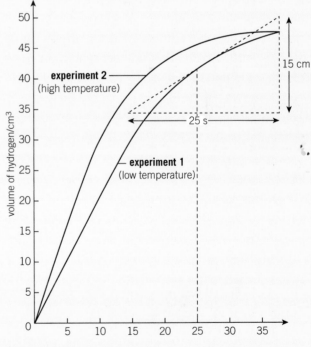

▲ **Figure 4** *Reaction rate graph*

Question

5 Find the rate for experiment 2 over the first 10 seconds.

Tangents

When you get to the part of the graph where the line starts to curve, the best you can do is to take a tangent at the point you are investigating and find the gradient of the tangent.

A tangent is a line drawn so that it just touches the graph line at one particular point.

The rate in experiment 1, after 25 seconds is
$$\frac{15\,cm^3}{25\,s} = 0.6\,cm^3\,s^{-1}$$

Questions

6 Find the rate after 30 s in experiment 2.

Logarithms

A logarithm, or log for short, is a mathematical function – the log of a number represents the power to which a base number (often ten) has to be raised to give the number. This is easy to do for numbers that are multiples of ten such as 100 or 10 000. 100 is 10^2, so the log to the base ten (written $\log_{10}$ or just log) of 100 is 2 and $\log_{10}$ of 10,000 is 4. Logs can also be negative numbers. The log of $\frac{1}{1000}$ is -3 as $\frac{1}{1000}$ is 10^{-3}. With other numbers you must use a calculator to find the log. $\log_{10} 72.33$ is 1.8593.

Make sure that you are confident using your calculator to find logs.

Question

7 What is the $\log_{10}$ of:
  a   1000
  b $\frac{1}{100}$
  c   0.0001
  d   48.2
  e   0.037

You will need to use a calculator for the last two you can go back from the log to the original number by using the antilog, or inverse log function, of the calculator. Log (10^{27}) is 27 (which doesn't need a calculator to work out) and log (21379.6) is 4.33.

Make sure that you are confident to use your calculator to find antilogs (inverse logs).

Question

8 What is the antilog (inverse log) of:
  a   3
  b -2
  c   14
  d   8.2
  e   0.37

You will need to use a calculator for the last two. The log function turns very large or very small numbers into more manageable numbers without losing the original number (which you can recover using the antilog function). So $\log_{10} 6 \times 10^{23} = 23.778$ and $\log_{10} 1.6 \times 10^{-19} = -18.79$.

This can be very useful in plotting graphs as it can allow numbers with a wide range in magnitudes to be fitted onto a reasonable size of graph. For example, the successive ionisation energies of sodium range from 496 to 159 079 whereas their logs range from just 2.695 to 5.201.

Another important use of logs in chemistry is the pH scale, which measures acidity, and depends on the concentration of hydrogen ions (H^+) in a solution. This can vary from around $5\,mol\,dm^{-3}$ to around $5 \times 10^{-15}\,mol\,dm^{-3}$ – an enormous range. If you use a log scale, this becomes 0.6989 to -14.301 – a much more manageable range.

When multiplying numbers you add their logs and when dividing you subtract them. This is easy to see with numbers that are multiples of ten.

$100 \times 10\,000 = 1\,000\,000$, that is, $10^2 \times 10^4 = 10^6$ and $\frac{10^4}{10^2} = 10^2$, but the same rules apply for more awkward numbers.

Geometry and trigonometry

Simple molecules adopt a variety of shapes. The most important of these are shown below with the relevant angles. When drawing representations of three dimensional shapes, the convention is to show bonds coming out of the paper as wedges which get thicker as they come towards you. Bonds going into the paper are usually drawn as dotted lines or reverse wedges.

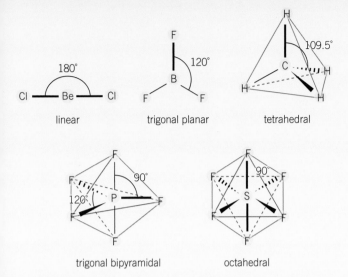

180°	120°	109.5°
Cl — Be — Cl	F–B(–F)(–F)	C with H atoms
linear	trigonal planar	tetrahedral

trigonal bipyramidal   octahedral

▲ **Figure 5** *Three-dimensional drawings of molecular shapes*

The three-dimensional shapes are based on
geometrical solid figures as shown

Data

Table A Infrared

Bond	Wavenumber / cm^{-1}
C—H	2850–3300
C—C	750–1100
C=C	1620–1680
C=O	1680–1750
C—O	1000–1300
O—H (alcohols)	3230–3550
O—H (acids)	2500–3000
N—H	3300–3500

Table B ^{1}H NMR chemical shift data

Type of proton	δ/ppm
ROH	0.5–5.0
RCH$_3$	0.7–1.2
RNH$_2$	1.0–4.5
R$_2$CH$_2$	1.2–1.4
R$_3$CH	1.4–1.6
R—C(=O)—C—H	2.1–2.6
R—O—C—H	3.1–3.9
RCH$_2$Cl or Br	3.1–4.2
R—C(=O)—O—C—H	3.7–4.1
R₂C=CH (C=C, R H)	4.5–6.0
R—C(=O)—H	9.0–10.0
R—C(=O)—O—H	10.0–12.0

Table C ^{13}C NMR chemical shift data

Type of carbon	δ/ppm
—C—C—	5–40
R—C—Cl or Br	10–70
R—C—C(=O)—	20–50
R—C—N	25–60
—C—O—	50–90
C=C	90–150
R—C≡N	110–125
(benzene ring)	110–160
R—C(=O)—	160–185
R—C(=O)—	190–220

Phosphate and sugars

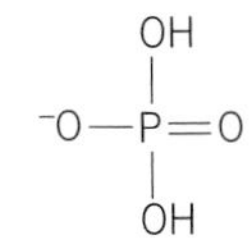

phosphate

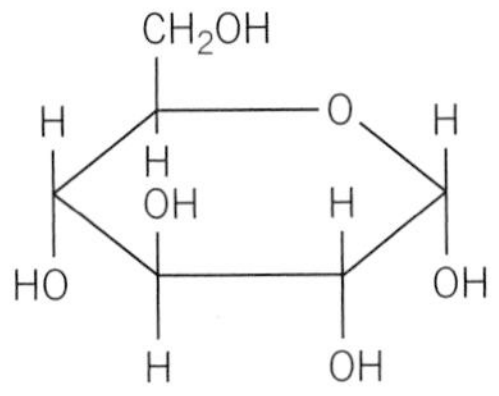

glucose

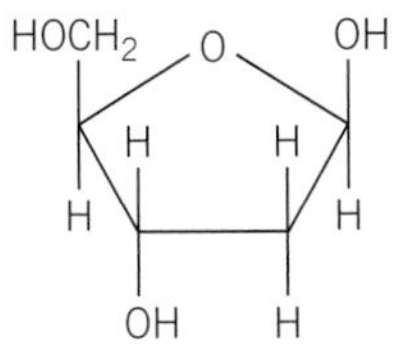

2-deoxyribose

Bases

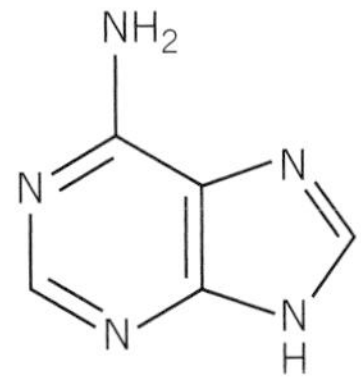

adenine

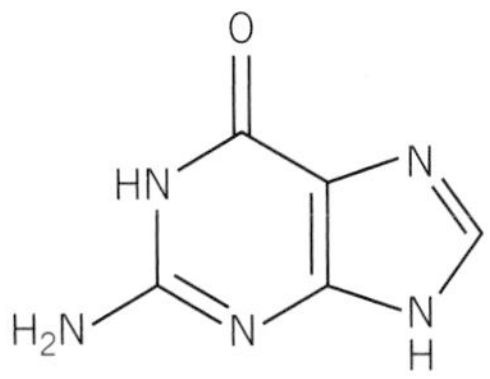

guanine

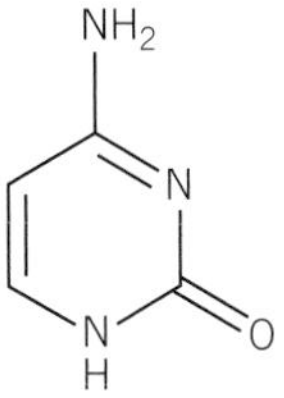

cytosine

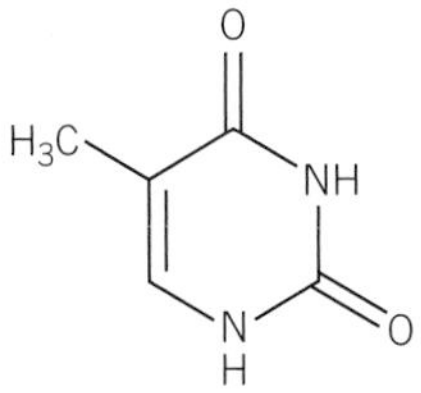

thymine

Amino acids

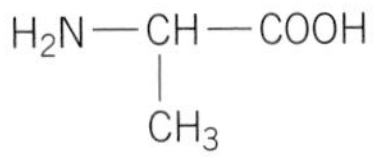

$H_2N-CH-COOH$
$\quad\quad\;\; |$
$\quad\quad\; CH_3$

alanine

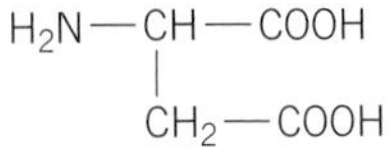

$H_2N-CH-COOH$
$\quad\quad\;\; |$
$\quad\quad\; CH_2-COOH$

aspartic acid

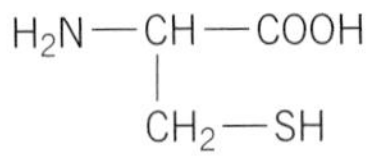

$H_2N-CH-COOH$
$\quad\quad\;\; |$
$\quad\quad\; CH_2-SH$

cysteine

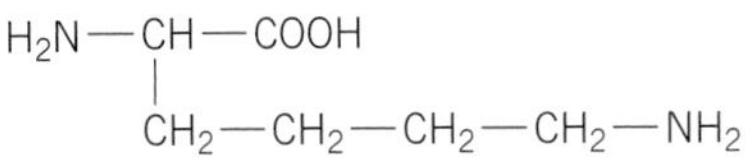

$H_2N-CH-COOH$
$\quad\quad\;\; |$
$\quad\quad\; CH_2-CH_2-CH_2-CH_2-NH_2$

lysine

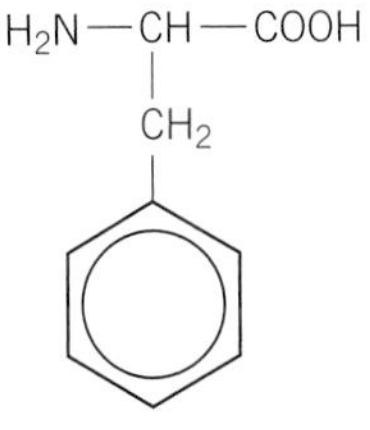

$H_2N-CH-COOH$
$\quad\quad\;\; |$
$\quad\quad\; CH_2$

phenylalanine

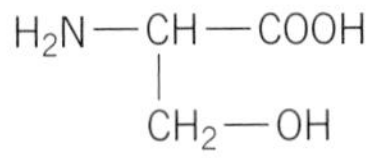

$H_2N-CH-COOH$
$\quad\quad\;\; |$
$\quad\quad\; CH_2-OH$

serine

Haem B

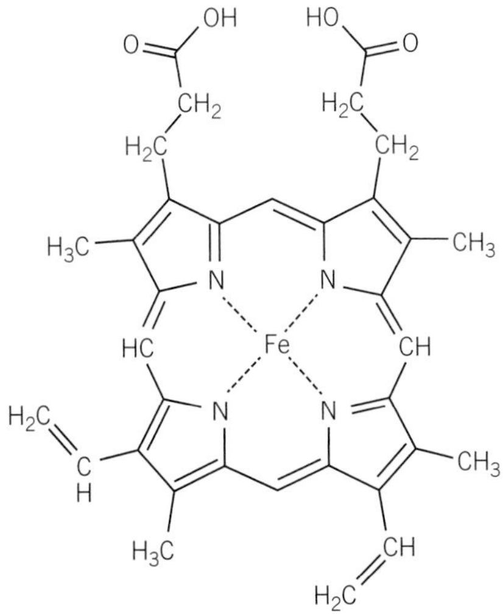

Periodic Table

The Periodic Table of the elements

Key

| relative atomic mass |
| **atomic symbol** |
| name |
| atomic (proton) number |

Example:
| 1.0 |
| **H** |
| hydrogen |
| 1 |

(1)	(2)											(13)	(14)	(15)	(16)	(17)	(18)
																	4.0 **He** helium 2
6.9 **Li** lithium 3	9.0 **Be** beryllium 4											10.8 **B** boron 5	12.0 **C** carbon 6	14.0 **N** nitrogen 7	16.0 **O** oxygen 8	19.0 **F** fluorine 9	20.2 **Ne** neon 10
23.0 **Na** sodium 11	24.3 **Mg** magnesium 12	(3)	(4)	(5)	(6)	(7)	(8)	(9)	(10)	(11)	(12)	27.0 **Al** aluminium 13	28.1 **Si** silicon 14	31.0 **P** phosphorus 15	32.1 **S** sulfur 16	35.5 **Cl** chlorine 17	39.9 **Ar** argon 18
39.1 **K** potassium 19	40.1 **Ca** calcium 20	45.0 **Sc** scandium 21	47.9 **Ti** titanium 22	50.9 **V** vanadium 23	52.0 **Cr** chromium 24	54.9 **Mn** manganese 25	55.8 **Fe** iron 26	58.9 **Co** cobalt 27	58.7 **Ni** nickel 28	63.5 **Cu** copper 29	65.4 **Zn** zinc 30	69.7 **Ga** gallium 31	72.6 **Ge** germanium 32	74.9 **As** arsenic 33	79.0 **Se** selenium 34	79.9 **Br** bromine 35	83.8 **Kr** krypton 36
85.5 **Rb** rubidium 37	87.6 **Sr** strontium 38	88.9 **Y** yttrium 39	91.2 **Zr** zirconium 40	92.9 **Nb** niobium 41	95.9 **Mo** molybdenum 42	[98] **Tc** technetium 43	101.1 **Ru** ruthenium 44	102.9 **Rh** rhodium 45	106.4 **Pd** palladium 46	107.9 **Ag** silver 47	112.4 **Cd** cadmium 48	114.8 **In** indium 49	118.7 **Sn** tin 50	121.8 **Sb** antimony 51	127.6 **Te** tellurium 52	126.9 **I** iodine 53	131.3 **Xe** xenon 54
132.9 **Cs** caesium 55	137.3 **Ba** barium 56	138.9 **La*** lanthanum 57	178.5 **Hf** hafnium 72	180.9 **Ta** tantalum 73	183.8 **W** tungsten 74	186.2 **Re** rhenium 75	190.2 **Os** osmium 76	192.2 **Ir** iridium 77	195.1 **Pt** platinum 78	197.0 **Au** gold 79	200.6 **Hg** mercury 80	204.4 **Tl** thallium 81	207.2 **Pb** lead 82	209.0 **Bi** bismuth 83	[209] **Po** polonium 84	[210] **At** astatine 85	[222] **Rn** radon 86
[223] **Fr** francium 87	[226] **Ra** radium 88	[227] **Ac†** actinium 89	[261] **Rf** rutherfordium 104	[262] **Db** dubnium 105	[266] **Sg** seaborgium 106	[264] **Bh** bohrium 107	[277] **Hs** hassium 108	[268] **Mt** meitnerium 109	[271] **Ds** darmstadtium 110	[272] **Rg** roentgenium 111							

Elements with atomic numbers 112–116 have been reported but not fully authenticated

* 58 – 71 Lanthanides

140.1 **Ce** cerium 58	140.9 **Pr** praseodymium 59	144.2 **Nd** neodymium 60	144.9 **Pm** promethium 61	150.4 **Sm** samarium 62	152.0 **Eu** europium 63	157.3 **Gd** gadolinium 64	158.9 **Tb** terbium 65	162.5 **Dy** dysprosium 66	164.9 **Ho** holmium 67	167.3 **Er** erbium 68	168.9 **Tm** thulium 69	173.0 **Yb** ytterbium 70	175.0 **Lu** lutetium 71

† 90 – 103 Actinides

232.0 **Th** thorium 90	231.0 **Pa** protactinium 91	238.0 **U** uranium 92	237.0 **Np** neptunium 93	239.1 **Pu** plutonium 94	243.1 **Am** americium 95	247.1 **Cm** curium 96	247.1 **Bk** berkelium 97	252.1 **Cf** californium 98	[252] **Es** einsteinium 99	[257] **Fm** fermium 100	[258] **Md** mendelevium 101	[259] **No** nobelium 102	[260] **Lr** lawrencium 103

Glossary

A

Acid Brønsted–Lowry: a proton donor; Lewis: an electron pair acceptor.

Acid derivative An organic compound related to a carboxylic acid of formula RCOZ, where Z = —Cl, —NHR, —OR or —OCOR.

B

Base Brønsted–Lowry: a proton acceptor; Lewis: an electron pair donor.

Base peak The peak representing the ion of greatest abundance (the tallest peak) in a mass spectrum.

Buffer A solution that resists change of pH when small amounts of acid or base are added or on dilution.

C

Chelation The process by which a multidentate ligand replaces a monodentate ligand in forming co-ordinate (dative) bonds to a transition metal ion.

Chiral This means 'handed'. A chiral molecule exists in two mirror image forms that are not superimposable.

Chiral centre An atom to which four different atoms or groups are bonded. The presence of such an atom causes the parent molecule to exist as a pair of nonsuperimposable mirror images.

Co-ordination number The number of ligand molecules bonded to a metal ion.

E

End point The point in a titration when the volume of reactant added just causes the colour of the indicator to change.

Entropy A numerical measure of disorder in a chemical system.

Equivalence point The point in a titration at which the reaction is just complete.

F

Fatty acid A long-chain carboxylic acid.

H

Hydration A reaction in which water is added.

Hydrolysis A reaction of a compound or ion with water.

I

Inductive effect The electronreleasing effect of alkyl groups such as —CH_3 or —C_2H_5.

L

Ligand An atom, ion or molecule that forms a co-ordinate (dative) bond with a transition metal ion using a lone pair of electrons.

O

Optical isomer Pairs of molecules that are non-superimposable mirror images.

Order of reaction In the rate expression, this is the sum of the powers to which the concentrations of all the species involved in the reaction are raised. If rate = $k[A]^a[B]^b$, the overall order of the reaction is a + b.

P

pH A scale for measuring acidity and alkalinity. pH = $-\log_{10} [H^+]$ in a solution.

Polarised This describes an atom or ion where the distribution of charge around it is distorted from the spherical.

Protonated Describes an atom, molecule or ion to which a proton (an H^+ ion) has been added.

R

Racemate A mixture of equal amounts of two optical isomers of a chiral compound. It is optically inactive.

Rate constant The constant of proportionality in the rate expression.

Rate-determining step The slowest step in the reaction mechanism. It governs the rate of the overall reaction.

Rate expression A mathematical expression showing how the rate of a chemical reaction depends on the concentrations of various chemical species involved.

Reaction mechanism The series of simple steps that lead from reactants to products in a chemical reaction.

S

Strong acid An acid that is fully dissociated into ions in solution.

T

Triglyceride An ester formed between glycerol (propane-1,2,3-triol) and three fatty acid molecules.

W

Weak acid An acid that is only slightly dissociated into ions in solution.

Answers to summary questions

17.1

1 a +788 kJ mol⁻¹

   b This is the reverse of the equation for lattice enthalpy formation, so the sign of ΔH is changed.

   c The enthalpy change of lattice dissociation.

2 a The electron is attracted by the sodium nucleus, so energy must be put in to remove it.

   b The electron is attracted by the chlorine nucleus so energy is given out during this process.

3 a i $Al(g) \rightarrow Al^+(g) + e^-$

    ii $Al^+(g) \rightarrow Al^{2+}(g) + e^-$

   b The sum of the enthalpies for **ai** and **ii**.

17.2

1 a

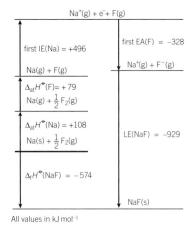

All values in kJ mol⁻¹

   b –929 kJ mol⁻¹

The first noble gas compound

1 $O(g) \rightarrow O^+(g) + e^-$

2 Xenon and krypton have lower ionisation energies than helium and neon, therefore it will be easier to form positive ions Xe⁺ and Kr⁺.

3 It is a compound of a metal and a non-metal – these are usually ionic and therefore expected to have high boiling points.

4 Xe is +1 as it has a single positive charge. F is always –1 in compounds and there are six of them. So the Pt must be +5 so that the sum of the oxidation numbers in the neutral compound is zero.

17.3

1

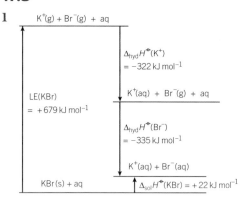

+22 kJ mol⁻¹.
Small because the energy put in to break the lattice is of similar size to that given out when the ions are hydrated.

2 They are small and highly charged positive ions so they strongly polarise negative ions. The calculated value would be greater because there is extra covalent bonding.

17.4

1 a i Approximately zero, two solids produce two solids.

    ii Significantly positive, a solid produces several moles of gases.

    iii Significantly negative, a gas turns into a solid.

    iv Significantly positive, a liquid turns into a gas.

   b i –7.8 J K⁻¹ mol⁻¹

    ii +876.4 J K⁻¹ mol⁻¹

    iii –174.8 J K⁻¹ mol⁻¹

    iv +119 J K⁻¹ mol⁻¹

   The predictions are upheld.

2 a i +493 kJ mol⁻¹   **ii** –52 kJ mol⁻¹

    iii It is feasible at 6000 K

   b 5523 K

3 –284 J K⁻¹ mol⁻¹

Determining an entropy change

1 240 kJ   **2** $M_r = 18.0$ 5.55 mol

3 43.2 kJ   43 200 J   **4** 115.8 J K⁻¹ mol⁻¹

5 2 s.f. as 2.4 kW has the smallest number of sf in the data, so $\Delta_{vap}S = 120$ J K⁻¹ mol⁻¹ to 2 s.f. (We cannot be sure of the number of s.f. for the two values of 100 g and 100 but they are probably 3 s.f.)

6 Heat loss from the sides of the kettle. The kettle should be insulated.

7 1%

8 There are hydrogen bonds between the molecules in water in the liquid state. This makes the liquid state more ordered that for non-hydrogen bonding liquids.

18.1

1 A reactant – its concentration decreases with time.

2 1.3×10^{-3} mol dm^{-3} s^{-1} to 2 sf

3 The rate of reaction after 300 seconds.

4 That at time 0 seconds, the gradient will be steeper than that at time 600 seconds.

5 The rate of reaction at the beginning is greatest as the concentration of reactants is greatest here and then decreases as reactants are used up.

Fast reactions
There are several possibilities, for example, two could combine to form an O_2 molecule or one could recombine with a ClO• radical in the reverse of the original reaction, and so on.

18.2

1 rate = k[A][B][C]^2

2 **a i** 1   **ii** 1   **iii** 2

   b i double   **ii** double   **iii** quadruple

   c i 1   **ii** 5   **iii** 6

    iv 3   **iv** 3

   d dm^9 mol^{-3} s^{-1}

3 **a** the rate constant

   b i 2   **ii** 0   **iii** 0

    iv 1

   c 3

   d dm^6 mol^{-2} s^{-1}

   e a catalyst

4 D It is impossible to tell without experimental data.

18.3

1 **a i** 1   **ii** 2

   b 3   **c** 27 mol dm^{-3} s^{-1} **d** rate = k[A][B]^2

   e No, other species might be involved that have not been investigated.

The iodine clock reaction

1 Iodide ions (oxidation state –1) are oxidised to iodine (oxidation state 0)

2 Hydrogen peroxide

3 The rate doubles

4 First order

5 Nothing

6 By making a stock solution and diluting it with water.

7 Temperature, the concentration of potassium iodide, the concentration of H$^+$ ions.

Reaction rate and temperature

1 The relationship is not linear (straight line). The rate rises rapidly with relatively small increases in temperature.

2 This is because as the temperature rises a much higher proportion of collisions between the reactants have enough energy to react, that is, they collide with more than the activation energy.

18.4

1 **a** 36 in 10^{19} molecules

   b 13 in 10^{18} molecules

2 103.75 kJ mol^{-1}

18.5

1 **a** B   **b** F   **c** Step (ii)

19.1

1 **a i** move right   **ii** move left

   b i move left   **ii** move right

   c i no effect   **ii** move right

2 **a** increase   **b** no change   **c** no change

3 **a i** move left   **ii** move left

    iii no change unit Pa−1

   b i move left   **ii** no change

    iii no change no unit

   c i move right   **ii** move right

    iii no change unit Pa

20.1

1 **a** Ni(s)|Ni^{2+}(aq) ‖ Ag$^+$(aq)|Ag(s)

    $E = + 1.05$ V

   b Electrons would flow from the nickel electrode to the silver electrode.

    Ni(s) → Ni^{2+}(aq) + 2e$^-$

    Ag$^+$(aq) + e$^-$ → Ag(s)

    Overall:

    Ni(s) + 2Ag$^+$(aq) → Ni^{2+}(aq) + 2Ag(s)

2 **a** +0.63 V   **b** −0.63 V

20.2

1 0.90 V zinc electrode positive

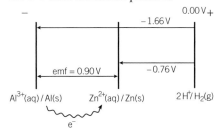

2 **a** +0.93 V **b** −0.53 V **c** +0.97 V **d** +0.26 V

3 Chlorine and bromine, but not iodine.

4 **a** no   **b** no

20.3

Answer from main text.

21.1

1 **a** acid: HNO_3
   base: OH^-

   b acid: CH_3COOH
   base: H_2O

2 1×10^{-10} $mol\,dm^{-3}$

3 **a** H_2O **b** NH_4^+ **c** H_3O^+ **d** HCl

21.2

1 2.00 **2** 1×10^{-6} $mol\,dm^{-3}$ **3** 1×10^{-5} $mol\,dm^{-3}$

4 1.70 **5** 13.30

Mixing bathroom cleaners

1 $CaCO_3(s) + 2HCl(aq) \rightarrow CaCl_2(aq) + CO_2(g) +$
   $H_2O(l)$

2 5×10^{-2} $mol\,HCl$ **3** 5×10^{-2} $mol\,Cl_2$ **4** $1200\,cm^3$

5 Some chlorine would remain dissolved in the
   water in the toilet bowl.

21.3

1 chloroethanoic acid   **2** They are the same.

3 **a** 1.94   **b** 3.10

21.4

1 **a** $2NaOH(aq) + H_2SO_4(aq) \rightarrow Na_2SO_4(aq) +$
   $2H_2O(l)$

   b 0.50 **c** 1.5×10^{-3} **d** $0.120\,mol\,dm^{-3}$

2 **a i** A   **ii** B

   b i B   **iv** A

   c Strong, as there is a rapid pH change in the
   alkaline region at the equivalence point.

21.5

1 b and d   **2** b and d

21.6

1 **a** 5.07   **b** 4.20

Making a buffer solution

1 **a** 122.1   **b** 144.1

2 **a** 1.53 g benzoic acid   **b** 3.60 g

3 Accurately weigh out the two compounds into
   separate weighing boats. Using a funnel in the
   mouth of the flask, transfer all of each solid into
   the flask using a wash bottle containing distilled
   (deionised) water. Fill the flask with water to
   a few centimetres below the graduation mark.
   Stopper the flask and shake until all the solid has
   dissolved. Now carefully fill the flask with water
   using the wash bottle until the meniscus of the
   solution sits on the graduation line. Stopper
   the flask and shake again to ensure that the
   concentration of the buffer solution is uniform.

22.1

1 They are malleable (can be beaten into sheets)
   and ductile (can be drawn into wires). They are
   good conductors of heat.

2 They tend to be brittle – they are poor
   conductors of heat.

3 **a** +1

   b −1. Oxidation state of oxygen is usually −2.

   c +2 −2 +1
   $Mg(OH)_2 = + 2 + 2(-2 + 1) = 0$

4 +6

22.2

1 **a** $Na_2O(s) + H_2O(l) \rightarrow 2NaOH(aq)$

   b i before +1, after +1   **ii** neither

2 **a** OH^- ions   **b** greater than 7

3 **a** acidic

   b It is a non-metal.

   c $P_4O_6(s) + 6H_2O(l) \longrightarrow 4H_3PO_3(aq)$

The structures of oxo-acids and their anions

1 109.5°

2 They are all the same (and intermediate between
   the P=O and P—O lengths).

3 10

4 0.150 nm (the average of S—O and S=O)

5 10

22.3

1 a $Na_2O(s) + 2HCl(aq) \rightarrow 2NaCl(aq) + H_2O(l)$

b $MgO(s) + H_2SO_4(aq) \rightarrow MgSO_4(aq) + H_2O(l)$

c $Al_2O_3(s) + 6HNO_3(aq) \rightarrow 3H_2O(l) +$
$$2Al(NO_3)_3(aq)$$

2 a   +4 −2   +1 −2 +1   +1 +4 −2   +1 −2

$SiO_2(s) + 2NaOH(aq) \rightarrow Na_2SiO_3(aq) + H_2O(l)$

b No

c No change in oxidation number of any of the elements occurs.

3 $P_4O_{10}(s) + 6H_2O(l) \rightarrow 4H_3PO_4(aq)$

23.1

1 a $1s^2\ 2s^2\ 2p^6\ 3s^2\ 3p^6\ 3d^5$

b $1s^2\ 2s^2\ 2p^6\ 3s^2\ 3p^6\ 3d^4$

2 a $1s^2\ 2s^2\ 2p^6\ 3s^2\ 3p^6$

b the two 4s electrons

c one of the 3d electrons

23.2

1 a i octahedral   **ii** octahedral   **iii** tetrahedral

b 6, 6, 4

c Cl^- is a larger ligand than either ammonia or water, so fewer ligands can fit around the metal ion.

2 a The two negatively charged oxygen atoms, because they have lone pairs of electrons.

b   **c** bidentate

Ionisation isomerism

1 They both a have lone pairs of electrons.

2 Compound 1 as it has most ions.

3 $AgNO_3(aq) + Cl^-(aq) \rightarrow AgCl(s) + NO_3^-(aq)$

4 Compound 1 as it has most free chloride ions.

5 *Cis trans (EZ)* isomerism

6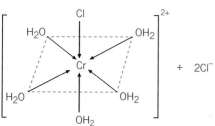

23.3

1 a The copper ion has part-filled d-orbitals, so electrons can move from one d-orbital to another and absorb light. Zinc has full d-orbitals.

b blue   **c** They are absorbed.

2 a 5 : 5   **b** 5 : 5   **c** $[NiEDTA]^{2-}$

23.4

1 a $Zn(s) + 2VO_2^+(aq) + 4H^+(aq) \rightarrow$
$$2H_2O(l) + 2VO^{2+}(aq) + Zn^{2+}(aq)$$
$Zn(s) + 2VO^{2+}(aq) + 4H^+(aq) \rightarrow$
$$Zn^{2+}(aq) + 2H_2O(l) + 2V^{3+}(aq)$$
$Zn(s) + 2V^{3+}(aq) \rightarrow Zn^{2+}(aq) + 2V^{2+}(aq)$

b This low oxidation state can be oxidised by air.

2 0.0698 g

3 $E^{\ominus}$ for the reaction:

$$Cl_2(g) + 2e^- \rightarrow 2Cl^-$$

is + 1.36 V, but $E^{\ominus}$ for the reaction of $Cr_2O_7^{2-}$ and Cl^- and is 0.03 V. Therefore the reaction is not feasible. Hence potassium dichromate(VI) will not oxidise chloride ions to chlorine.

23.5

1 a Homogeneous catalysts are in the same phase as the reactants, heterogeneous catalysts are in a different phase from the reactants.

b i heterogeneous   **ii** heterogeneous

iii homogeneous

2 It lowers the activation energy. This means that reactions can be carried out at a lower temperature than without the catalyst, so saving energy and money.

3 First, iron(III) oxidises iodide to iodine, itself being reduced to iron(II):

$$2Fe^{3+}(aq) + 2I^-(aq) \rightarrow 2Fe^{2+}(aq) + I_2(aq)$$

Then the peroxodisulfate ions oxidise the iron(II) back to iron(III):

$$S_2O_8^{2-}(aq) + 2Fe^{2+}(aq) \rightarrow 2SO_4^{2-}(aq) + 2Fe^{3+}(aq)$$

Catalyst is needed since the reaction is slow. The reaction is slow since both reactants are negative ions. These ions will repel each other, reducing the chance of effective collisions.

4 $4Mn^{2+}(aq) + MnO_4^-(aq) + 8H^+(aq) \rightarrow$
$$5Mn^{3+}(aq) + 4H_2O(l)$$
$2Mn^{3+}(aq) + C_2O_4^{2-}(aq) \rightarrow 2CO_2(g) + 2Mn^{2+}(aq)$

Multiply the top equation by 2 and the lower one by 5 to give the same number of Mn^{3+} ions:

$$8Mn^{2+}(aq) + 2MnO_4^-(aq) + 16H^+(aq) \rightarrow 10Mn^{3+}(aq) + 8H_2O(l)$$

$$10Mn^{3+}(aq) + 5C_2O_4^{2-}(aq) \rightarrow 10CO_2(g) + 10Mn^{2+}(aq)$$

Then combine the two equations and cancel any species that appear on both sides:

$$2MnO_4^-(aq) + 16H^+(aq) + 5C_2O_4^{2-}(aq) \rightarrow 2Mn^{2+}(aq) + 8H_2O(l) + 10CO_2(g)$$

24.1

1 **a** +7 **b** +6 **c** +6

2 Acid–base. It is not redox as the chromium is in the +6 state before and after. One of the oxygen atoms accepts two protons (H^+ ions).

24.2

1 $Cu(H_2O)_4Cl_2$ or $Cu(H_2O)_2Cl_2$

2 **a**

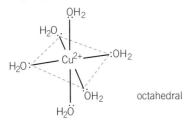

octahedral

b Bromide ions are bigger than water molecules, so fewer can fit around the copper ion. The shape is tetrahedral.

3 $[Cu(H_2O)_6]^{2+}(aq) + en \rightarrow$
$\qquad [Cu(H_2O)_4(en)]^{2+}(aq) + 2H_2O(l)$
$[CuH_2O_4(en)]^{2+}(aq) + en \rightarrow$
$\qquad [Cu(H_2O)_2(en)_2]^{2+}(aq) + 2H_2O(l)$
$[Cu(H_2O)_2(en)_2]^{2+}(aq) + en \rightarrow$
$\qquad [Cu(en)_3]^{2+}(aq) + 2H_2O(l)$

At each step two entities produce three, therefore the entropy change of each step is likely to be positive.

4 **a** No, it remains at +2.

b $[Co(H_2O)_6]^{2+}$ is octahedral and $[CoCl_4]^{2-}$ is tetrahedral.

c The ligands have changed as has the co-ordination number.

24.3

1 They are smaller and more highly charged and therefore more strongly polarising. They can thus weaken one of the O—H bonds in one of

the water molecules that surround them, so releasing a H^+ ion.

2 Aluminium is not a transition metal. It has no part-filled d-orbitals. Most transition metal compounds are coloured because of the electrons moving between part-filled d-orbitals that absorb light.

3 Both NH_3 and H_2O are of similar size and are neutral ligands.

4 **a** **i** $[Fe(H_2O)_6]^{3+}(aq) + 3OH^-(aq) \rightarrow$
$\qquad Fe(H_2O)_3(OH)_3(s) + 3H_2O(l)$

ii $[Cu(H_2O)_6]^{2+}(aq) + 4NH_3(aq) \rightarrow$
$\qquad [Cu(NH_3)_4(H_2O)_2]^{2+}(aq) + 4H_2O(l)$

b **i** The orange brown solution changes to a reddish brown solid.

ii The pale blue solution turns to dark blue.

25.1

1 **a**

H—C—H
H—C—H
H—C—C—C—C—C—C—H (with branches)
H—C—H

b

H—C—H
H—C—C—C—C—C—H (with branches)
H—C—H

2 **a** propan-1-ol **b** 2-chloropropane

c hex-3-ene **d** ethanol

e pentanoic acid

25.2

1 **b**

2 $CH_3 - C^* - C$ (with Cl, H, and O, OH)

3 $HO_2C - C^* - C^* - CO_2H$ (with OH, OH, H, H)

25.3

1 **a** 2-hydroxybutanoic acid

H—C—C—C—COOH (with H, H, OH and H, H, H)

b Yes, the carbon to which the –OH group is bonded has four different groups attached to it.

2 a 2-methyl-2-hydroxypropanoic acid

b No, it has no carbon to which four different groups are attached.

3 This carbon atom does not have four different groups attached to it.

26.1

1 a pentan-3-one   **b** propanal

2 a A ketone must have a C=O group with two carbon-containing groups attached to it, so at least three carbon atoms are required.

b The C=O group can only be on carbon 2, otherwise the compound would be an aldehyde.

c The C=O group must always be on the end of the chain.

3 A hydrogen bond requires a molecule with a hydrogen atom covalently bonded to an atom of fluorine, oxygen, or nitrogen. There is no such carbon atom in propanone.

4 The hydrogen atom in an –OH group in a water molecule can hydrogen bond with the oxygen atom in the C=O bond of propanone.

26.2

1 Cl^-

2 a no

b A negatively charged nucleophile will be repelled by the high electron density in the C=C.

c

3

4 The CN^- ions can attack the planar C=O group from above or below. In the case of CH_3CHO this will produce a compound with a chiral centre (four different groups attached to it). This will be a pair of optical isomers. As attack from above or below is equally likely it will be a 50:50 mixture (racemic). In the case of CH_3COCH_3, the product will not have a carbon atom with four groups

attached to it. It will not therefore be chiral and will be a single compound.

26.3

1 3-bromobutanoic acid   **2**

3 The carboxylic acid must be at the end of a chain and therefore in the 1 position.

4 ethyl ethanoate; methyl propanoate

26.4

1 They react with carbonates and hydrogencarbonates to give carbon dioxide, metal oxides to give salts, and alkalis to give salts.

2 methanol and ethanoic acid

3 ethanol and methanoic acid

4 They have the same molecular formula but have a different arrangement of their atoms in space.

26.5

1 The carbon of the C=O group is strongly $C^{\delta+}$ because, as well as oxygen, it is also bonded to an electronegative chlorine atom which draws electrons away from it.

2 OH^-, because it is negatively charged and contains a lone pair of electrons.

3 Because the nucleophile and the acylating agent join together and then a small molecule is lost.

4

27.1

1 CH   **2** 3

3 Six electrons are spread out over all six carbon atoms in the ring rather than being localised in three distinct double bonds.

4

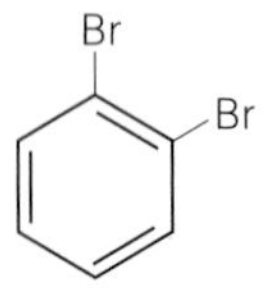

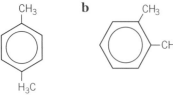

27.2

1 van der Waals forces
2 **a** 1,3-dichlorobenzene
   b 1-bromo-4-chlorobenzene
3 **a** [structure: benzene ring with CH₃ at top and H₃C at bottom]   **b** [structure: benzene ring with two CH₃ groups]
4 R⁺

27.3

1 **a** A electrophilic substitution
   b A electrophilic substitution
2 1,2-dinitrobenzene and 1,4-dinitrobenzene
3 Addition reactions would involve the loss of aromatic stability.
4 [reaction scheme: benzene + CH₃CH₂COCl → ketone product + HCl, with CH₃CH₂CO⁺ shown below]

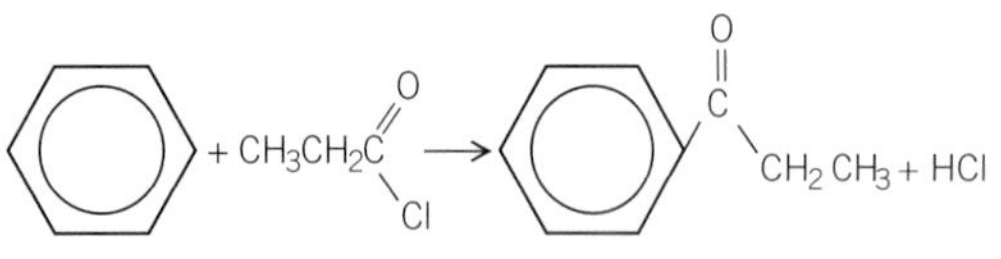

28.1

1 secondary   2 ethylpropylamine
3 $$H_3C-N-CH_3$$ with CH₃ above   4 a gas
5 It has the same value of M_r as ethylamine, which is a gas.

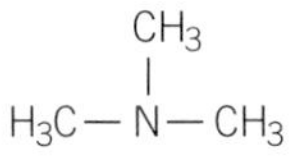

28.2

1 **a** $(CH_3)_2NH + HCl \rightarrow (CH_3)_2NH_2^+ + Cl^-$
   b dimethylammonium chloride
2 **a** It will dissolve, that is, the oily drops will disappear to give a colourless solution.
   b The oily drops will re-appear.
3 Stronger, as it has two electron-releasing alkyl groups.

Solubility of drugs

1 secondary
2 [structure: phenyl ring with OH, and quaternary ammonium group N⁺ with two H, CH₃, and CH₃ branches, with Cl⁻]
3 It is an ionic compound.
4 Solid, because it is ionically bonded.
5 –OH, alcohol, and an aromatic ring.
6 [structure: phenyl ring with OH, chiral centres marked with *, quaternary ammonium group N⁺ with H, H, CH₃, CH₃, with Cl⁻]
7 It is possible that only one of the optical isomers is active as a drug.

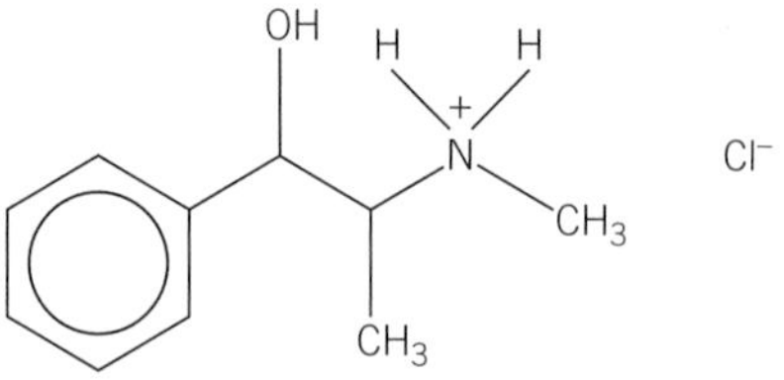

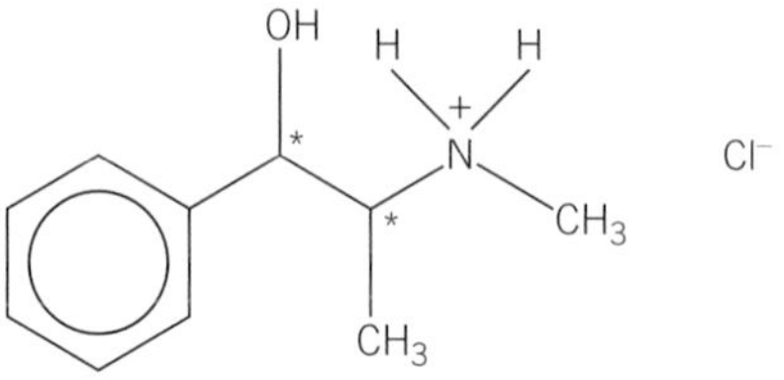

28.3

1 Because secondary and tertiary amines (and quaternary ammonium salts) may be formed as well.
2 **a** $CH_3CH_2Cl + 2NH_3 \rightarrow CH_3CH_2NH_2 + NH_4Cl$

   $CH_3CH_2Cl + 2NH_3 \longrightarrow CH_3CH_2NH_2 + NH_4Cl$

   [mechanism diagram showing NH₃ attacking CH₃–CH₂–Cl forming CH₃–C–N⁺–H + Cl⁻]

   [mechanism diagram showing CH₃–C–N⁺–H with :NH₃ → CH₃–C–N + NH₄⁺]

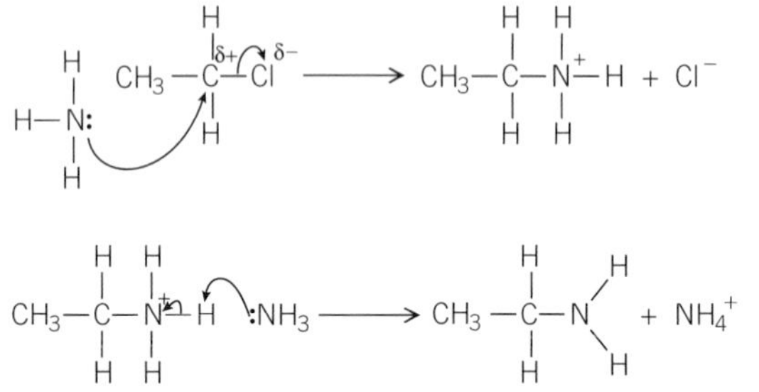

   b $(C_2H_5)_2NH$; $(C_2H_5)_3N$; $(C_2H_5)_4N^+Cl^-$

29.1

1 **a** the number of carbon atoms in each monomer
   b 1,10-diaminodecane
      $NH_2(CH_2)_{10}NH_2$
2 They also have CONH linkages.
3 any diol, for example, propane-1,3,-diol
4 **a** amide   **b** ester
5 $\{CO-R-CO-O-R'-O\}_n + nH_2O \rightarrow$
   $n\{CO-R-COOH + HO-R'-O\}$

Hermann Staudinger

1 2-methylbuta-1,3-diene

2 Addition since it contains a —C=C— in the structure.

30.1

1 2-aminopropanoic acid

2 Carbon number 2 in alanine has four different groups attached to it.

30.2

1 a amine and carboxylic acid

   b The carboxylic acid is acidic; the amine is basic.

2 two

3 They will be protonated, *ie* there will be a H^+ ion attached to the $-NH_2$. $^+H_3NCH(CH_3)\,CO_2H$

4

2-dimensional TLC

(approximate values) Solvent 1: orange = 0.85
blue = 0.60

Solvent 2: orange = 0.52   blue = 0.65

30.3

1 Ionic bonding and hydrogen bonding.

2 van der Waals forces and dipole–dipole forces

30.4

1 GGTCAACTGG

2 The phosphate groups which form the backbone of the DNA have P—O—H groups. These can donate the hydrogen atom as a proton, ie P—O—H → P—O⁻ + H⁺

DNA, amino acids, and proteins

30.5

1 The two chloride ligands are too far apart to effectively bond to two adjacent guanine bases.

31.1

1 a React with HCN.

   b React with methanol (with an acid catalyst).

   c Addition of water (by reaction with concentrated sulfuric acid followed by water).

   d Dehydrate with P_2O_5, for example.

2 a Add water (by reaction with concentrated sulfuric acid followed by water) to produce ethanol, then oxidise with $K_2Cr_2O_7/H_2SO_4$.

   b Reduce (using $NaBH_4$, for example) to propan-2-ol, then react with KBr/H_2SO_4.

3 Step 1: Friedel–Crafts acylation using ethanoyl chloride and aluminium chloride.

   Step 2: Nitration electrophilic substitution using a mixture of concentrated nitric and sulfuric acids.

   Step 3: Reduction (hydrogenation) using Sn/ concentrated HCl.

31.2

1 React with sodium hydroxide, acidify and then add silver nitrate solution. A precipitate will form – white for chlorine, cream for bromine, and pale yellow for iodine. To more clearly distinguish the cream and pale yellow precipitates, add concentrated ammonia – the precipitate will dissolve in the case of bromine, but not in the case of iodine. The white precipitate formed with chlorine will dissolve in aqueous ammonia.

2 a Because OH^- ions will form a precipitate with silver nitrate.

   b Using HCl adds Cl^- ions, which will form a precipitate with silver nitrate.

3 a alkene and carboxylic acid

   b CH_2=CHCOOH

   c CH_2=CHCOOH + Br_2 → CH_2BrCHBrCOOH

   CH_2=CHCOOH + $NaHCO_3$ →
   $\qquad\qquad CH_2$=CHCOONa + CO_2 + H_2O

32.1

1 a a single peak

   b Both carbon atoms are in exactly the same environment.

2 The upper one is propan-1-ol, as it three peaks because all three carbon atoms are in different environments. The lower one is propan-2-ol, as it has two peaks because there are carbon atoms in just two different environments.

A brief theory of NMR

They are all odd numbers.

32.2

1 **a** Isomers have the same molecular formula but different arrangements of atoms in space.

   b

               A   B   C   D

   propan-1-ol: $CH_3CH_2CH_2OH$

               A   B   C   A

   propan-2-ol: $CH_3CH(OH)CH_3$

   c **i** 4   **ii** 3

   d **i** A3; B2; C2; D1   **ii** A6; B1; C1

   e **i** D>C>B>A

      ii C>B>A

32.3

1 **a** A is ethanol, B is methoxymethane.

   b A: 4.5 is R—O—H, 3.7 is R—CH_2—O-, 1.2 is R—CH_3

      B 3.3 is —O—CH_3

   c A: 4.5, 1; 3.7, 2; 1.2, 3.

      B: It is not possible to tell.

2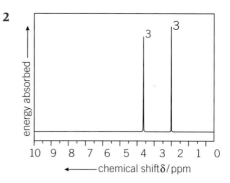

There is no spin-spin coupling because the two CH_3 groups are not on adjacent carbon atoms.

33.1

1 In column chromatography the eluent is a liquid. In gas–liquid chromatography it is a gas.

2 It can separate and help to identify minute traces of substances. For example (among others) substances used in making explosives. [Many other suggestions are also acceptable.]

3 **a** A   **b** C   **c** A   **d** C

Index

acid–base chemistry
    amino acids 204–205
    aqueous transition metal ions 124–128
    titrations 71–76
acid chlorides 169–171
acid derivatives of esters 165–166
acid dissociation constant K_a 68–70
acidic buffers 77–78
acids
    additions to buffers 80–81
    definition 62–63
    dissociation constants 68–70
    inorganic chemistry 124–126
    Lewis-type 125
    metal ion oxidation states 124–125
    pH scale 64–67
    pK_a 70, 76
    strong 66–67
    titrations 71–76
    weak 68–70
activation energy 34
    Arrhenius equation 35–38
active sites 212
acylations 168–172, 181–182
addition–elimination reactions 169–172
addition reactions, nucleophilic 158–159,
    168–172
alcohols
    aldehyde/ketone reductions 160
    chemical tests 225
    dehydrations 222
    naming 145
    nucleophilic addition to carbonyls 169–170
    oxidations 221
aldehydes 156–160
    chemical tests 159, 225
    naming 145
    nucleophilic additions 158–160
    oxidations 221
    reductions 160
alkaline batteries 56–59
alkalis
    buffers 78
    pH 66–67
    transition metal ions 117
    see also bases
alkanes, naming 144–145
alkenes
    amine synthesis 223
    chemical tests 225
    naming 144–145
alkyl amines, basicity 188–189
alkynes, naming 145
amide bonds 198–199, 206–207
amides
    from carbonyls 169–171
    naming 145
    peptides 198–199, 206–207
    polymers 197–198
    synthesis 192
amines 186–195
    amide formation 192
    as bases 188–189
    economic importance 192–193
    from nitriles 191

inductive effect 188–189
    naming 145, 186
    as nucleophiles 169–171, 190
    polyamides 197–198
    reactivity 169–171, 187–192
    synthesis 190–192, 223
amino acids 204–211
    acid–base chemistry 204–205
    isomerism 151
    naturally occurring 205
    peptide links 206–207
    thin-layer chromatography 210–211
α-amino acids 148
ammonia 186
    halogenoalkane reactions 190
    ligand substitution 129–130
    nucleophilic additions to carbonyls 169–171
    production 120
    in solution 68
    transition metal ions 126–127
anhydrides 145, 169–171
anti-cancer drugs 217
aqua ions 106, 124–125, 132–133
aqueous solutions
    inorganic compounds 124–139
    pH 64–67
    protons 62–63
arenes 174–185
    naming 145
    see also aromatic compounds
aromatic compounds 174–185
    amines 186, 188–189, 191–193
    combustion 179
    electrophilic substitutions 178–182
    Friedel–crafts acylations 181–182
    naming 177–178
    nitrations 179–180
    organic synthesis 223
    physical properties 177
    reactivity 178–182
aromatic stability 174–175
Arrhenius equation 35–38
asymmetric carbon atoms 148

base pairs, DNA 214–217
bases
    additions to buffers 81
    amines 188–189
    aqueous metal ions 126–133
    definition 62
    ester hydrolysis 165–166
    Lewis-type 125
    pH 64–67
    titrations 71–76
    weak 68–70
basic buffers 78
basic oxides of period 3 elements 97–98
batteries 56–59
benzene 174–176, 191, 223
bidentate ligands 106, 131
bond enthalpies 5–14, 157
bonding
    benzene 174–176
    hydrogen bonding 13, 162, 207–209,
    214–215

lattice enthalpies 13–14
    metal oxides 95
    non-metal oxides 95–97
    oxo acids 96–97
    polarisation 13–14
    proteins 206–209
    strengths 157
    transition metal ions 106–109
bond lengths, carbon–carbon 174
Born–Haber cycles 7–12
Brønsted–Lowry theory of acidity
    62, 126
buffers 77–81
    acid/base additions 80–81
    acidic 77–78
    basic 78
    calculations 78–79

cadmium/nickel batteries 58
calculation of entropy changes 16, 18–19, 21
carbon-13 NMR 229–231
carbonates, transition metals 127
carbon–carbon bonds 157, 174–176
carbon–carbon double bonds 157, 174
carbon–carbon triple bonds 174
carbon–oxygen bonds 157
carbon–oxygen double bonds 157
carbonyl compounds
    hydrolysis 169–170
    nucleophilic additions 150–151, 158–159,
    168–172
    physical properties 157
carbon/zinc cells 56–57
carboxylic acids 161–162, 164–165
    aldehyde oxidations 159
    chemical tests 225
    from carbonyls 169–170
    naming 145, 161
    physical properties 162
    polyamides 197–198
    reactivity 164–165
    synthesis 222
catalysis
    auto 122
    enzymes 207, 212
    heterogeneous catalysts 119–120
    homogeneous catalysts 121
    transition metals 105, 119–122
cells
    division 216–217
    electrochemical 56–59
    representation 51
chelation 106–107, 129, 131
chemical shifts 229–230, 232
chemical tests
    functional groups 159, 225
    iron 127–128
chirality 109, 148–153
chromatography 241–249
    columns 242
    GCMS 243
    HPLCS 243
    phases 241
    thin-layer 210–211, 241
cis-platin 217

colorimetry 111–112
coloured ions 105, 110–112
column chromatography 242
combinatorial chemistry 193, 221
combustion 179, 200
complexes of transition metal ions 105–109
complex ions
    colorimetry 111–112
    isomerism 108–109
    ligand substitution 129–131
    transition metal elements 106–109
concentration
    colorimetry 111–112
    water 63
condensation polymers 196–203
    monomers 199–200
    polyamides 197–199
    polyesters 196–197
    repeat units 199
condensation reactions 196
Contact process 120
co-ordinate bonds 106–109, 125, 129–131
co-ordination numbers 106–107
covalent bonding 14, 95–97

Daniell cells 49, 56
dative bonding 106–109, 125, 129–131
d-block elements 104–123
    catalysis 105, 119–122
    chemical properties 105
    coloured ions 110–112
    complex ions 106–109
    complex isomerism 108–109
    electronic configuration 104–105
    ions 104–112
    oxidation states 113–118
dehydrations 222
delocalisation 96–97, 164, 174–176, 179
deprotonated amino acids 204
2-dimensional thin-layer chromatography 211
dipeptides 206
diprotic acids 73
disposal of polymers 200
dissociation
    buffers 77–78
    enthalpy 5, 157
    strong acids 66–67
    weak acids 68–69
DNA 213–217
double bonds 157, 174

EDTA see ethylenediaminetetraacetate
efficiency of heterogeneous catalysts 119–120
electrochemical cells 56–59
electrochemical series 50–51
electrode potentials 48–61
    batteries 56–59
    electrochemical series 50–51
    half cells 48–49
    hydrogen electrodes 49–50
    redox reactions 52–55
    reducing agents 49–50
electrodes 48–55
    hydrogen 49–50
    redox reactions 52–55
    voltage 50–51

electronic configuration of d-block
        elements 104–105
electrons
    aromatic compounds 174–176, 178–179
    excitation 110
electrophilic substitutions 178–182
elements, periodicity 92–103
enantiomers 148–153
endothermic reactions 15
end points 74–76
enthalpy changes ΔH 4–14
    of atomisation 4, 6–10
    benzene 175
    Born–Haber cycles 7–12
    of combustion 4
    of dissociation 5, 157
    of formation 4, 7–12
    Gibbs free energy 17–21
    Hess's law 4–6
    of hydration 5
    hydration 13
    ionic bonding 5–9
    of solution 5, 13
entropy changes ΔS 15–21
    calculation 16, 18–19, 21
    chelate effect 107
    melting 18
    vaporisation 15, 18–19
enzymes 151, 207, 212
equivalence points 72–73
esters 162–163, 165–167
    formation 165
    from carbonyls 169–170
    hydrolysis 165–166
    naming 145, 163
    polymers 196–197
    uses 166–167
extraction, metals 19–20

fatty acids 166–167
feasible reactions 15–21
    entropy 15–16
    Gibbs free energy 16–21
Fehling's test 159
first electron affinity 4–10
first ionisation energy (first IE) 4, 6–12
first order reactions 28–29
flash photolysis 26
frequencies of light 110
Friedel–Crafts acylation reactions 181–182
fuel cells 58–59
fuels 167
functional groups 145–146, 221, 225

gas chromatography–mass spectrometry
        (GCMS) 243
gas–liquid chromatography (GC) 242–243
GCMS (gas chromatography–mass
        spectrometry) 243
Gibbs free energy changes ΔG 16–21

Haber process 120
haemoglobin 107
half cells 48–49
half equations 114–116
half-neutralisation points 76
half reactions 49–50

halogenoalkanes 145, 190, 222, 225
α-helices, proteins 207–209
Hess's law 4–6
heterogeneous catalysts 119–120
high-performance liquid chromatography
        (HPLC) 243
H⁺ ions 62–67
homogeneous catalysts 121
HPLC (high performance liquid
        chromatography) 243
hydration, enthalpy changes 13
hydrocyanic acid 69–70
hydrogen bonding 13, 162, 207–209, 214–215
hydrogen cyanide 158–159
hydrogen economy 59
hydrogen electrodes 49–50
hydrogen ions 64–67
    see also protons
hydrogen–oxygen fuel cells 58–59
hydrogen storage 59
hydrolysis
    carbonyl compounds 169–170
    enzymes 207
    esters 165–166
    peptides 206–207
    polyamides 200
    transition metal ions 125
hydroxide ions 66–67

indicators for titrations 74–76
inductive effect, amines 188–189
inhibition of enzymes 212
initial rate method 31–32
inorganic chemistry 90–141
    acidity 124–126
    amphoteric hydroxides 128
    aqueous solutions 124–139
    catalysis 119–122
    chelation 129, 131
    ligand substitution 129–131
    metal carbonates 127
    period 3 elements 92–103
    periodicity 92–103
    transition metals 104–123
insoluble oxides of period 3 elements 97–98
integration traces, proton NMR 232–233
International Union of Pure and Applied
        Chemistry (IUPAC) 146
ionic bonding 5–9, 14, 95, 208
ionic salts 164–165
ionisation isomerism 109
ionisation of water 63
ions
    acid–base chemistry 66–69, 77–78,
        124–128
    chelation 106–107, 129, 131
    coloured 105, 110–112
    complex 106–109, 111–112, 129–131
    d-block elements 104–112, 127–128
    enthalpy 4, 6–12
    hydrogen 64–67
    hydroxide 66–67
    iodide 32, 121
    iron 112, 124–125, 127–128
    manganese 112
    oxo acids 96–97
    zwitterions 204–205

isomers 108–109, 149
  ions 109
  naming 147–148
  optical 147–153
  organic compounds 147–153
  racemates 151–153
  synthesis 150–153
IUPAC (International Union of Pure and
    Applied Chemistry) 146

K_a see acid dissociation constant
ketones 156–160
  chemical tests 159
  naming 145
  nucleophilic additions 158–160
  reductions 160
kinetic energy, temperature 34
kinetics 20–21, 24–43
  activation energy 34
  order of reactions 28–32, 40–41
  rate constants 27–29, 33–34
  rate–determining steps 39–41
  rate expressions 27–29
  rates of reactions 24–26, 33–41

lattice formation enthalpy 5–10, 13–14
lead-acid batteries 57
Leclanché cells 56–57
Lewis acids and bases 125
ligands 106–108
  coloured ions 110–112
  substitution 129–131
light 110, 149
linear complexes 108
lithium ion batteries 58

macromolecules 201
magnetic resonance imaging (MRI) 238
mean bond enthalpy 5, 157
measurement
  pH 65
  rates of reactions 24–26, 31–32
melting points
  benzene 177
  carboxylic acids 162
  entropy 18
  period 3 oxides 96
metal ions
  ammonia ligands 129–130
  aqueous acid–base reactions 126–127, 132
  chelation 129, 131
  hydrolysis 125
  ligand substitution 129–131
  oxidation states 124–125
  see also transition metal ions
metal oxides of period 3 elements 95, 97–98
metals, extraction 19–20
mobile phases 241
monomers 199–200, 213–215
monoprotic acids 73
MRI (magnetic resonance imaging) 238
multidentate ligands 106–107, 131

naming see nomenclature
naturally occurring amino acids 205
nickel/cadmium batteries 58
nitrated arenes 179–180

nitriles 145, 191
nitryl cations 179–180
NMR see nuclear magnetic resonance
    spectroscopy
noble gases 11–12, 92
nomenclature
  aldehydes 156–157
  amines 186
  aromatic compounds 177–178
  carboxylic acids 161
  esters 163
  isomers 147–148
  ketones 156–157
  organic compounds 144–148
non-metal oxides of period 3 elements 95–97
non-rechargeable cells 56–57
$n + 1$ rule 234
nuclear magnetic resonance (NMR)
    spectroscopy 228–240
  carbon-13 229–231
  development 237
  interpretation 229–230, 234–236
  magnetic resonance imaging 238
  proton 232–240
  spin–spin coupling 234
  theory 228
nuclei
  resonance 228
  shielding 229
nucleophiles 169–171, 190
nucleophilic addition reactions 150–151,
    158–159, 168–172
nucleophilic substitution reactions 190, 192
nucleotides 213–215

octahedral complexes 108
oils 166–167
optical isomerism 109, 147–153
order of reactions 28–32, 40–41
ores, metal extraction 19–20
organic analysis 225
  carbon-13 NMR 229–231
  column chromatography 242
  gas–liquid chromatography 242–243
  GCMS 243
  HPLC 243
  NMR theory 228
  proton NMR 232–240
  thin-layer chromatography 210–211, 241
organic chemistry 142–251
  acylations 168–172
  aldehydes 156–160
  amines 186–195
  amino acids 204–211
  analysis 225–249
  aromatic compounds 174–185
  carbonyl compounds 156–173
  carboxylic acids 161–162, 164–165
  chirality 148–153
  chromatography 241–249
  condensation polymers 196–203
  DNA 213–217
  enzymes 212
  esters 162–163, 165–167
  isomerism 147–153
  ketones 156–160
  naming 144–148

NMR spectroscopy 228–240
  optical isomerism 147–153
  polyamides 197–199
  polyesters 196–197
  proteins 198–199, 204–212
  racemates 151–153
  stereoisomerism 147–148
  structure determination 228–240
  synthesis 150–153, 220–224
oxidations
  alcohols 221
  aldehydes 159, 221
  cobalt ions 107
  ethanedioic acid 122
  ketones 159
  organic synthesis 221
  period 3 elements 93–94
  transition metal ions 117
oxidation states 92, 105, 110–111, 113–117,
    124–125
oxides of period 3 elements 95–100
oxidising agents 221
oxo acids 96–98

peptide bonds 206–207
Period 3 elements 92–103
  oxides 95–100
  reactions with oxygen 93–94
  reaction with water 92–93
Period 3 oxides
  acidic/basic nature 99–100
  bonding 95–97
  physical properties 96
  reactions with water 97–98
periodicity 92–103
pH
  acid–base titrations 71–76
  buffers 77–81
  measurement 65
  scale 64–67
  strong acids 66–67
  temperature 65
  weak acids 69–70
phase boundaries 51
phases 51
photolysis 30
physical chemistry 2–89
  acids, bases and buffers 62–83
  activation energy 34
  electrode potentials 48–61
  kinetics 24–43
  thermodynamics 4–23
physical properties
  carbonyl compounds 157
  carboxylic acids 162
  period 3 oxides 96
  primary amines 187
  proteins 208–209
pK_a 70, 76
pleated sheets, proteins 207–209
polarimeters 149
polarisation
  bonding 14
  carbonyl groups 164, 168–171
  carboxylic acids 164
  hydration 13
  isomers 149

polarity, carbonyl compounds 157
polyamides 197–199, 200
polyesters 196–197
polymerisation/polymers 196–203
    disposal 200
    DNA 213–214
    monomers 199–200
    peptides/proteins 198–199
    polyamides 197–199
    polyesters 196–197
polypeptides 198–199
potential difference 48–51
prefixes, systematic naming 145–146
primary alcohols 160
primary amines 169–171, 186, 187, 190–191
primary structure of proteins 206, 208
proteins 198–199
    enzymes 212
    hydrogen bonding 207–209
    hydrolysis 206–207
    ionic bonding 208–209
    peptide bonds 206–207
    primary structure 206, 208
    secondary structure 208–209
    structure 207–209
    sulfur–sulfur bonds 208–209
    tertiary structure 209
    thin-layer chromatography 210–211
proton NMR 232–240
    chemical shifts 232
    interpretation 234–236
    predicting spectra 237
    solvents 236
    spin–spin coupling 234
protons
    acidity 126
    aqueous solutions 62–63
    Lewis acids and bases 125
    pH scale 64–67
    see also hydrogen ions
proton transfers 62

quaternary ammonium salts 190, 192

racemates 151–153
randomness 15–21
    see also entropy
rate–concentration graphs 30
rate constants k 27–29, 33–34
    Arrhenius equation 35–38
rate-determining steps 39–41
rate expressions 27–29
rate of the reaction 24–43
    measurement 24–26, 31–32
    rate-determining steps 39–41
    temperature 33–34
reaction mechanisms 39–41, 220, 222–223
reactions
    initial rates 31–32
    kinetics 24–47
    order 28–32
    rate constants/expressions 27–29, 33–34
    temperature 33–34
reaction schemes 39–41, 220, 222–223
reagents for organic synthesis 221–224

rechargeable batteries 57–59
redox
    aldehydes and ketones 159
    carbonyl compounds 159
    electrode potentials 50, 52–55
    half equations 114–116
    Period 3 elements 92–94
    titrations 114–116
    transition metals 113–117
reductions/reducing agents
    aldehydes 160
    electrode potentials 49–50
    ketones 160
    nitriles to amines 191
    organic synthesis 221–222
repeat units of condensation polymers 199
replication of DNA 216–217
resonance 176, 228
retention times 242–243

salt bridges 48–49
salts
    carboxylic acids 164–165
    quaternary ammonium 190, 192
secondary alcohols 160
secondary amines 186
secondary structure of proteins 208–209
second ionisation energy (second IE) 4, 8–9
second order reactions 28–29
shape of complex ions 108–109
β-sheets, proteins 207–209
shielding of nuclei 229
silver mirror test 159
solubility of primary amines 187
solvents
    esters 163
    glycerol 167
    proton NMR 236
spin–spin coupling 234
spontaneous reactions 15–21
    entropy 15–16
    Gibbs free energy 16–21
square-planar complexes 84, 108
standard enthalpy of atomisation 4, 6–10
standard hydrogen electrodes 49–50
standard molar enthalpy change of combustion 4
standard molar enthalpy of formation 4, 7–12
stationary phases 241
stereoisomerism 147–148
stereospecificity of enzymes 212
strong acids 66–67
structures
    complex ions 108–109
    period 3 oxides 95–97
substrates of enzymes 212
suffixes, systematic naming 145–146
systematic naming of organic compounds 144–147

target molecules 220
temperature
    Gibbs free energy 17–20
    pH 65
    rates of reactions 33–34
tertiary amines 186
tertiary structure of proteins 209

tests
    aldehydes/ketones 159
    functional groups 225
    iron 127–128
tetrahedral complexes 108
tetramethylsilane (TMS) 233
thermodynamics 4–23
    Born–Haber cycles 7–12
    enthalpy changes 4–14
    entropy 15–21
    Gibbs free energy changes 16–21
    Hess's law 4–6
    kinetics 20–21
thin-layer chromatography (TLC) 210–211, 241
third order reactions 28–29
titration curves 72–73
titrations 71–76
    equivalence points 72–73
    half-neutralisation points 76
    indicators 74–76
    transition metal ions 114–116
TLC (thin-layer chromatography) 210–211, 241
TMS (tetramethylsilane) 233
TNT (trinitrotoluene) 180
Tollen's reagent 159
transition elements 105
transition metal ions
    alkaline solutions 117
    aqueous acid–base chemistry 124–128, 132–133
    autocatalysis 122
    carbonates 127
    chelation 131
    chloride ion complexes 130–131
    colorimetry 111–112
transition metals 104–123
    catalysis 105, 119–122
    chemical properties 105
    coloured ions 110–112
    complex ions 106–109
    oxidation states 113–118
triple bonds 174

units, rate constants 28–29
universal indicator 74

vaporisation of water 15, 18–19
variable oxidation states 113–118
voltage 48–51
voltmeters 49

water
    as acid/base 62
    aqua ions 106, 124–125, 132–133
    ionisation 63
    nucleophilic additions to carbonyls 169–170
    reactions with period 3 elements 92–93
    reactions with period 3 oxides 97–98
    vaporisation 15, 18–19
weak acids 68–70, 76
weak bases 68–70

zero order reactions 29, 30
zinc/carbon cells 56–57
zinc chloride cells 57
zinc/copper cells 56
zwitterions 204–205